THE COMPLETE GUIDE TO CONTRACTING YOUR HOME

DAVE MCGUERTY & KENT LESTER

BETTERWAY BOOKS
CINCINNATI, OHIO

METRIC CONVERSION CHART		
TO CONVERT	TO	MULTIPLY BY
Inches	Centimeters	2.54
Centimeters	Inches	0.4
Feet	Centimeters	30.5
Centimeters	Feet	0.03
Yards	Meters	0.9
Meters	Yards	1.1
Sq. Inches	Sq. Centimeters	6.45
Sq. Centimeters	Sq. Inches	0.16
Sq. Feet	Sq. Meters	0.09
Sq. Meters	Sq. Feet	10.8
Sq. Yards	Sq. Meters	0.8
Sq. Meters	Sq. Yards	1.2
Pounds	Kilograms	0.45
Kilograms	Pounds	2.2
Ounces	Grams	28.4
Grams	Ounces	0.04

 Published by Betterway Books, an imprint of F&W Publications, Inc., 1507 Dana Avenue, Cincinnati, Ohio 45207. (800) 289-0963. Third edition.

Other fine Betterway Books are available from your local bookstore or direct from the publisher.

02 01 7 6

Library of Congress Cataloging-in-Publication Data

McGuerty, Dave
The complete guide to contracting your home / Dave McGuerty and Kent Lester.—3rd Ed.
p. cm.
Includes index.
ISBN 1-55870-465-5 (pb)
1. House construction—Amateurs' manuals. 2. Building—Superintendence—Amateurs' manuals. 3. Contractors—Selection and appointment—Amateurs' manuals. I. Lester, Kent, 1953-. II. Title.
TH4815.M27 1997
692′.8—dc21 97-853
CIP

Editor: R. Adam Blake
Production editor: Bob Beckstead
Cover designers: Angela Wilcox and Clare Finney
Cover photograph: D. Altman Fleischer

Contents

SECTION III: APPENDICES

Introduction

Welcome to the exciting world of home building. Whether you are a homebuilder or an individual planning to build your own home, this manual will be a great step in helping you to achieve your goal. The fact that you have come this far indicates that you have the desire to do things in an efficient, organized manner and to save money on your project.

Homes have existed as long as man himself. Over centuries, homes have evolved from simple dwellings to complex structures combining state-of-the-art skills and materials from many industries. Such rapid advancement has made the task of building a home increasingly difficult and complex. What the residential builder needs is a simple and complete method covering all of the activities involved in completing a construction project—from beginning to end. That's what this book is all about.

WHO CAN BE A HOMEBUILDER?

A homebuilder is anyone who wants to be one. Many states have no examinations or licensing requirements for builders. Check with local building authorities in your area to determine whether you must meet any special requirements. The homebuilder, or contractor, is the person in charge of overseeing the entire construction project, coordinating the efforts of professional subcontractors and labor, and managing materials. The term "homebuilder" is somewhat a misnomer, since a homebuilder normally does not even have to raise a hammer to build a home. His efforts are best concentrated on managing the project. A professional homebuilder is anyone who builds for a living.

BEING YOUR OWN CONTRACTOR

Many would-be homebuilders are discouraged from attempting to build their own homes because of fear due to the "mystique" of modern construction practices and the profusion of available reference materials. Although many steps covered in this book provide you with information on what is happening at a specific point in time, you are usually not the person doing it. The steps you choose to perform are up to you. This book provides concise step-by-step instructions for residential construction. Building a home requires a sound knowledge of contracting. Contracting is the process of coordinating the efforts of skilled tradesmen (carpenters, masons, electricians, plumbers, etc.) and managing materials in an efficient manner. This book will show you how to work with subcontractors and material suppliers to become a successful homebuilder, whether you plan to build one house or a hundred.

CONTRACTING OBJECTIVES

- Save time
- Save money
- Reduce frustration

Save Time

This book provides you with a complete, detailed masterplan that will help keep your project organized by dividing the work into discrete, manageable tasks. Efficiency will be gained by only doing necessary tasks—and doing them only once.

Save Money

This book can easily help save you from 25 to 42 percent on the home you build over a comparable home purchased through a real estate agent. Here are the major ways:

Real estate commission	up to 7%
Builder markup	up to 20%
Savings on material purchases	up to 4%
Cost-saving construction	up to 2%
Doing work yourself	up to 9%
Total	**up to 42%**

This book will provide you with a wealth of instantly usable knowledge equivalent to many other books on construction. This book can also save you from numerous, costly pitfalls associated with residential construction. After all, learning from bad experiences doesn't mean they have to be your own!

Reduce Frustration

This book will:

- EQUIP you with the knowledge to effectively manage a residential construction project.
- HELP you avoid major problems and pitfalls.
- PROVIDE you with the benefit of years of experience.
- ADVISE you when to avoid tackling tasks that are best left to specialists.
- SHOW you how to specify and inspect the work performed by others.

HOW THIS BOOK DIFFERS FROM OTHER CONSTRUCTION MANUALS

There are more than enough books on the market covering construction techniques. However, most fail to cover the important, practical, day-to-day details of managing a construction project. Remember, you are not building the home per se; you are *managing* the building of the home. Also, many other publications fail to provide the necessary forms and sequenced, step-by-step instructions helpful in guiding you through the day-to-day details of a construction project.

The Complete Guide to Contracting Your Home is different. It provides you with the information and skills to *manage* and *execute* a successful project, whether you do the work yourself or you hire professional help. Your ultimate success depends much more on your project management skills than on your personal skills as a do-it-yourselfer.

This Book Is Filled With Valuable Reference Materials

- A listing of more than 100 home plan books
- Listings of other books about architecture and construction
- More than 50 Internet Web sites filled with up-to-date construction information to further your construction knowledge
- Listings of all regional government mortgage loan offices
- An address listing of major construction suppliers and manufacturers (in the appendix)
- Detailed steps at the end of each chapter
- A Masterplan of every construction step, to guide you through the entire project
- Inspection checklists in each chapter, to help ensure that each job is done right the first time
- Sample specifications in each chapter that you can include in agreements with subcontractors
- Every form and contract you need to deal with subcontractors and suppliers
- A materials list for estimating

OVERVIEW

The Complete Guide to Contracting Your Home is composed of two major sections:

- Concepts
- Project Management

Concepts

Certain aspects of construction do not deal with specific construction steps, but are important to understanding successful home building. Concepts such as lot and plan selection, financial and legal issues, dealing with members of the construction industry and other topics are important in gaining a proper orientation to this environment. Much of the Concepts material will be helpful throughout the Project Management section of the manual.

Project Management

Residential construction involves the orderly coordination of some two dozen specialty trades each with its own special instructions, specifications, inspection requirements, detailed steps and special terms. The Project Management section will cover each trade, one at a time, in the order in which the project normally proceeds.

HOW TO USE THIS BOOK

This book is designed to be easy to use and straightforward. It has been carefully sequenced to provide you with the right information at the right time. Before beginning your project:

- READ this book thoroughly.
- FAMILIARIZE yourself with all material in the Project Management section.

Before Your Project:

- READ the glossary section to familiarize yourself with common construction terms. Check back to the glossary whenever you hear a new term or phrase.
- CHECK for names, addresses, phone numbers and websites of important companies and organizations. The Planning Your House Design and Home Building and Computers sections have extensive lists of books, Internet websites and phone numbers that will help you find exactly the information you need.
- REFER to the Sources section in the Appendix for additional names and addresses of construction-oriented manufacturers, associations and suppliers.

During Your Project:

- FOLLOW the steps outlined in the Project Management section. Certain steps may not pertain to your project. For example, if you are building on a slab foundation, excavating a basement will not be applicable. However, be very careful not to skip steps that do apply to your project. Some common sense and judgment are obviously essential in this respect. Interpretation of the execution of many steps will naturally vary from project to project. Do not even think of starting your project until you are familiar with the steps covered in this book.
- REFER to the sample specifications for ideas on what to include in your specifications when dealing with tradesmen. These are primarily intended to remind you of items easily overlooked and to illustrate the level of detail you should use.
- REFER to the Inspection checklists for ideas on how and what to inspect to help ensure you are getting good workmanship as your project progresses. It may take ten years to learn how to lay bricks properly, but only ten minutes to check the work for quality craftsmanship.
- USE the forms provided when applicable. Most of these forms are included in the appendices:
  - House plan evaluation checklist
  - Lot evaluation checklist
  - Lien waivers
  - Home construction contract
  - Material and labor take-off sheets
  - Purchase orders
  - Subcontractor specification form
  - Change order

EXCUSES FOR NOT BUILDING

If you decide not to contract your own home, here are a few excuses you can use:

- I have plenty of money. I don't need to save money by acting as contractor myself.
- I'm not particular about getting exactly what I want in a home, representing the single largest financial commitment of my life.
- I don't know how.
- It seems too complicated—I don't know where to begin.
- I can't even hammer a nail straight.
- It's not worth the trouble.
- I don't have the time.
- It's not my bag.

The last excuse is the only valid one. Building isn't for everyone. Many will use one of the above excuses or create their own. To make up an excuse is to rationalize why one pays "retail" for a home. Sadly enough, the money saved by contracting your own home can make all the difference in affording one or not.

The second most valid excuse is:

"I Don't Have the Time"

Most of us who can afford to contemplate building a home have full-time jobs. So how can one build a home in the evenings and on weekends? This is resolved in the How to Be a Builder section.

HELPFUL PERIODICALS

The magazines listed below are some of many you may wish to use as a source of ideas:

Fine Home-Building
Better Homes & Gardens
New Shelter
Architectural Digest
Southern Homes
Southern Living
Country Living
House & Garden
Fine Homes
Town & Country
Southern Accents
Home

ADDITIONAL HELPFUL MATERIAL

In conjunction with this manual, several other documents will provide you with a wealth of information while minimizing your expenditures on reference material. They are:

YOUR LOCAL BUILDING CODE—These are the rules you must abide by when building in your area. Obtainable from your county building inspector.

MODERN CARPENTRY—A general-purpose text with more than 1400 excellent illustrations covering the actual techniques of construction. By Willis H. Wagner. 1973. 480 pages. The Goodheart-Willcox Co., Inc. South Holland, IL ISBN 0-87006-208-5.

MINIMUM PROPERTY STANDARDS FOR SINGLE FAMILY HOUSING—HUD guidelines that FHA and VA officials use to evaluate homes. Document number 999999. May 1979. 325 pages. U.S. Department of Housing and Urban Development. Available from the Superintendent of Documents, U.S. Government Printing Office, Washington, DC 20402.

MANUAL OF ACCEPTABLE PRACTICES—Extensive manual covering most aspects of residential construction. Document number 4390.1. U.S. Department of Housing and Urban Development. 1973. 420 pages. Available from the Superintendent of Documents, U.S. Government Printing Office, Washington, DC 20402.

ARCHITECTURAL AND BUILDING TRADES DICTIONARY—R.E. Putnam and G.E. Carlson. Chicago: American Technical Society, 1974.

BUILDER'S COMPREHENSIVE DICTIONARY—Robert Putnam. Reston, VA: Reston Publishing Co., 1984.

BUILDING TERMINOLOGY—Peter Brett. Oxford: Heinemann Newnes, 1989. British terminology.

BUILDING TRADES DICTIONARY—Leonard Toenjes. Homewood, IL: American Technical Publishers, 1989.

CONSTRUCTION DICTIONARY—Phoenix: Greater Phoenix, Arizona, Chapter of the National Association of Women in Construction, 1973.

CONSTRUCTION GLOSSARY—J. Stewart Stein. New York: John Wiley & Sons, 1980.

CONSTRUCTION MATERIALS: TYPES, USES, AND APPLICATIONS—Caleb Hornbostel. New York: John Wiley & Sons, 1991. Comprehensive descriptive information, including tabular data and metric equivalents.

CONSTRUCTION: PRINCIPLES, MATERIALS, AND METHODS—Harold Olin. Chicago: Institute of Financial Education, 1983. Highly recommended.

DICTIONARY OF ARCHITECTURAL ABBREVIATIONS—New York: Odyssey Press, 1965.

DICTIONARY OF ARCHITECTURAL SCIENCE—Henry Cowan. New York: Halsted Press, 1973.

DICTIONARY OF ARCHITECTURE AND CONSTRUCTION—Cyril Harris. New York: McGraw-Hill, 1993. Very good, combining historical and technical information.

DICTIONARY OF BUILDING—Randall McMullan. New York: Nichols, 1988.

DICTIONARY OF BUILDING—John Scott. London: Granada, 1984.

ENCYCLOPEDIA OF ARCHITECTURE: DESIGN, ENGINEERING, AND CONSTRUCTION—Joseph Wilkes and Robert Packard. New York: John Wiley & Sons, 1988. Five-volume, comprehensive encyclopedia of historical and technical issues.

ENCYCLOPEDIA OF BUILDING AND CONSTRUCTION TERMS—Hugh Brooks. Englewood Cliffs, NJ: Prentice-Hall, 1983.

ENCYCLOPEDIA OF BUILDING TECHNOLOGY—Henry Cowan. Englewood Cliffs, NJ: Prentice-Hall, 1988.

HANDBOOK OF BUILDING TERMS AND DEFINITIONS —Herbert Waugh and Nelson Burbank. New York: Simmons-Boardman, 1954. Older terms, useful for historical applications.

ILLUSTRATED DICTIONARY OF BUILDING MATERIALS AND TECHNIQUES—Paul Bianchina. Blue Ridge Summit, PA: Tab Books, 1986.

MEANS ILLUSTRATED CONSTRUCTION DICTIONARY—Kornelis Smit and Howard Chandler. Kingston, MA: R.S. Means Co., 1991.

SECTION I: CONCEPTS

How to Be a Builder

INTRODUCTION

A lot of people incorrectly believe that builders must know everything about building a house and must pass several tests to become licensed. Many states do not require small homebuilders to be licensed, and in states that do, building your own home is exempt from licensing requirements.

Few working people can afford the time to build their houses themselves by hand. This book is directed at the person wanting to function as the contractor. The contractor can be thought of as manager of the project, making sure that subcontractors and materials show up at the site when needed. Leave the difficult and tedious tasks to the experts: They have spent years learning their trades.

Home building may challenge your abilities in ways yet untested. It will test your ability to:

- Get people to do what you want, when you want it done.
- Be fair, yet tough.
- Save money in creative ways.
- Manage an exciting, major project. You are the Boss.

Before you embark on this major project, you must become educated. Just as important, after you have studied this manual, you must decide whether you can handle all the details, people, surprises, insanity and periodic frustration associated with home building. Building is not for the squeamish. This book will not necessarily make home building a bed of roses for you. Home building favors hard work, perseverance, guts, the ability to "horse trade," common sense and anything else you can muster to get the job done. To get yourself on the right path to being a builder, get your mind set for doing three things right from the beginning:

- Acting like a builder
- Keeping good records
- Keeping a builder's attitude

Act Like a Builder

Remember: Unless special licensing is required in your area, you are just as much a builder as the next guy. But check with your county Building Inspection Department for a copy of the local building code (see Planning Your House Design section) and information on licensing. If you are building your own house, even areas with licensing requirements will probably not require you to have a license.

Get comfortable using building terms, and present yourself in a self-confident and professional manner. Trade-specific terms are covered in the Project Management section. The respect of your subs, labor and suppliers is essential to your effectiveness as a builder. The better you relate to those you work with, the easier your job will be. As good practice, get in your jeans, construction boots and your favorite plaid shirt and walk around a few construction sites. Get some mud on your shoes. This is a dress rehearsal. What you see is probably similar to what you'll have when you break ground. Get used to the smell of lumber, lots of mud and nails by the box. Talk to subs and imagine they're working for you. Figure out what makes them tick. And keep telling yourself you're a builder!

How you are perceived by others in this industry can make a difference. Much of a contractor's job is human relations, so acting like a builder will help you in getting along with others involved with your project. Order a set of business cards. It is amazing how many doors a simple business card will open. The business card will help convince suppliers that you are indeed a professional. The only time you should wear a suit while working on this project is when you're at the bank polishing apples.

Keep Good Records

If this manual accomplishes any one thing, it should be to emphasize the need for keeping track of costs and expenses. You will be at a distinct advantage over the common builder if you keep good records. Some of the primary records are:

COST TAKE-OFF—The single most important tool you can use to save money on your project is the cost take-off. The cost take-off will tell you approximately what your house will cost before starting the project. It also gives you that checklist of materials necessary to make sure that all essential materials are ordered at the proper time. The take-off has a further advantage of giving you a cost goal to strive for when purchasing materials. When bargaining with materials suppliers, don't hesitate to point out that "I've got to stay within my cost take-off budget!"

PURCHASE ORDERS—Make sure to use purchase orders when ordering materials. This will eliminate possible future confusion over whether certain items were ordered and from whom. Purchase orders and invoices are essential when figuring the house's cost basis for the IRS.

INVOICES—Most of the checks you write will be initiated by invoices. It is your responsibility to make sure they are accurate and that you pay them on time, especially if discounts apply. Match your purchase orders to invoices to control payments.

RECEIPTS AND CANCELED CHECKS—Lots of money is going to flow out on invoices. Keep them in a book. You'll need to compare them against purchase orders to avoid paying for things you didn't order and paying for things twice. Compare what was delivered with the invoice. Keep all proofs of payments together in an organized manner. Try to get a receipt—not just a canceled check. Your canceled check may take a month to get back to you as proof of payment.

PHOTOGRAPHS—Consider taking photographs of your home at various times during construction. This allows you to see changes take place and record them visually. Use a tripod and mark three spots on the ground. This way, you will always be taking your pictures from exactly the same location.

CONTRACTS—Nothing will make or break the project like being able or unable to lay your hands on a good, sound, written contract when a dispute arises. A man's memory is only as good as the paper it's written on. Contracts and their related specifications can almost never be too detailed. You can specify anything you want; but if you don't specify it, don't count on it being done.

WORKER'S COMPENSATION RECORDS—You must keep track of which subs have their own worker's compensation policies. For all those who don't, you should retain approximately 6 percent of the total charges. For subs with their own worker's compensation, you should make note of their policy numbers and the expiration dates.

A SMALL PILE OF BANK LOAN PAPERS—These papers will vary from one lender to another, but will include all or some of the following:

- Loan documents
- Draw schedules
- Funds drawn
- Original take-off
- Surveys and plats
- Title search and insurance

Keep a Builder's Attitude

Your attitude and that of others working on your project is important. You may not always be able to see it, but a poor attitude can cause:

- Delays
- Liens
- Poor workmanship
- Additional costs due to carelessness
- Lots of frustration and headaches

Undertaking such a bold project takes a special frame of mind. And once you break ground, there is no turning back. Here are some important attitudes you'll need to build successfully:

- PERSEVERE. Don't let anything get you down. When problems arise, tackle them immediately. Keep the project moving forward as close to schedule as possible. Realize that time is money and that you are responsible.
- BE FIRM. When you have a dispute with a sub over sloppy work or a high bid, stick to your guns. You are the *boss*. Don't let anyone talk you up

against a wall. Remember the golden rule: "He who has the gold makes the rules."

- DON'T BE A PERFECTIONIST. You want the work done on your home to be of high quality, but don't be upset every time you see a bent nail (provided you don't see too many of them). Chances are, the flaws you see in your home will not be noticeable by others when it's completed.
- BE FRUGAL. Always look for new and creative ways to save money on anything pertaining to the project. Small savings can really add up in a project of this size.
- BE THOROUGH. Think your ideas through and be meticulous. Pay attention to detail. Have the desire to keep detailed records organized and up to date.
- DON'T GET MAD. Avoid chiding your subs on the job, as this will reduce their desire to cooperate with you.
- DON'T WORRY TOO MUCH. Instead of worrying, do something about the problem.

THE BUILDER'S RESPONSIBILITIES

The laborers and subcontractors on your project are responsible for performing quality work and providing materials as specified and when specified by written contracts. That is their single most important responsibility. As the contractor, your responsibilities include the following:

- PROVIDE clear, detailed specifications for those who work for you.
- PAY laborers and subcontractors promptly.
- SCHEDULE material to arrive when or just before needed.
- SCHEDULE labor and subs to work when needed.
- MINIMIZE costs while maximizing quality.
- PROVIDE a safe site at which to work.
- MAINTAIN control of the project and the subs.
- INSPECT work as it proceeds and is completed.
- MAINTAIN accurate, up-to-date records on all project matters such as:
    - Insurance
    - Drawings and plans
    - Inspections
    - Bank papers
    - Purchase orders and invoices
    - Legal documents such as variances, permits and licenses
    - Payments, receipts, bank documents and other financial matters

HOW TO BE A PART-TIME BUILDER

Most people have a full-time job and cannot take time off to construct a house. How can you have your cake and eat it too? There are two good options available.

Option I—Hire a Supervisor

Check the want ads for retired carpenters or builders who would like some light work in the construction industry. They have the time and experience to do things right and, chances are, they've been in the building industry most of their lives. Even if they are too old to do physical work, they can spot good work—and bad. This is also a great source for subs and the occasional use of a pickup truck.

Arrangements can probably be made on an hourly, daily or even monthly basis. The supervisor can be around at three o'clock in the afternoon when the concrete is about to be poured, or when a load of bricks needs to be delivered and you are at work or out of town. These individuals can play an important part in your project. Without them, you may not be able to build at all. If this route is chosen, involve the individual early in the process, when designs and the project are being planned. This will allow you time to determine the individual's credibility, skill and honesty. If you would like more control over your project, consider getting a pager so you can reach the supervisor at all times.

Most subs can work unsupervised when not being paid by the hour. Your "supervisor" only needs to confirm that they are doing the proper work, answer questions and assist in inspecting the work. He's there to keep the job rolling along and answer questions that would otherwise bring work to a halt.

Option II—Pay a Builder for His Assistance

Another option is to pay another builder to help run your project. This allows you:

- Use of his subs. This saves you the time and hassle of finding reputable subs. In return, let him put up his building sign at your site.
- Assistance during your work hours. If you can be at the site before and after work, this can be a helpful bridge during those unattended hours.
- Assistance in securing a construction loan.

In return, there may be a number of favors you can do for the builder when you don't have anything to do at your site. Also, your site may attract another prospective buyer for your builder. Let him keep a copy of your plans if he likes them.

THE HOME-BUILDING PROCESS

Regardless of what size home you are building or where you build, your project will have certain inevitable characteristics.

Your Project Will Proceed With Decreasing Degrees of Risk

For example, the first steps, such as digging the foundation and pouring footings, will have more impact if done improperly than if a door is hung to swing in the wrong direction. Hence, your utmost care should be taken at the beginning of the project when every step is critical—particularly through the framing stage.

Your Project Will Proceed Erratically

As the result of unforeseen circumstances, many weeks may pass with no progress on the project. Material shortages, poor weather and subcontractor scheduling can all contribute to delays. Then, for no apparent reason, three different subs and all your materials will show up at the site at one time.

You Will Make Changes and Unplanned, Impromptu Decisions

You will probably change the position of a door, window, wall or major appliance. You may put in a dormer, a skylight or a screened-in porch. A prefab fireplace may be substituted for a masonry one at the last minute. There is no limit to the things that can change once a project is started. Don't let this frighten you. Details not necessarily shown on your plans will unfold before your very eyes. These trivial issues are normal and always occur when an idea on paper is converted to reality.

You Will Spend Lots of Money and Use It as a Tool

Until you have paid a sub, you have his undivided attention. If he doesn't fix those squeaky stairs, that leaky pipe or that crooked window, you simply won't pay him. Once you have paid too much, too early, you have killed his incentive to return and you have failed to use a builder's most important tool.

You Will Make the Difference

Small jobs will appear with no one assigned to do them. Replacing a bad piece of plywood or sturdying up the floor with a few nails—the list is endless. You make the difference. You may sacrifice a lazy morning in bed to pick up some brick, test out paint samples or get a free load of slate somewhere. If you are determined to get the job done, you will get up at 5:00 A.M. if that's what it takes. You will crawl in bed some nights aching—but knowing you did what had to be done to make a lasting contribution to a structure that may well last a hundred years.

COMMON PROBLEMS FOR BUILDERS

Being a builder means handling problems. Every project is different and has its peculiar quirks, but there are several problems that occur to some degree on just about all construction projects. Although you may not be able to avoid these problems altogether, it may be helpful for you to know what some of them are now so that you can take precautions to minimize their adverse impact on your project. The most common problems and suggestions for handling them are:

Subs Are Often Late

- Let subs tell you when they'll be there.
- Call the sub the day before he is scheduled to arrive at the site.
- Meet the sub at the site at a certain time.
- Give incentives for early completion. Cash talks.

Subs Don't Show at All

- Let them know up front that you expect prompt attendance and that you will spread the word around if they cancel with no warning.
- Have other subs as backups.

Subs Proceed Slowly

- Give subs a bonus for finishing work on or before schedule. A cash bonus makes a great incentive.

Subs Do Work Incorrectly or Disagree on Work to Be Done

- Ask them to repeat instructions back to you.
- Have detailed specifications and drawings where needed.
- Don't let too much work get done without checking on it. Visit the site daily.
- Discuss detailed drawings with sub and make sure he understands them.
- Have a site phone so that the sub can call you when questions arise.

Dispute Arises Over Amount to Be Paid, Especially When Changes Are Involved

- Put payment amount and payment schedule in writing and have it signed by both parties.
- All changes to the original plans should be written on a change order form and signed by both parties.
- Use lien waivers to control your liability.
- Pay with checks to ensure proper records.

Materials Not Available When Needed

- Order custom pieces early to ensure that they are on-site when needed.
- Confirm delivery dates with suppliers and have them notify you of delays several days in advance.
- Determine which suppliers deliver after hours and on weekends.
- Get friendly with the delivery men and schedulers at your supply houses.

Wrong Material or Wrong Quantity of Material is Delivered

- Use detailed descriptions or part numbers on all purchase order items.
- Print legibly.
- Don't order over the phone unless extremely urgent.
- Have materials delivered early so they can be exchanged or corrected before they are needed.
- Have your subs help you determine actual quantities and materials.
- Get to know your salespeople and work with the same ones every time.

Bad Weather

- Don't begin project during or just before the rainy season.
- Get roof up as soon as possible.
- Make the most of good weather.
- Provide a drive area of #57 crusher run to avoid muddy areas.
- Get a roll of poly (plastic sheet) to cover materials.

Theft and Vandalism

- Visit site each day.
- Call the local police department and have your site put on surveillance.
- Don't always visit site at a routine time. Make unannounced visits, especially near quitting time.
- Don't keep too much material on-site. This is especially true for windows, doors, millwork, good plywood, and expensive tubs and appliances. Some builders even put a 3″ screw in the top corners of a stack of plywood.
- Get builder's risk and theft insurance with a low deductible.
- Install locks as soon as possible and use them.

RULES FOR SUCCESSFUL BUILDERS

- PUT EVERYTHING IN WRITING—*not written, not said.*

- BE PREPARED *before* breaking ground and don't get behind.
- KEEP PROJECT BINDER current with all project details.
- BE FAIR with money. Pay subs what was agreed, when agreed.
- ALWAYS GET REFERENCES and check out three on each major sub. You must see their work.
- DON'T TRY TO DO TOO MUCH of the actual physical work yourself. Your time and skills are often better used managing the job.
- SHOP CAREFULLY for the best material and labor prices. Don't hesitate to insist on builder discounts.
- USE CASH to your maximum advantage.
- RELY ON YOUR CONTRACTORS for advice in their fields. But above all, use common sense.
- DON'T ASSUME anything. Material is not on-site unless you see it. Subs are not on-site unless you see them there. Don't expect subs to know what you want unless it's in writing.
- USE AS FEW SUBS AS POSSIBLE without sacrificing cost or quality.
- USE THE BEST MATERIALS you can afford.
- KEEP CHANGES TO A MINIMUM and budget funds for them carefully.
- ESTIMATE ON THE HIGH SIDE and have a contingency fund.
- KEEP MONEY UNDER CONTROL and track progress vs. your original budget.
- RELY ON YOUR LOCAL BUILDING CODE as a primary reference for specifications, inspection criteria, material guidelines and settling of disputes. Learn to use it comfortably.
- STAY AT LEAST A WEEK OR TWO AHEAD of the game. Staying ahead will allow you time to compensate for material delays or time needed to replace unreliable or unsatisfactory subs.
- USE THE BEST MATERIALS you can afford.
- LET YOUR FINGERS DO THE WALKING. If it looks like rain, call the roofer. If something can't be delivered, advise your subs.

PROVERBS FOR BUILDERS

Below are a few sayings very appropriate to the building industry. As a builder, you are bound to encounter times when one or more of them apply.

MURPHY'S LAW: If anything can go wrong, it will.

If there is a possibility of several things going wrong, the one that will cause the most damage will be the one to go wrong.

If you perceive that there are four possible ways in which a procedure can go wrong, circumvent it, then a fifth way will promptly develop.

It is impossible to make anything foolproof, because fools are so ingenious.

DAVE'S CONSTANT: Matter will be damaged in direct proportion to its value.

If you explain so clearly that nobody can misunderstand, somebody will.

Do not believe in miracles—rely on them.

LESTER'S FIRST LAW: Everything put together falls apart sooner or later.

LESTER'S SECOND LAW: No matter what goes wrong, it will probably look right.

If only one bid can be secured on any project, the price will be unreasonable.

KING'S LAW: No matter how long or how hard you shop for an item, after you've bought it, it will be on sale somewhere, for less.

Everything takes longer than you think.

THE GOLDEN RULE: He who has the gold makes the rules.

THE EIGHTY-EIGHTY RULE OF PROJECT SCHEDULES: The first 20 percent of the task takes 80 percent of the time, and the last 20 percent of the task takes another 80 percent.

Once you have exhausted all possibilities and fail, there will be one solution, simple and obvious, that is highly visible to everyone else.

All delivery promises must be multiplied by a factor of 2.0.

HARRISON'S LAW: The most vital dimension of any plan or drawing stands the greatest chance of being omitted.

COMMONER'S LAW: Nothing ever goes away. When things just can't get any worse, they will. Left to themselves, things tend to go from bad to worse.

Nothing is as easy as it looks.

WHY CONSTRUCTION PROJECTS FAIL

As a builder, it is up to you to make your project a success, but along with all the glory comes all the responsibility! Listed below are a number of common reasons why construction projects fail and/or encounter severe difficulties:

- Costs are underestimated.
- Lack of budget control. Excessive design changes are made late in the game.
- Project manager is not organized, lets things get out of control.
- Lack of understanding of contracting hampers efforts.
- Acts of vandalism, arson or natural catastrophes are not adequately covered by insurance.
- Lack of enthusiasm slows progress.
- Workers are paid before careful inspection is made of work and materials.
- Legal complications develop due to careless handling of business transactions, such as double invoice payment, payment for work uncompleted or undelivered, or payment for damaged or inferior goods.
- Conflicts arise with spouse or partner.

BUILDING FOR YOURSELF VS. BUILDING FOR OTHERS

One of the first questions that any of your workers are going to ask you is whether you are building this home for yourself or to sell. The difference can be subtle, but more likely the difference will change the house dramatically in terms of workmanship and materials used.

Those who build for a living learn quickly that the only way to make a fair profit is to build into a home *only* those features and the level of quality necessary to sell the house—and nothing more. Quality materials such as hardwood floors, marble, custom millwork, brickwork, fancy lighting and fancy bath fixtures are used sparingly or not at all.

However, you may defeat at least part of the purpose of building your own home if you don't outspend the neighborhood builder in a few areas. Perhaps you want a skylight, nice brass faucets, fancy wallpaper or better grade appliances. This is your prerogative. You should add a few special touches that will make your house stand out and easier to sell.

WHEN YOU GET NERVOUS

Nervous because you've never done anything so ambitious? Nervous because the carpenter hasn't returned your call yet? Or maybe you just can't put a finger on the source of your uneasiness. Relax.

Remember:

Fear is a natural response to the uncertain and the unknown. A goal of this book is to replace uncertainty with clear, concise knowledge—to remove those cloudy uncertainties. Some individuals maintain a certain level of fear as a natural defense mechanism—to always remain cautious and alert to the situation and possible problems.

Individuals with little or no formal education build houses every day—and they do it for a living!

A certain degree of fear is typical when doing anything for the first time. Chances are, after you have built your first home, you'll want to do it again!

Your construction project is staffed by knowledgeable, professional subcontractors, each specializing in his area. It is their job to do the work properly. Don't worry if you don't know everything they do. That is not a requirement of a builder. If the job is not done right, you can refuse to pay until the items in question are fixed.

The construction approach presented in this book has been time tested. Many homes have been built using information in this book. Using this book puts you way ahead of many individuals who may either try to brave it alone or otherwise push ahead without adequate knowledge and/or preparation.

Assure yourself that you are doing everything you can to protect your project against the pitfalls described in the Why Construction Projects Fail section.

Assure yourself that you are using the steps, specifications and inspection checklists provided.

WHEN YOU GET DISCOURAGED

You're into the project and things aren't going along perfectly. Your framing crew made some mistakes, you've had four straight days of rain, your roofing sub won't return your call and the price just went up on storm windows. So what else is new? Getting discouraged?

Did you ever hear of any project proceeding from beginning to end without a hitch or two—or twenty? Below are a few things to ponder when you wonder why you didn't just throw this book away and cut your losses short.

LOOK at the money you're saving. You probably have a good idea how much. Just look at that figure for a moment. That's the money you would have paid someone else to do the work you're doing right now. If someone else were performing your job, they'd run into the same kinds of problems. When the interest costs are figured in, you will probably be saving one or more year's salary by contracting yourself. Aren't a few months of challenge worth it?

LOOK at all the problems and solutions you're encountering. One day, when your home is completed, you'll look back at them and laugh. You'll tell of some of your experiences for the rest of your life.

LOOK at your accomplishment! Your friends, associates and family look upon you with admiration. You can honestly say that you built your home. Only you will know how much physical work and skill you didn't put in.

REMEMBER—bad days are always followed by better ones. Hang in there!

Legal and Financial Information

FINANCIAL ISSUES

Obtaining financing for your home will be one of the most critical and demanding phases of your building project. You must be bookkeeper, financier and salesman. The lender will want evidence that you know what you are doing and that you can finish your project within budget.

As the contractor and future owner of your home, you will be dealing with two loans: the construction loan and the home mortgage. Each one has different requirements, lending sources and durations. You will be wearing two different hats when you visit the bank. If you are lucky, you will be able to obtain both loans from the same bank and the same loan officer. If not, you may have to work with two different loan officers. Many mortgage companies do not handle construction loans and vice versa.

The first loan that must be addressed is the home mortgage. You must qualify for long-term financing before any bank will consider a construction loan. When choosing the appropriate plan for your dream home, you calculated the maximum loan you could afford. To satisfy the loan officer, you must go into greater detail.

THE HOME MORTGAGE

Prequalification

Prequalification enables you and the lender to calculate how much home you can afford to build. This step takes the guesswork out of the whole process of knowing what you can afford, giving you the knowledge of what your construction budget can be. The lender will have you fill out a Mortgage Credit Analysis Worksheet to calculate your purchasing power and your creditworthiness. The lender will depend on certain formulas to calculate the amount of loan you can afford:

- Mortgage Payment-to-Income Ratio (MR). The ratio of your proposed PITI (Principal, Interest, Taxes and Insurance) payment to your gross monthly income. This varies from 28 to 30 percent, depending on the lender.
- Total Debt-to-Income Ratio (DR). The ratio of your total monthly debt payments to your gross monthly income. This varies from 36 to 41 percent.
- Loan-to-Value (LTV). The ratio of the total value of the house to the loan amount. Most conventional loans require an LTV of 90 percent or less. This means that the loan you request is less than 90 percent of the appraised value of the house. Exceptions to this include government guaranteed loans (covered later).

The lender may use one or a combination of these formulas to determine a maximum monthly payment amount. Here's how it usually works:

- The lender will calculate the *MR* first. This will determine the absolute maximum house payment you can make.
- The lender will then add in your other debt obligations and calculate the *DR*. If it is less than the *DR* maximum ratio, then your maximum payment will be the *MR* amount. If it exceeds the maximum *DR* percentage allowed, this will drive down the maximum PITI payment you can make. By eliminating some of your other debt payments (by paying off debts), you may be able to push the maximum back up to the *MR* ratio.
- Once your maximum payment is known, the lender will calculate the total loan amount that will generate this monthly payment. This amount will vary according to the type of loan and interest rate you choose: fixed rate, adjustable, graduated payment, etc.
- The resulting loan amount cannot exceed 90 percent of the house's value.

However, in the case of a homeowner building his or her own home, the bank may require a larger down payment. A lender will be nervous about your ability to complete the project. If you contribute more money to the project, the lender will feel more comfortable. If your handiwork falls short of producing a home with the expected appraisal value, the bank will have less of its money on the line. Your challenge will be to convince the lender that you can finish the job on time and on budget. This shouldn't be too much of a problem for you because of sweat equity.

Sweat Equity

This refers to the labor "sweat" and materials you contribute to the value of your house. Most lenders will consider sweat equity as part of your down payment. As the contractor for your house, this becomes your secret weapon. It allows you to build a home with a value much higher than the loan amount, and also allows you to have a much lower Loan-to-Value (LTV) ratio. Most lenders will give you credit for value added through sweat equity.

Improving Your Chances

It's unfortunate that the banks seem to focus in on one item: your monthly debt payments vs. income. They don't consider your assets or savings to be quite as important. They are more interested in your credit payment history and your current debt obligations. There are several things you can do to improve your "score on paper" if you have additional funds:

- Pay off as many credit card debts as possible. This will incur no prepayment penalties and will reduce your monthly debt payments.
- Try to keep your debt payments timely. Most credit companies will report a late payment to a credit bureau after it is more than one month behind.
- Obtain a credit report on yourself before visiting the lender. These usually cost $15 to $20 and are available from all the credit reporting companies. (See the names and addresses of credit bureaus below.) By studying your own credit history first, you can make adjustments or file a letter of explanation for certain problem areas. This letter will then be available to any bank that requests the report.
- If you've had payment troubles with one credit company in particular, pay off that balance completely. This will show your determination to fix the problem permanently.

Obviously, fixing prior credit problems, buying a house and paying for construction all require cash, which you may not have available. To make matters worse, most loans will not allow you to use borrowed funds to make up your down payment. But you *can* use money from friends and relatives to pay off credit debts and pay construction costs.

Credit Bureaus

TRW Complimentary Credit Report Request
P.O. Box 2350
Chatsworth, CA 91313

Equifax Information Service Center
P.O. Box 740241
Atlanta, GA 30374

Trans Union Consumer Relations
P.O. Box 390
Springfield, PA 19064

How Flexible Can the Lender Be

Unfortunately, not very. His lending sources (whether government or independent) will have stringent requirements that must be followed in order to qualify. The lender will have very little leeway in making judgment calls. That's why you'll find most lenders focused on the "paper details" instead of common sense. Use this knowledge to your advantage. Make sure you qualify on paper before contacting the lender. With his requirements met, the lender will be free to help you structure a workable lending package.

Applying for the Mortgage

This mainly involves filling out lots of forms. Most lenders use a generic form called the Uniform Residential Loan Application. Make sure you bring all your background information (listed below) with you to expedite the process.

Things to Bring

- Bank and credit statements
- Current employer information: address, your supervisor, your title and length of employment. A letter from your employer verifying these facts is

useful. If you are self-employed, you will need two or three years of professionally prepared financial statements or tax returns for your business.

- Neighborhood or Homeowners Association covenants and bylaws
- Information on investments and pension plans
- List of other assets; you may be asked to provide verification.
- Social Security number
- Two years of tax returns and W-2s.

Additional Loan Costs

There are numerous costs and fees involved in originating and processing the loan. You need to add these costs to your down payment and points when calculating your monetary limits. RESPA (Real Estate Settlement Procedures Act) is a government law that requires the lender to provide you with an accurate estimate of these costs. He will also give you a copy of the government's *Settlement Cost* booklet, which explains your rights. In most cases, the lender will overestimate, just to be safe. He would rather you get a pleasant surprise at closing time.

Loan Costs

- Cancellation fee
- City/county taxes and stamps
- Credit report
- Discount points
- Escrow fees
- Flood report
- Loan origination fee
- Loan processing fee
- Notary fees
- Plan appraisal
- Recording fees
- Title insurance
- Title search

Prepaid Reserve Fees

- Homeowner's insurance
- Mortgage insurance
- Property taxes

Things Lenders Like To See

- A good credit history—no bankruptcies, foreclosures or late payments
- A healthy net worth
- A large down payment or substantial equity in the project; the more you have invested, the less likely you are to default on the mortgage
- A standard or popular house plan; they want to know that the house will sell easily in the future
- Free and clear title to the building lot with no loans due—lenders don't like to be second in line for their money
- Minimal debt payments; this shows a responsible attitude
- No outstanding lawsuits or claims—a divorce can complicate loan approval
- Stable employment: Most banks want to see two to three years of steady employment
- Two-paycheck households
- Your money—they don't want to see borrowed money used for your down payment

What the Lender Will Do

Evaluate You

In order to do this, he will:

- Perform a credit check, obtaining a credit report. This will show all loans you have opened and paid off, and any adverse payment history, such as delinquent and no-payment situations.
- Perform a Loan Analysis, looking at your monthly income and obligations.
- Submit a Verification of Employment to your employer, confirming that you really work there at the job you stated in the application.
- Submit a Request for Verification of Loan to lenders of any loan you may have.
- Submit a Request for Verification of Rent or Mortgage Account to lenders or landlords to assess your previous payment history.

Evaluate the Home You Plan to Build

- Evaluate your blueprints and your construction estimate to ensure that the budget is reasonable.
- Consider your experience and education and ability to complete the project. Some lenders like it when you hire a contractor part time to help you with the project.

LOAN TYPES

Fixed-Rate Mortgage

This is your most desirable loan, if you can afford the payments. The interest rate and the monthly payment stay the same for the life of the loan. If you take out the loan when interest rates are low, you have "inflation insurance." This is especially desirable if you plan to stay in the home for more than seven years.

Adjustable-Rate Mortgage (ARM)

This mortgage has lower interest rates and payments because the lender can raise the interest rate if rates skyrocket. Since the bank can adjust to changing times, they are more willing to offer a lower rate initially. Fixed-rate loans force the lender to add "insurance" points to the loan interest rate. The monthly payment of an ARM is tied to some form of index, such as the prime rate, one-year treasury bills (T-bills) or some other standardized cost of funds. Most programs have caps on the maximum amount of interest increase allowed over the life of the loan. Payments may be adjusted quarterly, semiannually or annually.

Graduated Payment Mortgage (GPM)

This mortgage is designed to allow first-time homebuyers to qualify for loans that would be too expensive as a conventional loan. The first few years of loan payments will be lower than the remaining term of the loan. The idea is to get the homeowner into the home with an acceptable payment amount. Payments go up only after the homeowner is settled and more able to pay the larger payment. You won't get something for nothing though. During the low-payment period, the principal of your loan actually goes *up*, not down (negative amortization). You will owe more money after three years than you did when you bought the home! This would not be very desirable in most circumstances, but for a home contractor, it might not be so bad. With the money you save building your own home, your house should be worth more than the loan amount virtually the day you move in. Even with an increased principal, you come out ahead. There are several other permutations of this loan. An FHA 245 loan is an example. Ask your lender to describe the different loan options.

Mortgage Loan Agencies

FNMA (Fannie Mae)

The FNMA (Federal National Mortgage Association) provides funds for FHA and VA loan guarantee programs and conventional loans. The largest single source of mortgage funds, they underwrite a large portion of the conventional loans that are provided by mortgage bankers. As secondary lenders, they purchase primary loans from other banking institutions. The actual funds come from private sources, mainly the sale of securities. Many of the loan requirements of mortgage companies are dictated by Fannie Mae rules. They publish a set of requirements for the loans they purchase. Most bankers want to make sure their loans meet these requirements so they have the option to sell their mortgages to Fannie Mae.

FHLMC (Freddie Mac)

The FHLMC (Federal Home Loan Mortgage Corporation), like Fannie Mae, purchases loans for FHA, VA and conventional loans.

GNMA (Ginnie Mae)

The GNMA (Government National Mortgage Association) deals exclusively with FHA and VA loans.

FHA (Federal Housing Administration)

An FHA loan is not really a loan, but instead a loan guarantee. The FHA insures the loan for the lender that actually provides the funds. The FHA is now a sub-agency of the Department of Housing and Urban Development. An FHA loan is designed to be "homeowner friendly." The terms are very liberal. You can choose from a variety of fixed- or adjustable-rate loan packages.

FHA Advantages

- It allows the buyer to finance up to 97 percent of the first $25,000 and 95 percent of the remaining loan amount, resulting in a much lower down payment.

- The down payment and closing costs can be a gift from relatives.
- The loan is assumable (if the new buyer qualifies).
- Some of the sale price of the house can be used to pay closing costs.
- There is no prepayment penalty.
- The Mortgage Payment-to-Income Ratio (MR) requirement is 29 percent.
- The Total Debt-to-Income Ratio (DR) is 41 percent.

Sound too good to be true? So, what's the catch? Homebuyers find out quickly that many builders hate FHA loans with a passion. Why? Since the FHA is guaranteeing the loan, the agency requires a stringent appraisal inspection. You will be required to fill out yet another set of forms: a complete description of materials. An FHA inspector will want to visit the building site several times during the project. Coordinating with both the municipal and FHA inspectors can be a real pain for builders who must finish their houses quickly. The home contractor, however, may find it to be a minor inconvenience. FHA building standards are higher than most building codes. You may want to build to this higher standard anyway. After all, the higher standards are an attempt to improve the value of the house. And you can't beat the loan terms!

FHA Disadvantages

- The seller is usually required to pay points and closing costs. Since a home contractor is both the seller and buyer, this isn't a problem.
- The mortgage insurance premium is higher than other loans.
- There is a 1 percent loan origination fee.
- Appraisals and inspections are more stringent than for conventional loans.

VA (Veterans Administration)

Like an FHA loan, the VA is a government loan guarantee that provides 100 percent financing to qualified veterans, National Guardsmen and reservists. They guarantee the loans provided by other lending sources. Like the FHA, VA loans are fixed-rate and assumable. The assumption policy is a little tricky, though. If a nonveteran assumes the loan, the veteran loses his eligibility for another VA loan until the assumed loan has been completely paid off. If another qualified veteran assumes the loan, the original mortgage holder can immediately qualify for a new VA loan. There is no down payment requirement. Obviously, this loan is not available to the nonveteran public.

THE CONSTRUCTION LOAN

Unless you are one of the few who build with cash, the construction loan is a necessity. It is typical for interest on a construction loan to be a few points higher than on a permanent loan. This gives you further incentive to finish the project as quickly as possible. Even though you have jumped through all the hoops for your mortgage, you'll have to do it all over again for the construction loan—unless you get a construction-permanent loan (see below).

WARNING: When you apply for your loan, apply for all the money you will need—and then some. Don't plan on getting extra money later. You are headed for a disaster of mammoth proportions if you run out of money before your project is over. Your lender will likely foreclose on your project and you'll end up with next to nothing. This also means that once you've done your take-off and are into construction, don't make expensive changes and upgrades unless you can pay for them out-of-pocket.

The Financial Package—Applying for Your Loan

Make sure that your financial package is complete before visiting the lender. As with the mortgage, the construction package should be neat, informative and well organized, and should comply with all of the lender's requirements. Your lender will want to be sure that you can keep good records. In addition to your mortgage information, you should include the following:

- Resume or personal biography
- Total cost estimate of the project
- Description of materials

- Survey of the proposed building site showing location of dwelling
- Set of blueprints

THE RESUME is designed to sell yourself to the lender. Start with a standard job resume, but be sure to emphasize any construction interests or experience. List any training courses you have taken (or even this book).

THE COST ESTIMATE will be very important, as your construction loan will be based on it. Make sure to include a 7 percent to 10 percent cost overrun factor in your estimates as a cushion for unexpected expenses. Many lenders can provide you with a standard cost sheet to fill out. If not, use the one included in this book.

THE DESCRIPTION OF MATERIALS form will aid the lender's appraiser in estimating the appraised value of your property. Many lenders use the standard FHA form for this purpose, but check to see if your lender uses its own special form.

THE SURVEY should be obtained from the seller of the property or from the county tax office. If the property is very old, you may have to request that a new survey be done. Indicate on a copy of the survey the approximate location of the house.

A SET OF BLUEPRINTS complete with all intended changes should be submitted. Give your lender the set without the coffee stains.

START-UP MONEY

You will need to put up a lot of your own money before you can expect to receive your first draw. Typically your initial outlay will include:

- Lot cost (to be paid before the construction begins)
- Closing costs (construction loan closing)
- Permits (building, water, sewer, etc.)
- Site preparation (clearing, grading and excavation)
- Foundation

If you can't do this out of your pocket before going to the lender, you aren't ready to begin building.

About Draws

Unlike a car loan, where you receive all your cash at once, construction loans are almost always handled on a draw basis. The lender will make you earn the money that you borrow—bit by bit. Each time the lender sees that you have completed, say, 5 percent of the project, you are allowed a "draw" of 5 percent of the construction loan. This nuisance is, in fact, a way of making sure that money isn't wasted carelessly up front. Whenever you would like a loan draw, you must contact the lender and request one. The lender has an inspector who will visit the building site to see what work has been completed. The lender will normally have a fixed draw schedule that allows a certain amount of money to be drawn for each major phase of construction. Many lenders inspect on Thursdays and distribute draws on Fridays. Builders are usually allowed only five draws free of charge. If you desire more draws than allowed per project, you are usually charged a nominal fee per draw.

Money is drawn after proof is shown that such money is deserved. Hence, you must go into the project with some start-up money of your own. The need for this "pocket money" can't be underestimated. It will be used to buy permits and licenses, dig your foundation and perhaps get your foundation built before money is drawn. It will also be used to cover unexpected expenses, such as price increases, last-minute additions you may want to make and underestimates. Be aware that draw schedules are normally back ended, meaning that you will spend more than you can draw up front, while receiving a 5 percent draw for painting your mailbox near the end.

The Conventional Construction Loan

These loans allow 80 to 90 percent LTV on fixed- or adjustable-rate loans. Borrower must have a minimum of 10 percent down payment of the combined land and construction costs. A portion of the loan may be used for land payoff/purchase. The time allowed for construction can range from 6 to 12 months. You may obtain a short-term construction-only loan or combine the construction and permanent loans together. Construction loans generally have a slightly higher interest rate and fees.

The Construction Permanent Loan

When possible, apply for a Construction Permanent Loan (also known as a convertible loan or rollover loan). This type of loan eliminates one of two closing costs related to separate loans for construction and permanent financing. This is done by putting both the construction loan and the permanent loan in your name. When the permanent loan is ready to close, the construction loan is converted to a permanent mortgage. Shop for loans carefully. Time here is well spent. Many institutions now offer 15-year payoffs that only cost a little extra each month over a standard 30-year loan. Charging points, a lump-sum cost when you originate a loan, is only a way for lenders to get a little bit more money out of you without raising the interest rate.

Dealing With the Construction Lender

When going to the lender, be prepared:

- Have your financial package ready.
- Know exactly how much money you need.
- Have all requested information typed.
- Dress up (preferably in a dark suit).
- Show up at the scheduled time.
- Don't mention how much pocket money you have on the side—the lender will want you to put it into the construction loan.
- Don't talk too much. Answer questions only when asked, and don't volunteer any information not asked for.
- Have a list of your subs and backups.
- Have all lender forms typed.

The Appraisal

The lending institution will have its capital at risk. As such, it will want to be assured that the house you are planning to build is worth more than the construction loan. The lender will also test your skill at estimating costs.

The lender will appraise your home in one of two ways; the Market Approach (most likely) or the Cost Approach.

MARKET APPROACH—This method of appraisal bases the value of the home on "comparables" in the area. The appraiser will look for houses comparable to yours in the surrounding neighborhood that have similar features, style, square footage and property values. Therefore, it is advantageous to build your house near a neighborhood with homes similar to but higher in price than yours. Don't get too far-fetched with your design. The appraiser may not be able to find suitable comparables and may have to guess at the value—usually to your disadvantage.

COST APPROACH—This is similar to the cost take-off, or a determination of value by square footage.

After the value is determined, the lender will then loan you a fixed percentage of the appraised value, usually up to 70 or 80 percent, to build the house. Very seldom will the lender give you a loan based totally on what you claim your construction costs to be. However, you should easily be able to build a house for less than 75 percent of its appraised value.

FHA Appraisal

This appraisal is completely different from a bank appraisal. Banks usually hire independent appraisers who follow the above guidelines. When the FHA is guaranteeing the loan, it will send its own inspector/appraiser to the site. Unlike the other appraisers, the FHA inspector will concentrate on structural construction details as well as market value. You should expect a few items on his list that must be corrected before the loan can close.

THE LEGAL ISSUES

Real estate laws have evolved over thousands of years. Needless to say, these issues are complex and are best left to the experts. Make sure to consult the appropriate authorities about any legal issue in order to avoid unpleasant surprises. The professionals you will need to deal with are attorneys, insurance agents and financial lenders. The key legal issues affecting your building project are:

- Filing in the public record
- Free and clear title to the property
- Land survey
- Easements
- Covenants
- Zoning
- Variances
- Liens
- Subcontractor contracts

- Builder's risk insurance
- Building permits
- Worker's compensation

FILING IN THE PUBLIC RECORD—Because of the importance and complexity of real estate law, all real estate transactions concerning the transfer of property, liens and loans are normally filed in the public records at your county's local courthouse. This way you are announcing to the public what property you own. Anyone has access to these records and may research the history of your property to see who owned it before you and if there are any mortgages, judgments or liens on it. This is for your protection, because you can do the same research on any property you intend to buy. Go to your local courthouse and familiarize yourself with the records. Ask one of the title lawyers in the room (there are always a few attorneys there) to help you learn how to search the records. This will give you an idea of the process of real estate transactions.

FREE AND CLEAR TITLE concerns your bundle of rights to a piece of property, whether it is land or improved property. Before closing on a property, your lawyer will do a title search of the property in the public record to determine its ownership history. Most real estate records are kept at the county courthouse and are accessible to anyone. If the lawyer finds someone with an "interest" in the property or a confusion over rights, this will create a "cloud on the title" which must be removed before title can transfer. A title search is required by all lenders and is usually done for a standard fee. *Never buy a piece of property without doing a title search!* Many so-called "bargains" are properties with title problems that the present owner is hoping to "pawn off" on some unsuspecting buyer. Don't be that buyer! You should also get title insurance, protecting your claim to the land.

A LAND SURVEY is another required item that must be done before title can be transferred. This survey can usually be arranged by your lawyer or financial lender. The survey will describe the exact location of your land in a written legal description and scale drawing of the property called a "plat map." These descriptions will be attached to the deed and filed in the public record.

EASEMENTS on the property will normally be discovered during the survey and title search. Easements give rights of traverse on your property to others, such as local governments (sidewalk easement), utility companies (sewer or electric company easements) or individuals. Make sure to check the location of your construction in relation to the easements to avoid any conflicts.

COVENANTS are building restrictions or requirements placed on construction in certain subdivisions, such as minimum square footage of the building or the type and style of construction. These restrictions are usually placed on the property by the developer of the subdivision in order to protect the value of the homes in the subdivision. Ask your lawyer to check for any building restrictions in the public record before buying a building lot. You may not be able to build the house of your dreams if the design is not allowed in the neighborhood! Watch out for any restriction stating that the home can only be built by a specific set of builders or in a specific time frame.

ZONING RESTRICTIONS, like covenants, control use of land in an area and the type of structures built. These restrictions are placed on property by the local government planning board in order to protect the value of land in the area. Check with your local planning office if you are unsure of the zoning of your property. A zoning map of the county can usually be purchased for a nominal fee and can be very helpful in searching for property. Zoning laws also define the area of the lot that the structure can actually occupy. The builder is required to place the structure a certain minimum distance from neighboring homes. This requirement is called the minimum setback. In many planned subdivisions, these boundaries are closely adhered to in order to make the homes in the area appear consistent and planned. You may also have requirements for a backyard of a certain depth and a minimum side yard of 20′. It is important to find out where your baselines are before you break ground. If you build outside a baseline and are caught, you may be in for a hefty fine.

VARIANCES allow you to "vary" from standard state and local zoning ordinances within prescribed, approved limits. For example, if you want to build your home 30′ from the front curb and the local zoning ordinance requires a 40′ setback, you must apply for a 10′ setback variance. Unless your request adversely affects neighborhood appearance or safety, your variance will usually get approval. To obtain a variance, you must normally pay an application fee and go before a panel of county zoning officials at a variance hearing, where you or your representative describe or illustrate your intentions and the reasons for them. Often, a notice of your variance request will be posted on your lot so that the public is aware of your intent to get a variance. They have a right to appear at the variance hearing to either support or fight against your variance.

LIENS can become extremely important to a homebuilder because of the effect they can have on the construction project and the subsequent closing of the property. A lien places an interest in a property that will create a cloud on the title, preventing the close of the sale until resolved. For example, if the plumbing contractor does not feel that he was fully paid for his work, he may file a lien in the public record on the property involved. This lien can stay attached to the property until the disagreement is resolved. Many states do not require that the owner be notified of the lien, but you will find out about it at closing! The lien will prevent selling or mortgaging of the property until the disagreement is resolved.

To avoid the pitfalls of liens to potential homebuyers, many states now require contractors to sign an affidavit guaranteeing to the purchaser that the contractor has paid all his bills due to suppliers and subcontractors. If any parties then file a lien, the contractor can become criminally or legally liable not only to the lien holder, but also to the purchaser. Builders or subs can be thrown in jail in many states for this offense!

If you are building for yourself or others, you can protect yourself from liens by requiring your suppliers and subcontractors to sign a subcontractor's affidavit. This agreement works like the contractor's affidavit. The subcontractor agrees that all bills have been paid and that the contractor has no legal claim against your property. This will prevent the subcontractor from filing any liens on the property. For your own protection, make everyone you deal with sign this agreement!

SUBCONTRACTOR CONTRACTS, like affidavits, protect you from surprises. Get everything important in writing, no matter how good your sub's memory or integrity. Subs may have contract amnesia if a problem comes up. Make sure to include the specifications for the work done, as well as a schedule of how payment will be made. It is wise to use a statement such as "Final payment will be made when work is satisfactorily completed."

BUILDER'S RISK INSURANCE is another "ounce of prevention" necessary before starting your project. This insurance will protect against liability if someone should be injured on the building site. Theft of building materials is sometimes covered also. Risk insurance may only cover theft after the dwelling can be locked. Be aware of very high deductibles.

BUILDING PERMITS must be obtained from your County Building Inspector's office before construction can begin. This will notify the city or county of your project so that the building inspector can schedule the required inspections of the property during construction. This permit must be displayed in a prominent location at the site. Check your local county for details. Normally, building permits cost a certain amount per $1,000 of construction value or per square foot. Obviously, this value is somewhat intangible. You may have to accept the authority's estimated value.

WORKER'S COMPENSATION is required in every state to provide workers with hospitalization insurance for job-related injuries. This accident insurance must be taken out by any person or company with hired employees. Most states also have special statutes that require builders, even when using subcontractors, to have this insurance protection for anyone working on the building site. At the end of each year, the insurance company does an audit to determine the amount of insurance to be paid. Each building trade is assessed an insurance rate depending on the relative risk involved in the trade. For instance, roofing trades will be charged a higher rate than trim carpenters. Since you are providing insurance to cover these workers on your site, make sure to deduct the amount of the insurance from the price paid to the subcontractor.

You are liable for any work-related injuries that occur on your property. Without this policy, a negligence lawsuit could take your life savings. Look for any situations that can void worker's compensation coverage. For example, in some states, drywall subs experiencing accidents while using drywall stilts are not covered. Better safe than sorry!

Your Lot

The lot you purchase will be a part of the home. It's a location and setting you had better be comfortable with. There's an old saying about the three most important factors when buying a lot: location, location and location. While this stretches the truth, it does make an important point.

CONSIDERATIONS

If you have not purchased your lot yet, you may wish to ask yourself the following questions:

Would you prefer to build in a planned development (subdivision), where home values will be more stable and controlled, or on a one-of-a-kind lot? If you are purchasing a lot in a subdivision, make sure that the building covenants work for you, not against you. Do the covenants allow you to build the home you want? When you want? Do the covenants allow you to subcontract the home yourself? Do they protect against low-priced homes which could affect your resale value? Don't build in an area where homes are more than 15 or 20 percent lower in value than your planned home.

Would you prefer a large lot, which may require lots of time and money in maintenance, or a smaller one that may cost less and involve less maintenance?

Are the lot and surrounding area suitable for the type of home you are planning to build? Consider this from both a physical and an esthetic point of view.

Will the lot support a basement, slab or crawl space if you plan to have one? A better location is almost always worth spending a little more money.

ADVICE

Don't attempt to save money by buying a less desirable lot. Curb appeal can affect the value of your home by 25 percent or more. Saving $2,000 in lot costs could cost you $10,000 in home resale value. (Don't step over a dollar to pick up a dime!) Don't overbuild for the area; build in an area of comparable value. Make sure that the lot and the neighboring homes are *worthy* of the home you plan to build.

Obtain a title search of the property before purchasing it to ensure free and clear title to the land. Make sure that there is public sewer access on the lot if the seller says it is available. Don't take his word for it.

It is wise to first determine the style and size home you plan to build before looking at lots. It would be a shame to find out that, after considering baselines, the home you want to build won't even fit on the lot. Also, you may want to build near homes of comparable value. To do this, you need to know what size home to build. This is the subject of the next chapter.

LOT EVALUATION

Review the following checklists of strengths and weaknesses, although many of them may already be obvious to you. Each weakness provides grounds for negotiating a lower offering price. Once you own the land, each weakness is your problem.

When evaluating a lot, you must walk the lot to check for soft spots, drain pipes, lay of the land, etc. Realize that all the trees on a lot may well disappear after the foundation area is cleared.

Corner Lots—Good or Bad?

Although some homeowners enjoy corner lots, they are generally not preferred by speculative homebuilders because they lack backyard privacy and often have a front setback, even on the side. The nonsquare shape complicates the design and placement of the house.

STRENGTHS: CHECKLIST

- ☐ Lot slopes to allow for basement (if desired).
- ☐ Lot has rear and side privacy.
- ☐ On sewer line. Check height of sewer line. Your lowest drain pipe should be slightly higher than the sewer line.
- ☐ On cul-de-sac or in other low-traffic area.
- ☐ Well-shaped square, rectangular or otherwise highly usable.
- ☐ Trees and woods—hardwoods and some larger trees.
- ☐ Good, firm soil base; important for a solid foundation.
- ☐ No large subterranean rocks that will require blasting.
- ☐ Not in flood plain.
- ☐ Safe, quiet atmosphere.
- ☐ Attractive surroundings, such as comparable or finer homes and attractive landscaping.
- ☐ Level or gradual slope up from road.
- ☐ Good drainage.
- ☐ Lot is located in a stable, respected neighborhood.
- ☐ Just outside of city limits. You may pay less in local taxes.
- ☐ Easy access to major roadways, highways and thoroughfares.
- ☐ Convenient to good schools, shopping centers, parks, swimming/tennis, firehouse and other desirable amenities.
- ☐ Situated in an area of active growth.
- ☐ Provides attractive view from dwelling site.
- ☐ Fully usable—no ditches, ruts or irregular surfaces.
- ☐ Seller will finance partially or in full.
- ☐ In an area zoned exclusively for single family dwellings.
- ☐ Underground utilities—phone, electrical, cable TV.
- ☐ Natural features in area will remain undisturbed.
- ☐ Private backyard area. If there are no trees to block the neighbor's house, check for room to plant trees or shrubs.

WEAKNESSES: CHECKLIST

- ☐ No sewer—you must install septic tank and must pass a percolation test.
- ☐ On or very near a major thoroughfare with yellow or white lines down the center of the road and a speed limit over 35 mph.
- ☐ Odd-shaped (too narrow or too shallow). Either will cause limited front or side yards.
- ☐ No trees or is barren.
- ☐ Rocky—you may have large rocks to excavate or dynamite.
- ☐ Sandy soil may require special foundation work and may not support a large foundation. Excavation is difficult.
- ☐ In flood plain. This normally involves expensive flood insurance and difficult resale. (Check local county plats).
- ☐ Near airport, railroad tracks, landfill, exposed electric power facilities, swamps, cliffs or other hazardous area, or near large power easements, commercial properties or radio towers.
- ☐ Slopes down from road and/or steep yard is hard to mow or presents difficult access by car when iced over. Water drains toward home. May require excessive or unusual excavation and/or fill dirt.
- ☐ Flat yard. Less than 2 percent slope may be difficult to drain.
- ☐ Water collects in spots. This may require expensive drainage landscaping and/or fill dirt, and may result in a wet/damp basement.
- ☐ Lot is located in or near unstable/declining neighborhood.
- ☐ Just inside city limits. You may have to pay more local tax.

- ☐ Relatively inaccessible by major roadways.
- ☐ Isolated from desirable facilities, such as shopping centers, swimming/tennis, parks, etc.
- ☐ Lot provides unattractive views, either from prospective dwelling site or from road.
- ☐ Shallow lots leave a small area for the backyard.
- ☐ Creek, gully or deep valley runs near center of lot.
- ☐ Area zoned commercial, duplex, quadruplex or other high-density, multifamily dwelling.
- ☐ Area is populated by less expensive homes.
- ☐ Utilities are above ground or not yet established, or there are unsightly electrical wires.

FINDING A LOT

Finding the right lot may involve approaching lots from any of the sources below and others as well:

- Local real estate agents and/or multiple listing services
- "For sale" ads in newspapers
- "For sale" signs on land
- Legal plats in county courthouses or other local government offices

PURCHASING THE LOT

Review the considerations and the checklists on the previous pages. Check for problems and weaknesses. *REMEMBER:* Once you buy it, it's *your* problem!

Ask the seller to provide a legal description of the lot, prepared by a registered surveyor, showing all legal easements and baselines (you may not be able to build just anywhere on the lot). Negotiate legal and other closing costs to be paid by or split with the seller. If you must pay all or part of the legal fees, *insist* on appointing an attorney of *your* choice.

Cash talks—especially when purchasing from an individual rather than an organization. Look for owner financing.

Insist on and pay for a title search and title insurance. Title insurance is important if there is ever a dispute over land ownership. Litigation could cost you lots of time and money. When applying for a construction loan at a local lending institution, you must have "free and clear title" to the land. This ensures that no dispute over title will interfere with the successful completion of a construction project. All financial institutions insist on you having free and clear title before negotiating a loan.

Planning Your House Design

Before you can build or purchase your dream home you must determine what features you can realistically afford. Many future homeowners start out by designing a dream home with all the features they have ever wanted. Unfortunately, reality tends to get in the way of many people's dreams. Before you spend valuable dollars on blueprints and architects' fees, determine the exact amount of home you can afford to purchase.

HOW MUCH HOME CAN YOU AFFORD?

The standard rule of thumb used by most banks is that your principal and interest (P&I) payments should not exceed 25 percent of your gross annual income. With interest rates fluctuating, many banks will now allow up to 35 percent of your income. To find the total amount of home you can afford, use the following equations. Look up the mortgage amount in Table 1 to determine the total amount of mortgage your monthly salary will allow you to cover financially.

Example

John earns $37,000 per year and has just bought a subdivision lot for $12,000. How much can he spend on house construction and still meet the bank's loan requirements? The bank is willing to finance 80 percent of the value of the house. The interest rate is 12 percent and the mortgage term is 30 years.

Since the bank will only approve a mortgage payment of about 25 percent of John's monthly income, divide John's yearly income by 12 to obtain the monthly income and then divide it by 4 (25 percent). This will give you the total payment amount available to put toward monthly mortgage payments.

$$\frac{\textit{Monthly Gross Income}}{4} = \textit{Monthly Mortgage Payment}$$

$$\frac{\$37{,}000}{12} = \frac{\$3{,}083.33}{4} = \$770.83$$

Look up the payment amount in Table 1 for a 30-year loan at 12 percent interest to determine the amount of the mortgage your monthly payments will cover. The table gives figures on a $100 per month payment. Multiply this by your available monthly payment to obtain the total mortgage amount.

$$\frac{\$770.83}{100} = 7.71 \times \$9{,}722 = \$74{,}956$$

Since the bank will finance this much mortgage at 80 percent, the total price of the house will be the mortgage amount plus the 20 percent down payment John must contribute to the purchase. Work backwards to obtain the total house value.

$$\frac{\textit{Mortgage Amount}}{\textit{Percentage Financing}} = \textit{Total Value of House}$$

$$\frac{\$74{,}956}{80\%} = \$93{,}700$$

Many times you will be able to secure 100 percent financing on your own construction once you have free and clear title to the land. In this example, however, we will use 80 percent.

Now take the total value of the house and subtract the cost of the land, construction loan cost and closing costs to determine the total amount of money you can afford to spend on the construction of your home. As a rule of thumb, the minimum cost of your home, excluding land, will be roughly $40 per square foot of heated space (excluding garage) if you contract the work yourself. For expensive brickwork, appliances, cabinets, etc., costs can exceed $55 per square foot. You should check with local building associations to find out the typical cost per square foot in your area.

Total cost of home	$93,700
Land cost	–12,000

Construction interest	–2,500
Closing costs	–1,800
Construction cost	$77,400

This amount should be the upper limit that you are willing to spend to secure a home of your own in this example. Next, determine what it costs builders to build in your area. One way of doing this is to ask a few builders directly what they would sell their homes for. Deduct about 15 to 25 percent profit for them and then divide by the square footage. This will give you the cost-per-square-foot. Another way is to ask experienced salesmen at a large building supply outlet. They always talk to builders and know what's going on. Remember, the cost-per-square-foot does not include the cost of the lot.

Back to our example: The total cost of the home less the cost of the lot, interest and closing is $77,400. Assuming that the cost-per-square-foot is $41.25, you could build a 1,935-square-foot home. This is quite a bit of house at your cost! Your actual cost-per-square-foot will obviously depend on the quality of your materials and the extras that you include. Only after your cost take-off will you know what your house will cost, but this figure will give you an idea of the size house you can afford. Keep your square footage figure in mind when deciding on a house design.

BUILDING VS. REMODELING

If you're looking to save money, you might consider remodeling an old home instead of building a new one. This can be especially attractive if a low interest loan can be assumed on the property. Examine this option carefully. Remodeling can have many pitfalls! Virtually all information in this manual, except for the discussion of buying land, is useful to remodelers.

Advantages of Remodeling

- Low interest rates can sometimes be assumed on older homes.
- There is potential for finding good bargains.
- The property can usually be occupied sooner than new construction can.
- Tax laws provide numerous incentives for remodeling, especially in areas designated as historical districts.

AMOUNT OF MORTGAGE PER $100 MONTHLY PAYMENT						
	10 YR. LOAN	15 YR. LOAN	20 YR. LOAN	25 YR. LOAN	30 YR. LOAN	35 YR. LOAN
5.0%	9,428	12,646	15,153	17,106	18,628	19,814
5.5%	9,214	12,239	14,537	16,284	17,612	18,621
6.0%	9,007	11,850	13,958	15,521	16,679	17,538
6.5%	8,807	11,480	13,413	14,810	15,821	16,552
7.0%	8,613	11,126	12,898	14,149	15,031	15,653
7.5%	8,424	10,787	12,413	13,532	14,302	14,831
8.0%	8,242	10,464	11,955	12,956	13,628	14,079
8.5%	8,065	10,155	11,523	12,419	13,005	13,389
9.0%	7,894	9,859	11,114	11,916	12,428	12,755
9.5%	7,728	9,576	10,728	11,446	11,893	12,171
10.0%	7,567	9,306	10,362	11,005	11,395	11,632
10.5%	7,411	9,047	10,016	10,591	10,932	11,134
11.0%	7,260	8,798	9,688	10,203	10,501	10,673
11.5%	7,113	8,560	9,377	9,838	10,098	10,245
12.0%	6,970	8,332	9,082	9,495	9,722	9,847
12.5%	6,832	8,113	8,802	9,171	9,370	9,476
13.0%	6,697	7,904	8,536	8,867	9,040	9,131
13.5%	6,567	7,702	8,282	8,579	8,730	8,808
14.0%	6,441	7,509	8,042	8,307	8,440	8,506
14.5%	6,318	7,323	7,813	8,050	8,166	8,223
15.0%	6,198	7,145	7,594	7,807	7,909	7,957
15.5%	6,082	6,974	7,386	7,577	7,666	7,707
16.0%	5,970	6,809	7,188	7,359	7,436	7,471
16.5%	5,860	6,650	6,998	7,152	7,219	7,249
17.0%	5,754	6,498	6,818	6,955	7,014	7,040
17.5%	5,650	6,351	6,645	6,768	6,820	6,841
18.0%	5,550	6,210	6,480	6,590	6,635	6,654
18.5%	5,452	6,073	6,322	6,421	6,460	6,476
19.0%	5,357	5,942	6,170	6,259	6,294	6,307
19.5%	5,264	5,816	6,025	6,105	6,135	6,147
20.0%	5,174	5,694	5,886	5,958	5,984	5,994

Table 1: Affordable mortgage amounts.

- Additional money can be saved by doing much of the work yourself.

Disadvantages of Remodeling

- Many subcontractors are hesitant to do remodeling because of the hassles involved. Good remodeling contractors are hard to find, and many require payment by the hour.
- The cost of remodeling may be hard to determine before starting the project due to hidden factors (rotten wood, air infiltration, foundation flaws, poor wiring, poor plumbing, etc.).
- The improvements on the house may raise its value above that of homes in the surrounding neighborhood. The value of neighboring homes may hold down the resale value of your home. Consider, though, that this can also work in your favor if you buy the smallest home in the area.
- Appraisers have a hard time determining the final value of a home before remodeling is done.
- Future utility costs are usually higher due to inferior insulation. This can be difficult or impossible to improve.
- Remodeling is more suited to someone who plans to do much of the work himself.
- If you plan to remodel an existing home, get several bids from remodeling subs and have the house inspected carefully before making your purchase offer. Then make your sales contract contingent on a good appraisal, approval of the remodeling loan and the discovery of no hidden defects.

Estimating Remodeling Costs

If you contract a "remodeling specialist," you are paying retail cost. He will be his own general contractor on your project. Often, bids will range from two to three times the sum of material and labor costs. By doing your own contracting, you can normally cut your cost to a fraction of this.

WHAT TYPE OF HOME SHOULD YOU BUILD?

Fit the Home to Your Lifestyle

Up front, some people choose their homes based on what they believe other people like. Of course, this defeats the purpose of "having your own castle." Make sure that you consider what you want based on the way you live.

Before deciding on a plan, make a list of those features you find most valuable to your own style of living. This will be your set of criteria for evaluating plans. Do you like a formal setting or a more rustic environment? Is a formal dining room or living room important to you? What are your family considerations? Listed below are some suggestions to keep in mind when making your planning decision.

- Eliminate as many hallways and walls as possible. Hallways waste building materials and add to heated space while making a house seem smaller.
- Combining living and dining rooms into one room will make a smaller home seem more spacious and will cut down on building materials needed for separating walls.
- Place the kitchen near the garage or driveway for easy access when bringing in groceries.
- Two-story and split-level homes are generally less expensive to build than ranch-style houses of the same square footage because there is less roof and foundation to build.
- Make sure the inside of the kitchen cannot be seen while entering or sitting in the living area.
- Use closets as sound buffers between rooms.
- Avoid house plans that have many different sizes of windows. During purchasing, delivery and installation, many mismatches can occur.

EVALUATING HOUSE PLANS

The pleasure of living in your new home will be greatly affected by its interior layout and appearance. When evaluating plans, examine the relationship and convenience of areas to each other, traffic circulation, privacy and room size. Many plans will not lend themselves to an ideal arrangement without excessive alterations.

Traffic Patterns

Study the circulation and traffic patterns of each room. A floor plan with good circulation should direct traffic flow to one side of the room rather than through its center. Some circulation problems can be improved simply by moving doors to the corners of rooms.

Kitchen Area

The location of the kitchen in relation to other areas of the house is critical. It should have direct access to the dining area and should be accessible to the garage or driveway for ease in unloading groceries. Being near the utility room is also convenient if you plan to have work in progress in the kitchen and utility room simultaneously. Traffic should not pass through the kitchen work area.

The size of the kitchen is important. There was a time when small kitchens were thought to be convenient, but with the advent of modern appliances, kitchens now require more space. Make sure your plan concentrates the main kitchen appliances (fridge, stove, sink) in a compact traffic triangle.

Private Areas

To ensure privacy, the bedroom and bathroom areas should be separated visually and acoustically from the living and work areas of the house.

Make sure you can reach the bathrooms without going through any other room, and that at least one bathroom is accessible to the work and relaxation areas. One of the basic rules of privacy is to avoid traffic through one bedroom to another. Check the size of the bedrooms. They should have a minimum floor area of 125 square feet for a double bed and 150 square feet for twin beds.

Living Areas

The living areas include the dining room, living room and den. In many older designs, these areas are broken into individual rooms. Modern designs combine small rooms into larger, multipurpose rooms. The larger rooms also add to the illusion of a larger home. The den area is usually located at the front of the house, but rooms at the side or rear may be desirable, particularly if this provides a view into a landscaped yard. The main entrance should usually be located at or near the living room. There should be a coat closet near this entrance and passage into the work area without going though the living room.

Energy Considerations

The energy efficiency of a home can be greatly improved with a little advance planning. Listed below are a few energy considerations:

- A window's insulating ability is much lower than

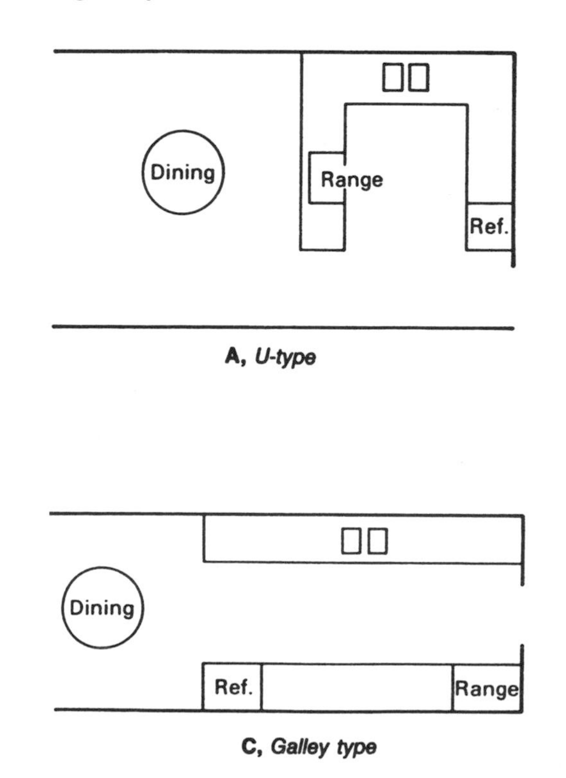

A, *U-type*

C, *Galley type*

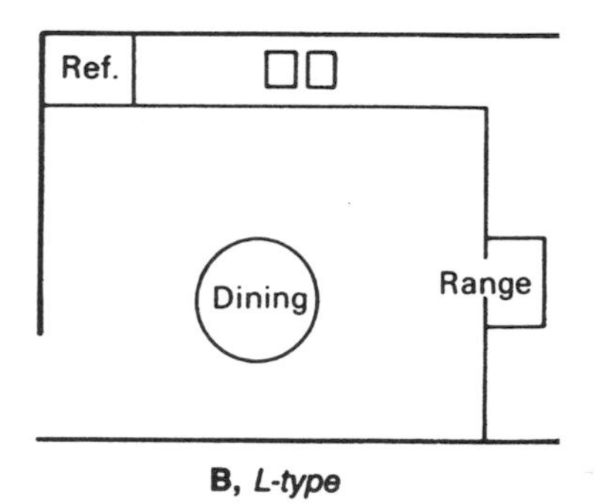

B, *L-type*

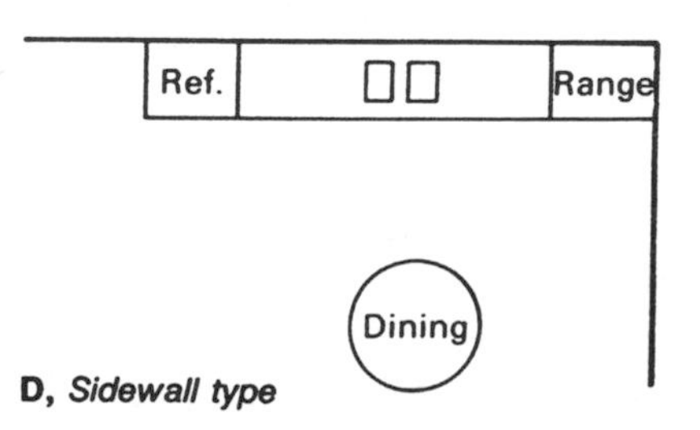

D, *Sidewall type*

Figure 1: Kitchen layouts.

that of a typical wall. Reducing the amount of window coverage will increase insulating efficiency. This can be done by reducing the number of windows (which also saves money) or the size of each window. Wooden windows, although more expensive, tend to insulate better than metal ones. Choose double- or triple-pane windows, or use storm windows, for the best insulation possible.

- Make sure insulation has a vapor barrier—either foil-backed insulation or a sheet of plastic placed between the interior drywall and the insulation. The vapor barrier prevents condensation from building up in the wall.
- Framing your exterior walls with 2×6s 24″ on center will allow 6″ batts of insulation in exterior walls, but will cost more in lumber costs. Also, framing doors and windows can be a problem with 6″ walls. Another way to achieve a higher insulation value is to use thinner walls but more efficient exterior sheathing, such as foil-backed urethane.
- Orient your building site so that the exterior walls with the smallest window area face north. Southern exposure to sunlight will allow more solar heat gain during the day.
- Planting hardwoods near the house will provide shade in the summer, but will allow the sun to shine through in the winter when the leaves fall. To prevent cracks, don't plant trees too close to foundations, driveways, patios or walkways.
- The two most efficient heating systems are gas heat and heat pumps. Gas heaters are cheaper to install and simpler to maintain; however, heat pumps use less energy to operate. As gas prices are deregulated, heat pumps will become an even greater bargain. Heat pumps lose their efficiency where winter temperatures stay below 30°. In this type of climate, gas heat may be better.
- The importance of caulking air leaks cannot be overemphasized. Air infiltration can contribute up to 40 percent of a home's heat loss. Buy a can of urethane spray insulation and caulk around doors, windows, electrical outlets, switch plates and pipes during construction.

SAVING ON FOUNDATIONS

Choosing the proper foundation can save money. The foundations listed below are the four major foundation types. They are listed in order from the least expensive to the most expensive. See the section on Foundations for further information.

All-Wood Foundation

An all-wood foundation can be installed at any time of the year and saves labor costs because only one subcontractor (carpenter) is used. Acceptable for crawl spaces and full basements.

Monolithic Slab

A monolithic slab is the least expensive concrete foundation and a good choice for quality and savings. Work can be completed quickly for additional savings. Care must be taken to position plumbing accurately before casting in concrete.

Cinder Block

Cinder block is somewhat cheaper than concrete but requires the use of several different subcontractors. This causes a longer completion period and more complications. Not as strong as concrete and more vulnerable to leaks.

Poured Concrete

Poured concrete is the best choice for full basements due to high strength and resistance to leakage. It is the most expensive type, but time and money can be saved because only one subcontractor is used.

COORDINATING SUBCONTRACTORS

The task of timing and coordinating subs can be difficult, but can yield savings by reducing construction time. In some instances, promoting cooperation between subs can make their jobs easier and give you a reason to bargain for a lower price. One example would be coordinating the painter and trim carpenter. If trim and walls are to be of different colors, then have the painter put on his primer coats before trim and doors are installed. Primer coats can also be put on trim and doors before they are set. This will save the painter considerable time and effort, and will speed up the construction process.

BUILDING CODES

Building codes were developed to provide minimum-quality construction guidelines, ensuring that structures are suitable, and safe, for their purposes.

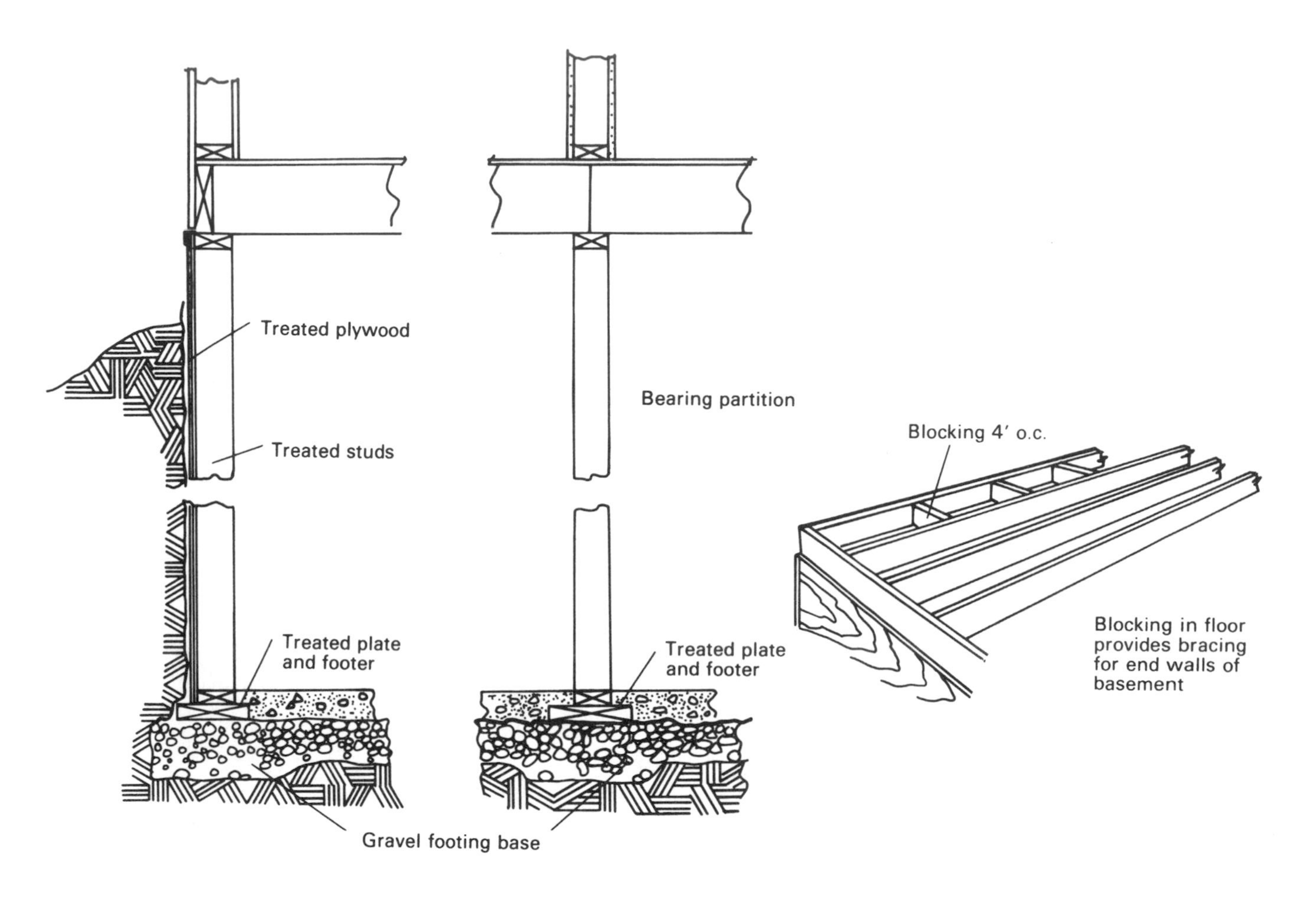

Figure 2: Pressure-treated wood basement footing and foundation wall. All-wood foundations are less expensive and easier to insulate.

The building codes were originally developed as reactions to major disasters: fires, earthquakes and hurricanes. Their requirements cover structural and fire safety items. If you are only changing the look or decor of a house, codes are usually unnecessary. But when you plan a new structural living space, you must obtain a building permit and comply with local code requirements.

Most localities use one of the three main building codes as a base for their local codes. Be aware that local codes may differ slightly from the three main codes. The adopting state or municipality may make changes to the code to suit its own purposes. You can order copies of the building codes for your area from the parent organization or your local inspector. Make sure to ask the local inspector about exemptions from the code standards.

Model codes are published on a three-year cycle, with supplementary information published in the interim. When you apply for a building permit, the building inspector will schedule several visits to the construction site to examine the work for code violations. The holder of the building permit (usually you or the building contractor) is responsible for fixing any shortcomings. All code violations must be corrected before the inspector will issue an occupancy permit.

You should study the building codes for your area during the design process to ensure that your plan meets code requirements. This is one area where a licensed architect can save a few headaches. Having your plan reviewed by a qualified architect or draftsman can save you countless headaches later. The three main building code organizations are:

The Uniform Building Code (ICBO)

International Conference of Building Officials
5360 S. Woekman Mill Rd.
Whittier, CA 90601
(310) 699-0541
http://www.icbo.org

The National Building Code (BOCA)

Building Officials and Code Administrators International
4051 W. Flossmore Rd.
Country Club Hills, IL 60478-5795
(708) 799-2300, Ext. 242
http://www.bocai.org

The Standard Building Code (SBCCI)

Southern Building Code Congress International
900 Montclair Rd.
Birmingham, AL 35213-1206
(205) 591-1853

The International Codes Council

This organization oversees the new consolidated residential building code known as the *One and Two Family Dwelling Code*. This code is brand new and attempts to consolidate all the strong features from the other three major code organizations. Check with your local municipality to see if it plans to adopt this code. Their Internet site can be found at http://www.intlcode.org.

HOUSE PLANS

How many copies? Regardless of how you arrived at a design, you will need plenty of copies of your plan. Plans refer to a standard set of blueprints showing basic elevations of the home and a layout of each floor. Special detail plans will be covered later in this section. The following is a representative list of the copies you are likely to need:

Personal copies	2
Bank copy	1
County inspector	1
HVAC	1
Plumbing	1
Electrical	1
Framing	1
Total	**8**

Sound crazy? Copies are cheap—normally $10 each after the first one. Don't scrimp here. Get what you need. It's always nice to have that extra clean copy when your originals become ragged from use. You cannot make a blueprint directly from a blueprint.

Finding the Right House Plan

There are five sources of house plans:

- Planning services (stock plans)
- Residential architects (custom plans)
- Draftsman (custom and stock plans)
- Hybrid plan (custom changes to a stock plan)
- Computer plan software (custom plans)

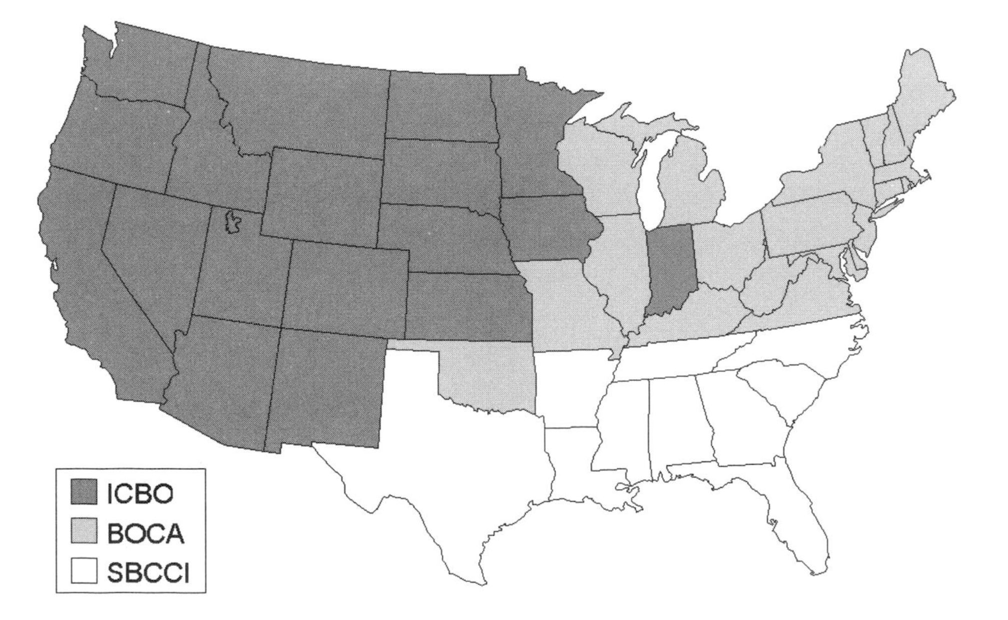

Figure 3: Location of states supporting one of three major building codes.

Stock Plans

House plan services specialize in catalogs of their standard plans. These catalogs usually specialize in a certain type of home of a particular size or price range. Other books will focus on certain styles of architecture, such as traditional, contemporary, southwestern or western European style homes. These books are readily available through bookstores and ads in construction magazines. After finding the plan you want, you can order the complete blueprints from the plan services. That is the plan service's main source of income. Since these plans have been refined over time, you benefit from experience and a low plan cost. Plans may be purchased by the set. Normally there is a discount after the first set. Many plan services offer custom design services to incorporate your design changes.

Residential Architects

Local architects provide design skill and knowledge of local architectural styles. They usually develop a portfolio of plans that will reflect a certain style, but their bread and butter is custom plan designs. Their portfolio plans will help lead you in the right direction. Look for a certified architect who carries AIA (American Institute of Architects) credentials. Architects can be expensive. If you choose an architect, make sure he has a general sense of your budget. He should keep an eye on your budget and caution you if your ideas are starting to get expensive.

Draftsman

A less expensive alternative to an architect is a plan draftsman. He is usually not a licensed architect, but is fully capable of preparing final working drawings. His knowledge of structural and design problems will be limited, however. He may be better suited to making minor changes to plans that are close to the final product.

Hybrid Plan

Probably the best method is to combine the two plan approaches. Find a plan that is close to the home of your dreams and have the architect or draftsman tune it for you. You can accumulate ideas from numerous plans and develop a composite plan. Take these plans or a sketch of your ideas to the architect or draftsman, or choose one of their base plans and sketch out your changes. This minimizes the time they have to invest in planning and should reduce costs considerably. Both architects and draftsmen charge by the hour.

Computer Plan Software

New computer software provides amateur designers with a new option. (See the Technology section for more details.) The new programs are amazingly powerful and easy to use, allowing you to play "what if" to your heart's content. Be careful though. These programs will do nothing to warn you of design or structural violations. Even if you design your own plan, it is best to pay an architect to review it for design flaws.

GENERAL ADVICE

When building your first home, consider using a standard plan, preferably a popular one designed by a reputable architect or plan company. Plans drawn by architects are generally engineered better, but are more expensive. If you are staying with a simple, standard layout, home planners' and draftsmen's plans are usually fine and much cheaper. Go to the architect if you plan any special layouts, such as special foundations or vaulted ceilings. Their expertise will be well worth the extra expense. Always use plans with a standard scale such as ¼″ to 1′.

If you build again, consider using the same plan. You'll learn from your mistakes and probably do an even better job. You will also know exactly how much material is needed. Your cost estimates will be more accurate.

Critically review the kitchen and kitchen cabinet layout. Cabinet layouts are normally there just to meet minimum requirements. You can probably do better. The kitchen design itself may not be the best, either.

For sources of house plans, check with builder plan services, architects, draftsmen and new home magazines. Architects and draftsmen usually charge a certain fee per square foot of house.

For an additional charge, your plan supplier may be able to provide you with a material take-off. But do not assume it will be accurate.

If you change interior partitions, be careful of any impact on load-bearing walls. Consult an architect if you are in doubt. If you want to finish the basement, opt for a 9′ basement ceiling to allow for a dropped ceiling to hide heating ducts.

PLAN CONTENTS

How do you know if your plans are complete? At a minimum, a set of plans should include the following diagrams:

- Front elevation
- Side elevations (one for each side of the home)
- Rear elevation
- Top view of each floor (floor plan)
- Detailed layout of kitchen cabinets
- Detailed foundation, footings and framing specifications

Special Detail Plans

Although you have a set of blueprints, they may not include the following:

- Mechanical layout (HVAC and ductwork)
- Electrical layout
- Plumbing layout

These are normally done by the respective subs, based in part on your specific lifestyle, tastes and ideas. Your blueprints may also include:

- Detail cabinetry layout
- Trim detail layout
- Detail tile plan
- Window and door schedule
- Detail roof plan

DETAIL CABINET LAYOUT—This should include a final layout of each specific cabinet you plan to have in the kitchen, including style, finish and placement. All cabinetry in bath areas, such as vanities, must also be spelled out.

TRIM DETAIL LAYOUT—This should include a room-by-room schedule of what trim is used where. Base, chair rail, crown and other decorative molding, as well as paneling, should be described in detail. Specific material finishes should also be covered. Trim detail layouts are not usually included with most standard plans.

DETAIL TILE PLAN—This should include diagrams showing areas to be covered with particular types of tile. All tiled shower stalls, bathtub areas, tiled walls and floors should be covered. Also, it should show all fixtures such as mirrors, soap dishes, safety rails, towel racks, tissue roll holders, toothbrush and tumbler holders, medicine cabinets and related fixtures. Grout types and colors also need to be determined.

WINDOW AND DOOR SCHEDULE—Itemized list of all window types and dimensions. Make sure that, if you plan to have shutters, you have adequate room between window sets for them (normally 12″ or 15″ per shutter). Also make sure you have symmetrical spacing between windows where applicable. Check before construction and during the construction process.

DETAIL ROOF PLAN—This plan is helpful, especially with complicated hip roofs. It consists of a framing diagram showing opening dimensions, hips, valleys, ridges, roof pitch, ridge venting and skylights.

PLAN CHANGES

After you have selected your plans, consider special changes such as zoned heating/AC, passive solar, special wall partitions, window modifications, special cabinetry, etc. Your original drawings should provide the basis for any details, refinements or other changes you wish to make. Be careful when planning changes. Some changes can have unforeseen effects. For example, changing the height of a ceiling on the first floor of a two-story home will change more than just the height of the studs; it will also add several risers and treads to every staircase, which will require more space for the stairs to spread out. When you decide to change a plan, mark the change on the plan with a red permanent marker. (This will keep the change from smearing if the plans get wet.) Make sure when making changes on plans that you mark *all* the sets of plans to eliminate costly problems later. Many lenders require that houses be built exactly as the plans indicate.

OTHER DETAILS

- Look for 42″ wide halls when possible.
- If you are installing a basement, specify a 9′ ceiling. This will make it easier to finish later.
- Make sure the room layout accents large furniture fixtures that will be present in the room, i.e., dining room table or master bed.

- Make sure the space left for stairs allows the proper number of steps. A 9′ ceiling needs several more steps in the stairway than an 8′ ceiling.
- Make sure that kitchens are designed to provide a compact and efficient work area.

STANDARD DIMENSIONAL SPECIFICATIONS

General

- Floor to ceiling: 7′6″ minimum with no beams projecting down more than 6″
- Smoke detectors in each sleeping area or in the hallway located centrally to the sleeping areas

Bedroom

- No dimension less than 7′ in any direction
- Two means of escape in case of fire: one door, and a window minimum 6 sq. ft. area and no more than 44″ above the floor
- Closet in all bedrooms
- Easy access to a bath
- Space on a wall for a queen-size bed (king-size for master bedroom)
- Space on another wall for a 6′ dresser
- At least one window
- Should not have to go through a bedroom to get to any other room
- Bed to closet: 36″ (minimum)
- Bed to dresser 36″ (minimum)
- Minimal space for dressing: 42″
- Minimal space between double beds: 22″
- Minimal space between bed and wall: 12″
- Minimal space between bed and dresser: 6″

Closet

- Depth: 24″
- Pole height: 5′ to 6′
- Pole depth: 1′ from closet rear wall
- Unsupported clothes rack (4′ maximum)
- Door clearance at bottom: 1″ from finished floor, for ventilation
- Shelving depth: 6″
- Shelving spacing: 8″ vertical clearance

Hallways

- Minimum width: 36″
- Minimize overall hallway length

Dining Room

- Eight-person dining room: 10′ × 12′ minimum
- Direct access to kitchen
- Closed-door access for formal DR (consider double swing doors)
- Should not have to walk through DR to access any other room
- One large free wall for a breakfront
- Consider space for extra furniture and guest seating

Kitchen

The following kitchen specifications represent the latest in kitchen standards and are provided by the National Kitchen and Bath Association.

- A clear walkway at least 32″ wide must be provided at all entrances to the kitchen.
- No entry or appliance door may interfere with work center appliances and/or counter space.
- Work aisles must be at least 42″ wide, and passageways must be at least 36″ wide for a one-cook kitchen.
- In kitchens 150 square feet or less, at least 144″ of wall cabinet frontage, with cabinets at least 12″ deep and a minimum of 30″ high (or equivalent), must be installed over countertops. In kitchens over 150 square feet, 186″ of wall cabinets must be included. Diagonal or pie-cut wall cabinets count as a total of 24″. Difficult-to-reach cabinets above the hood, oven or refrigerator do not count unless specialized storage devices are installed within the case to improve accessibility.

- At least 60″ of wall cabinet frontage with cabinets that are at least 12″ deep and a minimum of 30″ high (or equivalent) must be included within 72″ of the primary sink centerline.
- In kitchens 150 square feet or less, at least 156″ of base cabinet frontage, with cabinets at least 21″ deep (or equivalent) must be part of the plan. In kitchens with more than 150 square feet, 192″ of base cabinets must be included. Pie-cut/lazy Susan cabinets count as a total of 30″. The first 24″ of a blind corner box do not count.
- In kitchens 150 square feet or less, at least 120″ of drawer frontage or roll-out shelf frontage must be planned. Kitchens with more than 150 square feet require at least 165″ of drawer/shelf frontage. (Measure cabinet width to determine frontage.)
- At least five storage items must be included in the kitchen to improve the accessibility and functionality of the plan. These items include, but are not limited to: wall cabinets with adjustable shelves, interior vertical dividers, pull-out drawers, swing-out pantries or drawer/roll out space greater than the minimum.
- At least one functional corner storage unit must be included. (Rule does not apply to a kitchen without corner cabinet arrangements.)
- Between 15″ and 18″ of clearance must exist between the countertop and the bottom of wall cabinets.
- In kitchens 150 square feet or less, at least 132″ of usable countertop frontage is required. For kitchens larger than 150 square feet, the countertop requirement increases to 198″. Counter must be 16″ deep to be counted; corner space does not count.
- No two primary work centers (the primary sink, refrigerator, preparation center or cook top/range center) can be separated by a full-height, full-depth tall tower, such as an oven cabinet, pantry cabinet or refrigerator.
- There must be at least 24″ of counter space to one side of the sink and 18″ on the other side. (Measure only countertop frontage—do not count corner space.) The 18″ and 24″ counter space sections may be a continuous surface or the total of two angled countertop sections. If a second sink is part of the plan, at least 3″ of counter space must be on one side and 18″ on the other side.
- At least 3″ of counter space must be allowed from the edge of the sink to the inside corner of the countertop if more than 21″ of counter space is available on the return. Or at least 18″ of counter space from the edge of the sink to the inside corner of the countertop if the return counter space is blocked by a full-height, full-depth cabinet or any appliance that is deeper than the countertop.
- At least two waste receptacles must be included in the plan, one for garbage and one for recyclables, or other recycling facilities should be planned.
- The dishwasher must be positioned within 36″ of one sink. Sufficient space (21″ of standing room) must be allowed between the dishwasher and adjacent counters, other appliances and cabinets.
- At least 36″ of continuous countertop is required for the preparation center and must be located close to a water source.
- The plan should allow at least 15″ of counter space on the latch side of a refrigerator or on either side of a side-by-side refrigerator. Or at least 15″ of landing space that is no more than 48″ across from the refrigerator. (Measure the 48″ walkway from the countertop adjacent to the refrigerator to the island countertop directly opposite.)
- For an open-ended kitchen configuration, at least 9″ of counter space is required on one side of the cook top/range top and 15″ on the other. For an enclosed configuration, at least 3″ of clearance space must be planned at an end wall protected by flame retardant surfacing material, and 15″ must be allowed on the other side of the appliance.
- The cooking surface cannot be placed below an operable window unless the window is 3″ or more behind the appliance and/or more than 24″ above it.
- There must be at least 15″ of landing space next to or above the oven if the appliance door opens into a primary family traffic pattern. A 15″ landing space that is no more than 48″ across from the oven is acceptable if the appliance does not open into a traffic area.
- At least 15″ of landing space must be planned above, below or adjacent to the microwave oven.

- The shelf on which the microwave is placed is to be between counter and eye level (36″ to 54″ off the floor).
- All cooking surface appliances are required to have a ventilation system, with a fan rated at 150 CFM minimum.
- At least 24″ of clearance is needed between the cooking surface and a protected surface above; at least 30″ of clearance is needed between the cooking surface and an unprotected surface above.
- The work triangle should total less than 26′. The triangle is defined as the shortest walking distance between the refrigerator, primary cooking surface and primary food preparation sink. It is measured from the center front of each appliance. The work triangle may not intersect an island or peninsula cabinet by more than 12″. No single leg of the triangle should be shorter than 4′ or longer than 9′.
- No major household traffic patterns should cross through the work triangle connecting the three primary centers (the primary sink, refrigerator and cook top/range center).
- A minimum of 12″×24″ counter/table space should be planned for each seated diner.
- At least 36″ of walkway space from a counter/table to any wall or obstacle behind it is required if the area is to be used to pass behind a seated diner. Or at least 24″ of space from the counter/table to any wall or obstacle behind it is needed if the area will not be used as a walk space.
- At least 10 percent of the total square footage of the separate kitchen, or of a total living space that includes a kitchen, should be appropriated for windows/skylights.
- Ground fault circuit interrupters must be specified on all receptacles that are within 6′ of a water source in the kitchen. A fire extinguisher should be located near tile cook top. Smoke alarms should be included near the kitchen.

Cabinets

- Base cabinet depth: 24″
- Floor-to-countertop: 36″
- Countertop to bottom of range vent: 24″
- Countertop to bottom of unprotected (wood) wall cabinets: 30″
- Minimum of 40 square feet of cabinet shelf space per household member

Refrigerator

- At least 15″ from an inside countertop corner

Range

- At least 9″ from an inside countertop corner
- Horizontal clearance from a wall cabinet: 12″

Bathroom

The following bathroom specifications represent the latest in bathroom standards and are provided by the National Kitchen and Bath Association.

- A clear walkway of at least 32″ must be provided at all entrances to the bathroom.
- No doors may interfere with fixtures.
- Mechanical ventilation system must be included in the plan.
- Ground fault circuit interrupters specified on all receptacles. No switches within 60″ of any water source. All light fixtures above tub/shower units are moisture-proof special-purpose fixtures.
- If floor space exists between two fixtures, at least 6″ of space should be provided for cleaning.
- At least 21″ of clear walkway space exists in front of lavatory.
- The minimum clearance from the lavatory centerline to any side wall is 12″.
- The minimum clearance between two bowls in the lavatory center is 30″, centerline to centerline.
- The minimum clearance from the center of the toilet to any obstruction, fixture or equipment on either side of toilet is 15″.
- At least 21″ of clear walkway space exists in front of toilet.
- Toilet paper holder is installed within reach of person seated on the toilet. Ideal location is slightly in front of the edge of the toilet bowl, the center of which is 26″ above the finished floor.

- The minimum clearance from the center of the bidet to any obstruction, fixture or equipment on either side of the bidet is 15″.
- At least 21″ of clear walkway space exists in front of bidet.
- Storage for soap and towels is installed within reach of person seated on the bidet.
- No more than one step leads to the tub. Step must be at least 10″ deep, and must not exceed 7¼″ in height.
- Bathtub faucets are accessible from outside the tub.
- Whirlpool motor access, if necessary, is included in plan.
- At least one grab bar is installed to facilitate bathtub or shower entry.
- Minimum usable shower interior dimension is 32″×32″.
- Bench or footrest is installed within shower enclosure.
- Minimum clear walkway of 21″ exists in front of tub/shower.
- Shower door swings into bathroom.
- All shower heads are protected by pressure balance/temperature regulator or temperature-limiting device.
- All flooring is of slip-resistant material.
- Adequate storage must be provided in plan, including: counter/shelf space around lavatory, adequate grooming equipment storage, convenient shampoo/soap storage in shower/tub area and hanging space for bathroom linens.
- Adequate heating system must be provided.
- General and task lighting must be provided.

DEALING WITH MATERIAL SUPPLIERS

DEALING WITH LARGER COMPANIES offers advantages due to potential savings passed on as a result of volume purchasing. Larger companies may offer wider selections than their smaller counterparts. Larger companies have a larger reputation to protect. Hence, your satisfaction is of great importance.

DEAL DIRECTLY WITH MANUFACTURERS when possible. Companies that specialize in windows, doors, carpet, bricks or other such items may provide you with merchandise at close to wholesale prices. Be prepared to pick many of these purchases up yourself, since delivery may be costly or unavailable.

LIMIT THE NUMBER OF VENDORS YOU DEAL WITH in order to reduce confusion, building accounts and persons you must deal with. Dealing with a few vendors may mean special discounts due to larger volume purchasing.

OPEN BUILDER ACCOUNTS at your suppliers. You are a builder! And builders get discounts. Also, builders don't deal with retail sales personnel. Most large material supply houses have special contract sales personnel in the back of the store. Builder accounts often imply special payment schedules. For example, in one case you may pay the open balance on the first of each month. In this case, it is to your advantage to purchase your materials early in the month to gain a full month's use of the money. The time value of money is on your side in this case. Paying accounts within a specified period is a surefire way to earn early payment discounts. Keep track of them and earn them. That's money in your pocket.

INSPECT MATERIAL as it arrives on-site. If something is damaged or just doesn't look right, insist on having it returned for exchange or credit. This is normal and expected; you're paying good money and you should expect good quality material in good condition. Indicate returned items on the bill of lading as proof. Generally, once installed, the materials are your responsibility. This includes items such as doors, tubs, sinks, etc.

GET SEVERAL BIDS on all expensive items, or whenever you feel you could do better. The old saying "Only one thing has one price: a postage stamp" is alive and well in the building material business. Shop around and, chances are, you'll find a better price. Check with the local builders in your area and ask them where they buy their materials.

DETERMINE WHAT IS RETURNABLE. In the normal course of construction, you will more than likely order too many of certain items accidentally or on purpose to avoid delays due to shortages. Find out what overages can be returned to suppliers. Ask for written return policies. Certain cardboard cartons, seals, wrapping or steel bands may need to be intact for an item to be returnable. There may be minimum returnable quantities for items such as bricks in full skids (1,000 units). Restocking charges may also be incurred when returning materials.

MANAGE YOUR PURCHASES WITH PURCHASE ORDERS. Explain to your suppliers up front that you plan to control purchases with "P.O.s" and that you will not pay any invoice without a purchase order number on it. Purchase orders help to ensure that you receive what you ordered and no more. Make sure to use two-part forms. This makes it impossible for someone to create or alter a form. Prenumbered purchase orders are available at most office supply stores.

MINIMIZE INVENTORY and schedule your material deliveries with the project schedule as best you can. This is easier said than done. Minimizing the materials on-site minimizes your investment and exposure to theft or damage. Most material suppliers will be glad to store purchased items for you and deliver them when ready for use. Fragile materials such as drywall and windows are good examples of items not to have lying around too long. Try to be on-site when material is delivered. Write special delivery instructions on purchase orders. You're the customer; you make the rules.

HAVE MATERIALS DROPPED CLOSE TO WHERE THEY WILL BE USED in order to save time. Bricks, loads of sand and gravel are good examples. Place bricks around the house to reduce handling. Have delivery men move plywood, drywall and other heavy items to the second floor if that is where they will be used. Make sure to write your delivery instructions and directions to the site on the purchase order.

KEEP MATERIALS PROTECTED from weather by covering them with plastic or keeping them inside whenever possible. The garage is a great place to keep millwork, doors, windows and other valuable items that shouldn't get wet. Lumber, especially plywood, is susceptible to water damage. Lumber should be covered with heavy-duty plastic and should be ordered shortly before it will be used. Bricks and mortar should be covered carefully with plastic. Wet mortar is ruined and is difficult to throw away. Many suppliers will allow you to purchase materials and keep them in their warehouses until you need them. Rocks, brick or scrap wood should be used to hold the plastic down on windy days.

PROTECT MATERIALS FROM THEFT. You should not have material suppliers deliver until shortly before the material is needed. With stacks of plywood, some builders put a long wood screw through the top corners of the top sheet. If someone wants a sheet, he'll have to haul off four at a time. This usually stops kids. Make sure to lock up all movable items in a storeroom or basement. You will want to install the locks on your house as soon as possible to protect interior items from theft. Instead of giving subs keys, unlock in the morning and lock up in the evening.

GET DISCOUNTS for early payments. Many suppliers give a percentage discount if full payment is made by the end of the month, the beginning of the month or within 30 days of purchase. Make sure you don't lose these discounts. Take advantage of them because they can really add up. Deal with lumber yards and other suppliers who provide payment discounts.

USE FLOAT as much as possible. Many suppliers will invoice you at a certain period each month and will only charge you a finance charge 30 days later. If you pick up materials at the beginning of the month and they invoice you for them at the end of the month, you have had 30 days' free use of their money.

ORDER EARLY FOR PROMPT DELIVERY. Orders are normally queued up for next-day delivery. The earlier in the day you get your order in, the earlier your shipment should arrive the next day. Ask each supplier for a reliable delivery time.

GIVE PLANS TO YOUR LUMBER SUPPLIER. In many areas, you can arrange for your supplier salesman to drop by the site every other day or so to keep your framing crew properly stocked.

Working With Labor and Subcontractors

Labor generally refers to employees who are working for you and have employment tax and withholding taken from their pay. If you use employees, get a good bookkeeping service to figure withholding taxes for you. The subcontractors you use are in business for themselves and withhold their own taxes. Your subs may consist of workers who supply labor only, such as framers and masons, or businesses that provide materials and labor, such as heating and air conditioning contractors. Use contractors whenever possible to avoid the paperwork headaches that go along with payroll accounting. Using contract labor will also save you money because you are not paying their withholding taxes. The IRS may require you to file a 1099 form on your subs. This form shows the amount you paid your subs so that the IRS can be sure that they are paying their share of taxes. Check with a certified public accountant to obtain the proper forms.

LOCATING WORKERS

Finding good subs and labor is an art. The following is a surefire list of sources for locating subs and getting you on the right track for determining which is the best combination of skill, honesty and price:

- JOB SITES—(An excellent source because you can see their work.) As your building site shows progress, it will act as a billboard, attracting all sorts of subs looking for work.
- REFERENCES FROM BUILDERS AND OTHER SUBCONTRACTORS
- THE YELLOW PAGES—Look under specific titles. Do this only when you are in a pinch, because these folks always seem to charge a bit more.
- CLASSIFIED ADS—Here you can normally bargain more than when you meet them at a job site.
- ON THE ROAD—Look for phone numbers on trucks—busy subs are normally good subs.
- MATERIAL SUPPLY HOUSES keep lists of approved subs. Many of them keep a bulletin board where subs leave their business cards.

References are essential in the home-building business. Nobody should earn a good reputation without showing good work and satisfied customers. Insist on at least three references, one of them being the last job. If the last reference is four months old, it may be their last good reference. Some feel that companies that use a person's name as the title, such as "A.J. Smith's Roofing" will tend to be a safer choice over "XYZ Roofing" under the premise that a man's name follows his reputation. But this obviously cannot be taken at face value. Contact references and inspect the work. You may even wish to ask the customer what he paid for the job, although this may sometimes be awkward. A good local standing is essential.

BACKGROUND INFORMATION

Acquiring reference information should be a rule. It is not a bad idea to have your candidate subcontractors fill out a brief form in order for you to find out a little about them. By asking the right questions, you can get a fairly good idea of what types of people you're dealing with.

ACCESSIBILITY

It is important to be able to reach your subs. Try not to hire subs who live too far away. If they need to make a quick visit to fix something, it may not be worth their time. Your subs must have a home phone where they can be reached.

PAYING SUBS

When you have your checkbook out to pay your subs, remember Lester's Law—"The quality of a man's work is directly proportional to the size of his investment." If you want your subs to defy human nature and do the best job possible, give them motivation. ***Do not pay any subcontractor until you are completely satisfied with his work and materials.*** Once you have paid him, you have killed all his incentive to perform for you. After you have paid him, do not expect to see him again in your life. Subs make money by getting lots of jobs, not by taking time on one job. Make each of your subs earn his pay. Below is a suggested schedule for payment:

- 45 percent after rough-in
- 45 percent after finish work complete and inspected
- 10 percent held for about two weeks after finish work (to protect yourself if something is detected a bit later)

Let them know your pay schedule in writing before you seal the deal.

PAYING CASH

Paying with cash gets results. Telling a sub that you "will pay him in cash upon satisfactory completion of his job" should be ample incentive for him to get the job done—and may motivate him to do a good job to boot! Make sure when obtaining cash for work that you keep documentation of the work done for the IRS. Write a check in the sub's name and have him cosign it. Then cash the check. This way you will have an "audit trail" to prove that you paid the sub for work done.

BE FAIR

Be fair with money. Pay what you promise. Be discretionary with payments. Pay your subs and labor as work is completed. Not everyone can go a week or two without a little money. It is customary to pay major subs, such as plumbers, HVAC subs and electricians, 40 percent after rough-in and the remaining 60 percent after final inspection and approval. This keeps the "pot sweet." If you pay subs too much too early, their incentive to return may diminish. If they don't return, you'll have to pay other subs healthy sums to come in and finish the job. Some subs may not even guarantee work they didn't do completely. *Never pay subs for more than the work that has already been done.* Give them a reason to come back and finish the job.

SPECIFICATIONS

Don't count on getting anything you don't ask for in writing. *Remember: Not written, not said.* A list of detailed specifications reduces confusion between you and your sub and will reduce callbacks and extra costs. Different subs have different ways of doing things. Make sure they do it *your* way.

PAPERWORK

Remember that many subs live from week to week and despise paperwork. You will have difficulty getting many subs to present bids or sign affidavits. Many are stubborn and believe that "their word is their bond." You must use discretion in requiring these items. Generally, your more skilled trades, such as HVAC, electrical and plumbing, will be more businesslike and will cooperate more fully with your accounting procedures. These trades are the most important to get affidavits from because they purchase goods from other suppliers that become part of your house. Labor-only subs need not sign affidavits. Make sure that all your subs provide you with an invoice for work done and have them sign it "paid in full." This practice will save you many headaches later.

LICENSES

Depending upon who you have work for you, you may want them to furnish a business license. If required, make sure that plumbers, electricians, HVAC subs and other major subs are licensed to work in your county; if not, have them pay to obtain the license or get another sub.

ARRANGEMENTS FOR MATERIALS

Material arrangements must be agreed upon in a signed, written specification. If subs are to pay for all materials or some of the materials, this must all be spelled out in detail. If a sub is supposed to supply materials, make sure he isn't billing the materials to your account, expecting you to pay for them later. (Refer to "Lien Waivers" in the Legal section.) If you are to supply the materials, make sure they are at the site ahead of time and that there is sufficient material to do the job. This will be appreciated by the subs and will generate additional cooperation in the future.

Ask your sub about any special materials he needs. For instance, many framers use nail guns, which require special nails. Make sure to take this into consideration when ordering materials.

KEEP YOUR OPTIONS OPEN

Don't count on one excavator, framing crew or plumber—or anyone for that matter. Have at least one or two backup subs who can fill in when the primary doesn't show up or does not please you. It is best to be up front with your primaries; let them know that you plan to count on them and expect them to live up to their word. If they can't be there to do the work, skip to your second-string subs.

BE FLEXIBLE WITH TIME

Because of variables in weather, subs and their work, expect some variation in work schedule. Chances are, they are doing four or five other jobs at the same time. Expect some problems in getting the right guy at exactly the right time.

EQUIPMENT

Your work specifications should specify that subs are to provide their own tools, including extension cords, ladders, scaffolding, power tools, saw horses, etc. Before hiring a sub, make sure he has all the tools. Don't plan on furnishing them. Defective tools you supply can make you liable for injuries incurred when they're used. If you don't have sawhorses available for your framing crew and if they didn't bring any, count on them spending their first ten minutes cutting up your best lumber to make a few.

WORKER'S COMPENSATION

Make sure your sub's worker's compensation policy is up to date before the job is started and get a copy for your records. If the sub doesn't have worker's compensation, make it clear to him that you intend to withhold a portion of his payment to cover the expense. This can be a sore spot later if you forget to arrange this in advance.

Worker's compensation should be deducted from each check when paying a sub more than once. Typically, you must set a deduction percentage that matches the rate for the particular sub. Check with your insurance agent to determine the proper percentage.

POOR WORKMANSHIP

This can be one of your worst headaches. A sub you've contracted is halfway through a job and you notice poor workmanship. The work will have to be completely redone, so you fire the sub. He may file a lien on the project, halting or hindering further progress. Make sure you inspect often and well, and know your subs!

STANDARD BIDDING PROCESS

Most of the steps below will apply to the effort of reaching final written agreements with the subs you will be using. These steps will be referred to in each of the individual trade sections.

BD1 REVIEW your project scrapbook and locate potential subcontractors.

BD2 FINALIZE all design and material requirements affecting subcontractor.

BD3 PREPARE standard specifications for job.

BD4 PREPARE subcontractor packages (Subcontractor Information Sheet, Standard Contract Terms and Specification Sheet).

BD5 CONTACT subcontractor and discuss plans. Mention any forms to be completed and specify deadlines (should be written on forms). Get a good understanding of who will provide what.

BD6 RECEIVE and evaluate completed bids.

BD7 SELECT the best three bids, based on price and your personal assessment of the sub.

BD8 COMPARE bids against budget.

BD9 NEGOTIATE with prospective sub. Ask him what his best "cash" price is. When you can't get the price any lower, get more service out of the same cash by asking him to do other items in lieu of a cash discount.

BD10 SELECT sub. Make it clear that he has been selected and that he needs to commit a specific period of time for the job. Explain that you are fair with money but expect good, timely work according to a written contract.

BD11 CONTACT subs and tell them when you expect them at the site. Discuss upcoming schedule and keep in touch with them or you will lose them. If the building schedule changes, let the sub know as soon as possible so that he can make other plans. This courtesy will be returned to you when you need the sub at the last minute.

Making Changes

Changes are second nature. Most people can't even order lunch without changing their minds. We all see things and perceive how they could be made better. That's where a good imagination and careful thought are important. The idea here is to *make changes on paper, then freeze the design.* The cost of ripping out walls, adding windows or "that breakfast counter" can be staggering. In many cases, the cost of making a change after construction makes the change economically impractical. Making changes late in the game is a good way to wreck your budget. So be good to yourself—plan ahead and make all those modifications and additions to your plans while you're still just using an eraser.

Let's look at three examples. Suppose you have decided to have a bay window in the kitchen as opposed to the original idea of a flat window. For the sake of simplicity, let's assume the two windows are of the same dimensions.

CHANGE DURING PLANNING

You think the change over and decide that the $110 material differential is well worth the money. You take out your eraser and update your drawings, take-off and contract specifications. Not bad.

CHANGE PRIOR TO INSTALLATION

The flat window has been delivered to the site but is uninstalled. Now you will have to negotiate with the subcontractor who already signed the contract to install the flat window. The sub may charge you a little extra; that's just part of the game. But now you hope that the material supplier will exchange a bay window for a flat one (provided it is still in new condition). You also pay the $110 differential and a restocking fee that was not in your original budget; this is an out-of-pocket expense. The change will cost you money, but at least you aren't undoing work.

CHANGE AFTER INSTALLATION

The flat window has been installed. *Now* you (or your spouse) decide to change to a bay window. This could be expensive for several reasons:

- You may have to damage the flat window to remove it, so it may not be returnable.
- You have to purchase the bay window, say $700, out of pocket. You (or the bank) didn't count on this.
- You will have to pay to have the flat window removed and the bay window installed. Pray that you didn't paint or wallpaper yet.

In addition, you have to prepare the change order form. Note that in this case you are at the mercy of the subcontractor. He knows you want the bay window and may charge you heavily for it. If his price is too high, you have three alternatives:

- Pay the man and try to forget this ever happened.
- Do it yourself. This can be dangerous if you don't know what you're doing. If the sub damages the bay window during installation, it's his problem. If you damage it, that bay window may cost you another $200 to $300.
- Live with the flat window. (This is frustrating. So long as you live in that home, that flat window will be a constant reminder of a careless mistake.)

ADVICE

Think changes over carefully on paper. Really take the time to picture what that drawing represents in real life. Is it what you want?

If you must make changes, decide to act as early as possible. If you make a change, pay for it out of pocket; that is, with funds you had not earmarked for the project.

COMMON PITFALLS WHEN MAKING CHANGES

- When changing ceiling heights, pay close attention to stairs. Added ceiling height changes the number of steps used in a staircase. Make sure you have room for the extended length of the staircase.
- When walls are moved, pay close attention to the effect on load bearing. Load-bearing walls cannot be moved easily without providing some other means of supporting the wall. Even small changes in walls can make a big difference. Most walls are located directly above and below each other. If you move a wall one or two feet, the floor between them may sag over time.
- Any changes to walls in the house can affect the roof framing. Check with a home designer or an architect when moving any wall.
- When adding shutters to windows, make sure that the window spacing will allow it without changing the balance or symmetry of the house.
- Examine the door swing of each door in the house. Moving a door may cause the door to block entrances or other doors when open.
- Adding brick to the outside of a house designed for siding will change the spacing and size of the foundation and roof overhang. Take this into consideration.

Saving Money

Aside from the obvious ways to save money on a home, such as building yourself and reduction of square footage, other measures can be taken to minimize the cost of your home without compromising on quality. Let's look at a few ways to save more money before, during and after construction.

SAVING MONEY BEFORE CONSTRUCTION

The most effective means here are:

- Shop for prices aggressively
- Plan your home to minimize waste (modular construction)
- Plan for extras later
- Plan your home without extras
- Build closer to the street
- Minimize the house's "footprint" ✓
- Eliminate walls
- Finish only what you need
- Avoid "big house" expenses
- Use panelized, modular or precut kit homes

Shop for Prices Aggressively

This means not accepting bids as being the best job for the best price until you are satisfied. Use your take-off to help keep prices down. Subs are out there to earn a living, and they will try to get the best price possible. If you say the price is right, it's a done deal. Don't be too ready to give in. The more you save over your estimate early in the game, the more cushion you'll have later for unexpected expenses.

Plan Your Home to Minimize Waste

This means going below the surface. Here is an example:

Carpet normally comes in 12′ rolls. In estimating your job, a carpet sub will charge you for any and all scrap necessary to carpet designated areas. When possible, plan rooms to be twelve feet wide or 24′ long. This reduces waste significantly.

You can also save up to 30 percent on building materials by using new "value engineering" concepts endorsed by the Department of Housing and Urban Development. These techniques aim to reduce excess lumber and other materials in residential construction. These techniques are discussed in more detail below.

Use Modular Construction

Blueprints are often designed with esthetic purposes in mind. That's OK, but remember that many materials, lumber especially, are designed around 16″, 24″ and 48″ units. To a reasonable extent, plan your home to be composed of 24″ or 48″ units, eliminating waste and cutting time. Refer to the Framing section, which describes this approach in more detail.

Plan for Extras Later

Almost anyone would like to have all or some of the following items:

- Built-in microwave ✗
- Intercom system ✗
- Central vacuum system ✗
- Electronic alarm/security system ✓
- Skylights ✓
- Finished bonus room(s) ✗
- Patio or deck ✗
- Swimming pool and hot tub ✗
- Electric garage door opener ✓
- Crown and other fancy molding ✓
- Wallcoverings ✗
- Sidewalk or other paved area ✓
- Expensive landscaping ✗
- Den paneling and built-in bookcases ✗
- Wet bars (stub-in plumbing) ✗

- Paved driveway/garage ✓
- Gas grill ✗

But these items, and many others, are not essential to getting a certificate of occupancy. Don't strain to pay for items you can easily add later. Minimize that mortgage! Smarter individuals learn to live without certain luxuries and add them later when a little cash is available. The trick is to prepare for extras by doing work that is difficult or impossible to add later. If you want a gas grill, run the gas line *now* and cap it. Gas lines, plumbing and wiring are fairly inexpensive to put in during construction and will wait for you until you are ready to add those extras.

Plan Your Home Without Extras

Still, certain other extras can be avoided altogether. These include such items as basements, bay windows, brick veneer, an extra room, a fireplace, expensive cabinetry, etc. While it may be nice to have them, they may be the difference between a lesser home and no home at all.

Basements, for example, can easily cost you 5 to 7 percent of the cost of your home. Do you want a basement? Do you need a basement? Many only have one for possible resale purposes. No basement is better than a leaky, damp and smelly one.

Build Closer to the Street

Moving the house closer to the street will reduce the cost of the driveway and front yard landscaping. The additional benefit will be a bigger backyard—an added plus with modern houses. Just make sure that you meet the minimum setback requirements.

Reduce the Footprint of the House

The smaller the footprint of the house, the smaller (and less expensive) the foundation. How do you reduce a house's footprint without reducing its size? By using cantilevers. Bay windows, breakfast areas and fireplaces can all be extended 4′ or more out from the house. These cantilevers also serve to break up the straight lines of the house—adding more stylistic variety.

Eliminate Walls

Modern houses often have a more open design, with good reason. Interior walls cost money. In addition to the materials used in the wall partition, there are additional finishing costs: trim, doors, drywall finishing, painting, etc. Eliminating extra walls also opens up the plan and gives the house the appearance of being much larger. Every separate room must have an access door and a path for foot traffic. When you eliminate a wall separating two rooms, you also eliminate the need for this additional area. A combined living room-dining room will share foot traffic area and can be smaller than two separate areas. The combined area will also seem much larger than the two areas. Air is the cheapest building material you can use.

Build Only What You Need

Can't afford to build the home of your dreams now, but you don't want to move again later? Plan for expansion space in your new home. The largest single cost in a new home is the finishing work. You can leave areas of the house unfinished and complete them as needed. The second story of a story-and-a-half design can become new bedrooms when new children (or extra funds) arrive. Make sure the stair access, plumbing and electrical rough-outs and insulation are in place. These are expensive to add later.

A basement is the perfect expansion area for an office, den or workshop. Make sure you design in windows and outdoor access into the basement to improve the quality of finishing work you complete here. Make sure your plumbing is already in and that the basement floor is high enough to allow proper sewer drainage. If not, you may have to install a "flush-up" toilet when finishing the space. Design a 9′ basement ceiling to leave space to frame in below the wiring and plumbing. Open web floor trusses are perfect for supporting the floor above the basement. They can span large distances without support, and the open webs allow the plumbing, electrical and HVAC to be placed within the truss itself.

There is another advantage to finishing these areas later. It is more convenient for you to do the work yourself in these areas, thereby saving even more money. There is no rush to complete this additional work, and you can work at a pace that finances allow.

Watch for Extra Costs of Larger Houses

Larger homes carry with them additional hidden costs. If a 2,000 square foot house built at $40 per square foot costs $80,000, then why doesn't a 4,000 square foot home cost over $160,000? Home costs aren't directly related only to square footage. The additional size of larger rooms in large houses is mostly made up of air and does not add costs linearly. However, many other factors contribute to adding extra costs to larger houses:

- Larger homes typically require larger rooms with corresponding longer spans for floor and ceiling joists. Longer pieces of lumber cost more per board foot than shorter pieces. Longer spans may also require thicker lumber in order to carry the weight over the longer spans.
- Larger homes tend to have higher ceilings—9′, 10′, 12′ and up. Longer studs cost more per square foot. Taller rooms use more drywall, paint, trim, etc.
- Larger homes tend to have more structural weight, which often means thicker joists, steel support and I-beams, all of which add to the cost.
- Larger homes have larger and fancier doors, windows and trim.
- Larger kitchens mean more and fancier cabinetry.

Use Panelized, Modular or Precut Kit Homes

Advances in factory-built home design have made these homes viable alternatives to "stick-built" homes. In most cases, these homes are less expensive, better built, better insulated and more efficiently designed than their stick-built cousins. You save money in several ways:

- These preconstructed homes go up faster, shortening the overall project duration and saving you interest payments.
- There are no material surprises. All the materials of the structure will be included in the sales price. You don't have to worry about misestimating your material costs. Material can't mysteriously disappear from the building site.
- Usually, the equivalent manufactured home is less expensive than stick-built. This is because the factory can use materials more efficiently.

Don't confuse these manufactured homes with mobile homes or shoddy construction. It's a shame that most people want their houses built the good old-fashioned way. Homes built in the factory are made from high-quality, straight lumber. Nailing and framing is done in jigs to a precise accuracy unmatched in the field. These more precise dimensions also lead to a much more airtight and well-insulated home. There are three basic types of factory construction.

Modular Homes

Modular homes are completely constructed in the factory in modules that are shipped to the building site and connected together. Virtually all the construction except for some finish work is done at the factory. They go up fast, sometimes "dried in" in less than a day. The individual modules usually have the wiring, plumbing and interior wall finishes already installed. The obvious limitation with modular construction is the size of the module. Since it must be shipped to the site via the public highways, the dimensions of each individual module can grow no larger than a typical mobile home. In the industry's infancy, houses built from modules had that blocky, square look. Modern advances in computer design and factory fabrication have changed this, however. Now, custom home designs can be produced in the factory, and they're indistinguishable from site-built homes once they are assembled. You must make sure that your foundation is as accurate as possible. There's nothing more upsetting than to get your new home to the site and realize your foundation is off by a foot!

Panelized Construction

This is a hybrid between modular and stick-built construction. The wall panels are constructed on a jig in the factory. The windows and doors, siding and insulation can be installed. The panels are then stacked in a truck along with the rest of the house and roof trusses. Since the "air" is eliminated, the entire house can be shipped on one or two trucks. These houses are not limited by modular shipping dimensions, so any house plan can be converted to a panelized house. In fact, many factories specialize in taking your plan, pricing it and constructing a custom design.

Once the panels reach the site, they are unloaded and assembled on the foundation. Small variations in the foundation can be fixed by a little sitework. Once the walls are up, the roof trusses are installed and covered with sheathing, and the siding is completed. Small pieces of siding will be installed to cover the seams between the panels. These siding pieces will be staggered so that, once they are installed, it will be impossible to distinguish this house from a normally constructed home. Panelized houses go up as fast as modular housing, sometimes two days or less. The panels do not require a crane. They can be handled by small work crews. It is truly amazing to see an entire house go up in a couple of days, complete with windows and doors and fully dried in! This allows you to cheat the weather and get inside quickly, where the finishing work can be completed in warm and dry surroundings. Your building materials can be locked inside to protect them from theft.

Make sure you find a building crew that is familiar with panelized construction. A rookie framing crew can take as long to construct a panelized house as they would to stick-build it. There go your savings. Many panelized dealers offer a list of framers familiar with the intricacies of panelized assembly.

Kit Homes

Kit homes are not preconstructed, but they offer the advantage of having all the necessary materials cataloged and priced. Kit homes are especially popular for certain types of tricky construction or homes made from special materials. Examples include log homes, post-and-beam construction, cedar homes, steel construction and solar homes. The materials may come precut, as in a post-and-beam home, where large beams are connected with mortise and tenon joints. Other kits, such as steel framing, just include all the materials, especially those that are exotic or special. You will still have to hire a framer to cut and assemble them, but at least you know that all the necessary materials are available for a fixed price.

SAVING MONEY DURING CONSTRUCTION

The most effective ways here are:

- Do work yourself
- Make prefabricated substitutions
- Downgrade
- Economize framing with "value engineering"

Do Work Yourself

This can be lots of fun. You'll feel (and be) helpful, while saving even more money. In most cases, every hour you spend productively on-site can save you $10 to $15. Great! But be careful. There are two main reasons why you shouldn't get carried away with doing too much of the work yourself:

- There are many jobs you shouldn't even consider doing. Remember, where special skills are needed, skilled tradesmen serve an important purpose: They get professional results and save you from injury, expensive mistakes, frustration and wasted time. Certain jobs may only appear easy. You can ruin your health sanding hardwood floors or drywall. You can ruin a den trying to put up raised paneling and crown molding.
- Work can distract you from your more important job of overseeing the work of others. Your most valuable skill is that of *boss*—scheduler, inspector, coordinator and referee. Saving $50 performing a small chore could cost you even more than that in terms of problems caused by oversight elsewhere on the site. Doing lots of work yourself invariably extends your project timetable. Let the pros do the job quickly.

Jobs to Consider Doing Yourself

- Paint, sand and fill
- Wallpaper
- Install door and light fixtures
- Install mailbox
- Light landscaping
- Light trimwork
- Cleanup
- Organizing materials and locating them where needed (stack bricks where masons will use them, etc.)

Jobs Better Left to Pros

- Framing
- Masonry
- Electrical
- HVAC
- Roofing
- Cabinetwork
- Plumbing
- All others not listed

Make Prefab Substitutions

Many items, such as fireplaces, can be purchased as prefabricated units. This saves you labor costs. When prefabricated items are used, make sure the sub knows of this and takes it into consideration in his bid. One-piece fiberglass shower and tub enclosures, attic staircases and prehung doors are other examples.

Downgrade If Necessary

Certain items can make a major impact on cost reduction without making a visible impact on your home. Consider hollow vs. solid core doors, prefabricated cabinets vs. custom cabinets, prefabricated doors and windows vs. custom doors and windows, to name a few of your options. Another example is trim. Trim that will be stained will show all imperfections, so it should be as high a grade as possible. But if you plan to paint the molding, use "finger-joint" grade. This is molding made of lots of little scrap pieces glued together. It costs less, but will serve the purpose.

Windows with true divided panes are more expensive than solid pane windows with "fake" wood pane inserts. In most cases the fake panes look just as good and have the added benefits of easier window cleaning and better insulation.

Brick provides a great look to the house but is very expensive. If you use brick, consider bricking only the front of the house—that part visible from the street. Use siding on the back side.

Economize Framing and Other Materials

Since framing is hidden, appearance is not an issue—only workmanship and safety are important. The Department of Housing and Urban Development (HUD) has worked closely with all the major building code associations and the NAHB Research Foundation to develop a series of construction methods that can reduce construction material by more than 15 percent. It is called Optimum Value Engineering (OVE). You can use all or some of these suggestions to reduce costs considerably.

Planning and Design

Use a rectangular or near rectangular floor plan. The most efficient plan is a basic rectangle, which allows a simple floor and roof structure.

Consider slab-on-grade designs to minimize cost and foundation problems.

Locate partitions to intersect exterior wall studs. This will save additional framing material at the intersections and simplify insulation.

Coordinate window and door openings with normal stud locations to minimize framing. Windows and doors can often be moved several inches without upsetting the design balance.

Plan for straight-run stairs. Orient stairwell openings parallel to floor framing so as to disrupt as few floor joists as possible.

Locate attic/crawl space access doors between framing members in closet, hallway and other appropriate areas. A 24″ O.C. spacing of framing members provides ample access.

Use in-line framing. Use the same spacing (24″ O.C.) for all roof, exterior wall and floor framing, and align the framing members, one under the other, to provide structural strength. This will also reduce the need for double top plates on walls.

Consider centralized "back-to-back" plumbing. Cluster bathroom, utility room and kitchen fixtures as close together as possible to reduce plumbing and allow sharing of vent stacks. In two-story plans, arrange upper-level plumbing over lower-level plumbing so they can connect to the same stack. Concentrate plumbing in the same wall to minimize framing for plumbing.

Lay out plumbing to avoid structural members in the floor and roof. Do not place plumbing in exterior walls because of structural and insulation difficulties.

Locate heating and cooling equipment centrally to minimize duct runs and sizes and to provide good distribution.

Floor Construction

Eliminate or reduce size of sill plate. If the foundation is level and accurate, the sill plate is unnecessary. The floor framing may be erected directly on top of the foundation and anchored by metal straps.

Use built-up wood beams. If a center beam is required to support floor joists, construct a beam from standard 2× lumber rather than more expensive steel or laminated wood beams. The joists can be fastened directly to the wood beam.

Space floor joists 24″ O.C. to correspond with the two-foot planning module. This will allow wall studs placed 24″ O.C. to be aligned directly above each

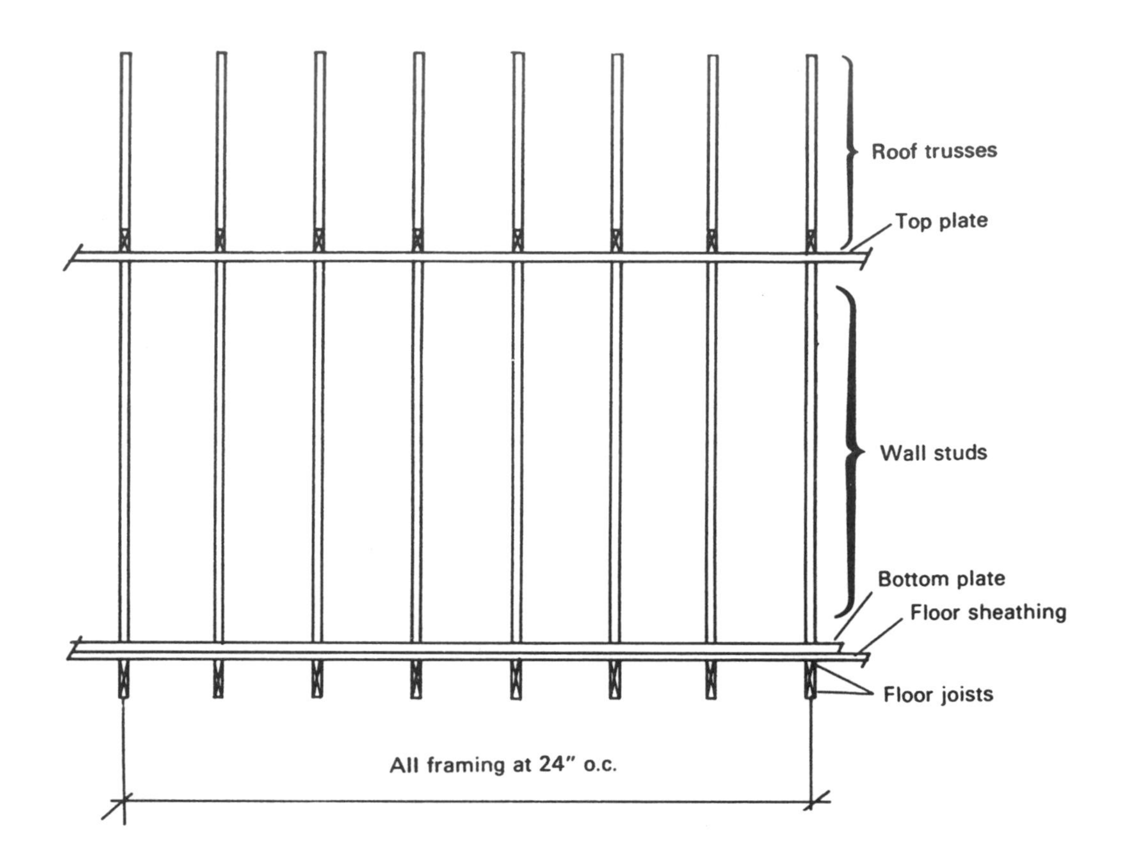

Figure 4: Vertical alignment of framing members simplifies framing and transmits loads directly down through structural members. This eliminates the need for double top wall plates and allows 2′ O.C. framing.

joist. Often, the same size floor joist used for 16″ O.C. can be used if the plywood floor is glue-nailed and the joists are spliced off center.

Use in-line, off-center spliced floor joists. The allowable span of floor joists may be increased by maintaining continuity over the center bearing beam. One way to do this is to splice unequal lengths of full-length floor joists at a less critical point away from the center support.

Use engineered floor trusses instead of standard lumber. These trusses will be engineered for the span of the floor and can be one solid piece for the entire width of the house. They can span greater distances, and work well when spaced 24″ O.C. Most floor trusses have a wider surface area for the subfloor adhesive. Open web trusses also provide a convenient space for running plumbing, wiring and HVAC vents. This eliminates the need to box in a utility duct.

Eliminate double floor joists. Don't use double floor joists under nonload-bearing interior partitions. In fact, there does not have to be a joist under nonload-bearing walls if a ⅝″ or thicker subfloor is used.

Reduce or eliminate the band joist. The band joist serves little or no structural function if wall studs are aligned directly over floor joists. If a band joist is necessary, use 1× lumber instead of 2× lumber.

Eliminate bridging between floor joists. Bridging does not contribute to the strength of the floor system and is no longer required by building codes, unless the joists are larger than 2×12. In fact, bridging becomes a source of squeaking in many floors.

Preplan floor sheathing to minimize scrap. If the floor is laid out on 2′ or 4′ modules, there should be very little waste.

Use a glued-and-screwed tongue-and-groove plywood floor to increase the allowable span of the floor system. (See the Framing section for information on the APA Sturdifloor system.) This system

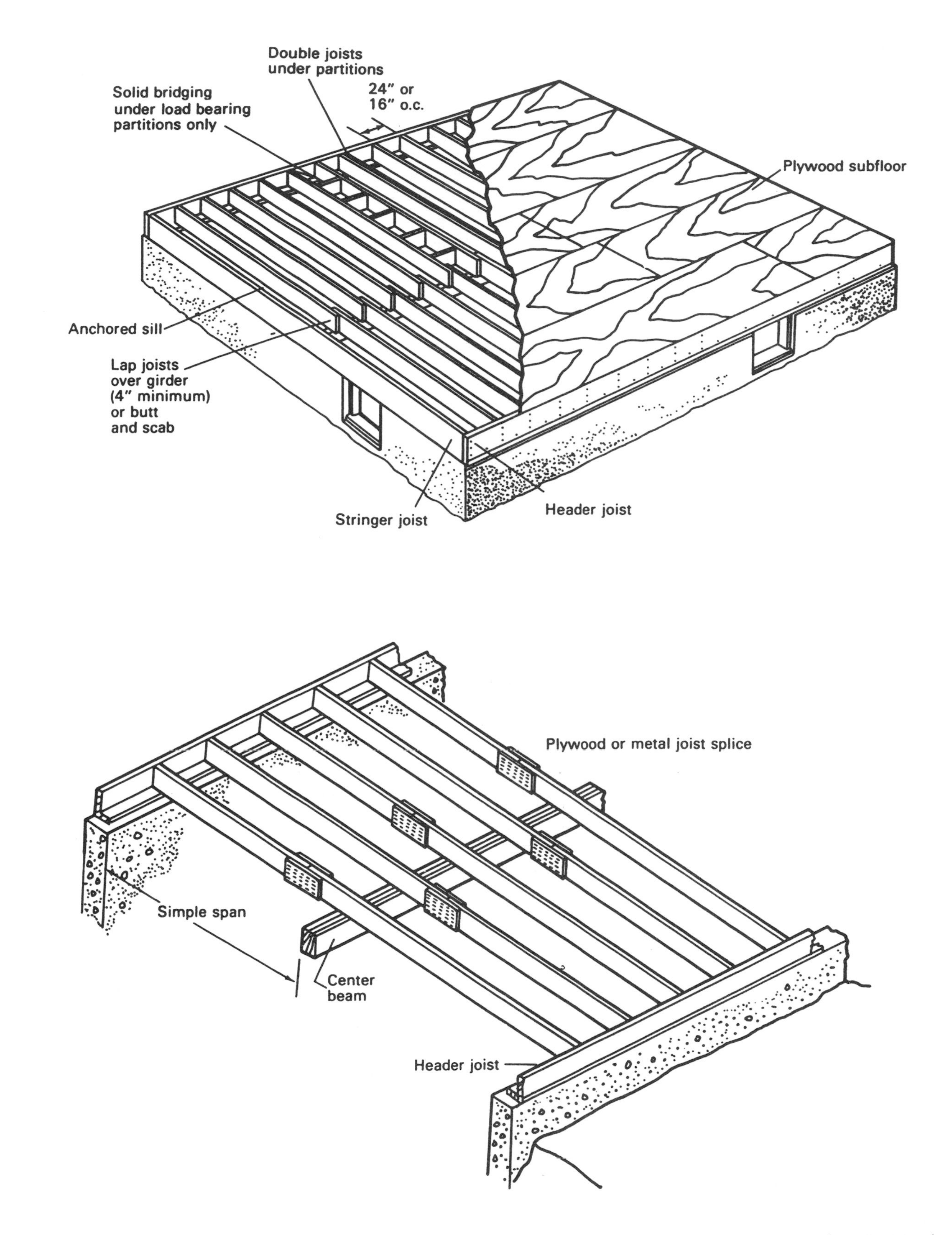

Figure 5: In-line joist system. Alternate extension of joists over the center support with plywood gusset allows the use of smaller joist size.

stiffens the floor and eliminates squeaks. It also eliminates the need for a separate underlayment.

Exterior Wall

Space studs 24″ O.C. to reduce the amount of framing lumber and labor in conjunction with the two-foot planning module. A 24″ stud spacing is acceptable in most areas of the country for one story and the second story of two story construction. If you are building a two-story house and want to preserve the two-foot planning module, use 2×6 lumber on the first floor and 2×4 lumber on the second. This also allows space for more insulation on the first floor.

Use a single top plate. Where all framing is vertically aligned (i.e. joists, studs and roof trusses are in line), the top plate serves no structural function. Special connections to splice joints in the top plate or to tie corners together are not required because the floor and roof system serves this function.

Consider a 1× bottom plate. When the framing is aligned, the bottom plate serves no structural function. A 1× bottom plate is adequate, and is easier to install with 8d nails.

Use two-stud corners. A three-stud corner post is not required and is used only to provide a backing for the drywall. By using metal drywall clips or wood cleats, you can eliminate the third stud. It also leaves more space for insulation.

Eliminate partition posts. A partition post is not required where an interior partition intersects an exterior wall. It serves only to back up the drywall. Drywall clips can be used here as well.

Eliminate mid-height blocking. Blocking is not required in exterior walls as a fire-stop or structural bracing. Standard platform framing provides all the fire-stopping that is necessary. The lack of blocking also makes the application of wall insulation much easier. Old habits die hard and many framers must be talked out of using blocking.

Eliminate structural headers in nonbearing walls. Structural headers are used to support loads over openings in load-bearing walls. However, in practice, they are often used over openings in walls that are not actually load-bearing. If the house uses roof trusses, only the walls that support the truss need headers over openings. The walls under the gable ends do not need headers.

Use 22½″-wide windows fitted between studs spaced 24″ O.C. and aligned with other framing members. Since the window fits between the studs, no header is necessary. Several windows can be placed side by side to provide the effect of a larger window.

Use a glued-and-screwed plywood header. Where a header is required, use a plywood box header. Make a box header by gluing and screwing a plywood face over the framing above an opening. The plywood face can be used in place of the exterior sheathing in the space above the door. Make sure to use a plywood thickness that matches the sheathing thickness.

Use single-layer panel siding. Siding products that do not require the additional support of sheathing may be applied directly to the studs. However, siding products with open joints or that might otherwise allow wind or rain to gain access to the interior should be installed over a suitable nonvapor barrier building paper, such as 15 lb. felt.

Minimize exterior trim. However, exterior trim often serves a valuable function by covering construction joints and providing tolerances at floor, wall and roof intersections, windows, corners, etc.

Eliminate interior window trim by returning the drywall into the opening.

Roof Construction

Use engineered roof trusses spaced 24″ O.C. in conjunction with the aligned modules. These trusses will be specifically engineered for the house, and often will span the entire interior space without load-bearing interior partitions. This saves headers in the interior walls. There are several truss designs available, including options that provide an attic space.

Use prefab gable end trusses. These are similar to the regular roof trusses, but have vertical members for nailing the gable siding.

Use ⅜″ Group 1 plywood roof sheathing with ply clips to support the edges of the plywood between the trusses.

Eliminate the rake overhang on a gable roof. The rake overhang is a costly detail and serves no real function except as partial shade for the windows.

Use an open-soffit overhang. If you use an overhang, use an open-soffit detail with blocking to fill the opening between the trusses at the wall. A dark stain used in the open soffit will conceal shingle nails or staples that penetrate the wood sheathing.

Interior Walls

Use 2×3 studs spaced 24″ O.C. for nonload-bearing partitions. The 2×3 studs offer ample strength, and the thinner walls add one inch to the dimensions of each room.

Use light-gauge steel studs for nonload-bearing walls if they offer a cost advantage. Steel studs install rapidly and are very common in commercial construction. Drywall must be installed with screws, however.

Use single frame openings in nonload-bearing partitions. It is not necessary to double the studs at either side of interior door openings, or to install headers or cripples over the opening.

Reduce blocking. Nonload-bearing partitions that are parallel to floor framing do not require blocking or other special support in the floor where ⅝″ or thicker plywood flooring is used. Where partitions are parallel with ceiling joists or roof framing overhead, use precut 2× blocks spaced at 24″ O.C. between the framing members to secure the top of the partition and to provide drywall backup.

Interior load-bearing walls should be aligned with the framing members in the floor and roof structures, just like the exterior walls.

Simplify closet framing by providing a full-width opening at the front, dimensioned to receive a standard-width bifold or sliding door.

Use 4×12-foot sheets of drywall installed across the framing on walls and ceilings. Drywall ½″ thick is approved for application over framing 24″ O.C. Blocking is not required behind tapered drywall edges.

Eliminate the bulkhead over kitchen cabinets. This is a costly detail and is difficult to insulate properly. Kitchen cabinets may be hung on the wall at the proper height with an open top. This area becomes additional storage space. Or you can use full-height cabinets to enclose the entire area.

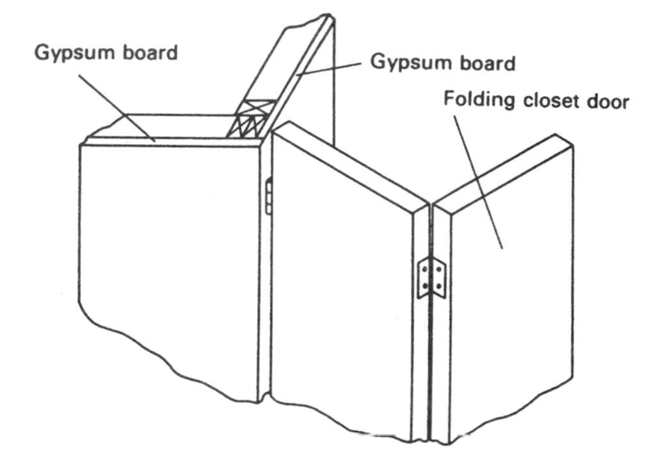

Figure 6: Folding closet door mounted directly on finished wall with hinges secured to wall framing.

Minimize interior trim. Eliminate the shoe molding; minimize window trim by framing the opening with drywall; install bifold closet doors that do not require trim.

Use metal drywall clips to eliminate studs and blocking for drywall backup. Metal clips have been proven cost-effective.

Plumbing

Cluster plumbing. This is a cost-saving design concept also referred to as back-to-back plumbing. The basic principle is to arrange typical plumbing groupings such as baths, kitchen and laundry on a common wall or, in multiple stories, vertically so that all fixtures can be attached to a common stack. This minimizes drain, water pipe, waste and vent materials and reduces installation labor. Make sure to bargain with the plumbing subcontractor, who often charges by the connection. Point out that your design will use considerably less material and labor, and ask for a substantial discount. Locate the water heater near this grouping. This will greatly reduce the time it takes for hot water to reach the faucet.

Use prefab shower and bath modules. These prefab units eliminate the cost and installation of ceramic tile, a significant expense. Other advantages include eliminating leaks and callbacks associated with tile installation, one of the trickiest installation steps. The material is also easier to clean.

Reduce venting. Most homes have one or more 3″ vent stacks plus several 1½″ and 2″ fixture vents. The vent's primary function is to equalize atmospheric pressure in the plumbing system in order to preserve the trap seal. The trap seal is the *S*-shaped pipe that prevents sewage gas from leaking into the house. Research shows that a single vent is adequate for several shared fixtures.

Use stack venting. Another simplified system is referred to as stack venting, which is recognized by most plumbing codes. In this system, each fixture drain is individually connected to the plumbing stack, eliminating the need for individual fixture vents.

Use CPVC (chlorinated polyvinyl chloride) plastic piping in place of copper. CPVC costs 20 percent to 40 percent less than copper. It installs quickly and easily and reduces the danger of fire from soldering the copper joints. The installation is so simple. The only disadvantage is its large expansion and contraction ratio when hot water is running through it. This can lead to popping and cracking noises in the wall when the hot water is turned on. To prevent this, do not anchor the plumbing to the wall. Leave additional area around the pipes to allow them to expand and contract freely.

Use an automatic vent. If you have a fixture that is located too far away to connect to the shared stack vent, consider an automatic vent. Automatic plumbing vents are located on the downstream side of a trap and admit room air through a pressure-activated diaphragm each time the fixture is used. Check to make sure these fixtures are allowed by the local building codes.

HVAC

Select the most efficient HVAC system. There are two costs to consider here: the installation cost and the operating cost. Choose the system that best fits the climate in your area. For instance, gas heaters are inexpensive to install and operate at a moderate cost. Heat pumps are more expensive to install but need very little electricity to operate. However, they lose efficiency as temperatures drop, so they don't work as well in northern climates with low winter temperatures. You should also consider the relative costs of electricity and gas in your area when making your decision.

Size the equipment to the house. Make sure the BTU rating of the HVAC unit matches the size and insulation efficiency of your house. A system that is too large will turn off and on much too often and will not operate efficiently. A system that is too small will be on all the time and will not be able to handle temperature extremes. If you insulate the house well, a smaller unit is required, saving money on installation and operation.

Downsize duct systems. If the house is very energy efficient where smaller units are installed, you can specify smaller ductwork that can be installed more easily and cheaply.

Locate the HVAC in the center of the house. Ducting is more efficient and less expensive when it radiates from a central point and is connected directly to the HVAC unit.

Use high-efficiency units with a high EER (energy efficiency rating). These units may be a little more expensive to install, but they pay back the expense in savings on monthly utility bills.

SAVING MONEY AFTER CONSTRUCTION

Cleaning Up

Cleaning up yourself can save you money. Normally, a cleaning crew works on a home for hours before it goes on sale. Burn or bury trash if you can. It is cheaper than hauling off the material. Check your local and county ordinances for rules on burning and burying.

MOVING IN

Most likely, you will build in the same city (or nearby) where you live during the project. Moving yourself can save hundreds of dollars. You can save yourself a little rent if you move in early and "finish up" while you're at home. Ask the local building inspector about minimum requirements for occupancy.

Homebuilding and Computers

COMPUTER SOFTWARE

You would have to be living in a cave to not notice the tremendous influence that computers and the Internet have had on our lives. The power of modern computers that can be purchased for under $2,000 far exceeds multimillion dollar mainframes of just a few years ago. This fact is important for you, the potential contractor, because building a home is just a large management project. And computers are perfect for managing the details for you.

Powerful computer software will help you build your house in a variety of ways:

- The Internet can provide a tremendous amount of information about construction suppliers, products and organizations. You can find detailed information about virtually every major supplier, or at least their address and phone.
- CAD design packages help you to visualize the home plan and produce the final drawings.
- Estimating software can produce accurate and complete materials estimates.
- Project management and accounting software follows the details of the project and keeps it on budget.

Remember, things change in the computer industry virtually overnight. The following information should be used as a guide only. By the time you read this, much of it will be obsolete. You can obtain a complete listing of the latest computer software for the construction industry by contacting the NAHB (National Association of Home Builders) or by visiting your local computer software stores.

The Internet

Covering everything you need to know about the Internet is beyond the scope of this book, but we'll cover a few of the basics to provide you with the tools you need to continue your construction education on the Web. We will assume that you have some familiarity with a computer and software; otherwise, most of this information will seem like gibberish to you. If you don't have a computer, but want to access some of the valuable information available, talk to a friend who has one, or visit your public library. Many libraries now offer free access to the Internet at public computer terminals.

The Internet is a wonderful and rapidly growing source of information related to just about everything—including homebuilding. The best information can be found on web pages—graphic pages of information available through a computer browser. This area is called the World Wide Web, better known as the web, WWW or W3. If you've seen or heard advertisements that say "visit us at www.something.com," that's what they're talking about—the company's web address, also known as a URL.

What You'll Need To Access the Internet

To access the Internet, you'll need two pieces of hardware and three pieces of software:

- **A computer.** This can be an IBM PC, Apple computer or UNIX-based system. If you have speakers connected to your computer, you can hear audio on web sites that feature sound. You may also wish to have word processing software on your computer so that you can save and view information you gather while on the Net.
- **A modem.** This is a small device, costing about $150 or less, used to communicate with other computers over the phone lines. The faster the modem, the better. Many construction pages have a lot of pictures, which take a long time to load. Higher speed access, such as ISDN and cable modems, are becoming available. These will greatly speed up access to the graphics-based web pages, such as product photographs. Because the modem uses your phone line, you

may prefer to install a second, separate phone line so that you can work on the Internet while still receiving phone calls.

- **Internet account**. Internet access can be obtained from several sources including the phone company, cable TV suppliers and online services such as CompuServe or America Online (AOL). Access normally costs less than $25 per month, based on a certain number of hours used. These services also allow you to send and receive electronic mail (e-mail). This is very useful for contacting construction manufacturers for information.
- **Access software**. This is usually supplied by your Internet provider. This software normally comes with simple instructions that walk you through the installation process.
- **Browser**. This is your window to the Internet. It's what you use to navigate ("surf" the web) and view text/graphical content. A browser has an area where you type a web address. Your service provider will probably furnish you with a browser along with your other software.

Finding Information (Surfing the Web)

Once you're connected, how do you find the information you're looking for? You will find a list of helpful websites at the end of this chapter, but be forewarned: The Internet changes drastically every week. Many (maybe most) of these sites may be obsolete by the time you read this. You need to learn how to use the search utilities on the Internet to find the information you want. This is only an overview. Read the documentation accompanying your browser software, or purchase one or more of the many third-party books available about the Internet.

Now what? Search . . .

If you know what you need, but don't know where to go, use one of the "search engines" available on the web. There are several. A search engine is just another website that allows you to type in several key words about the subject you're interested in. A popular search site is Yahoo. (But there are many others, such as Altavista, Excite and Infoseek. Check a recent book about the Internet for the addresses of others.) To access it, you simply enter:

http://www.yahoo.com

in the Address or Location window on your browser and press the Enter or Return key.

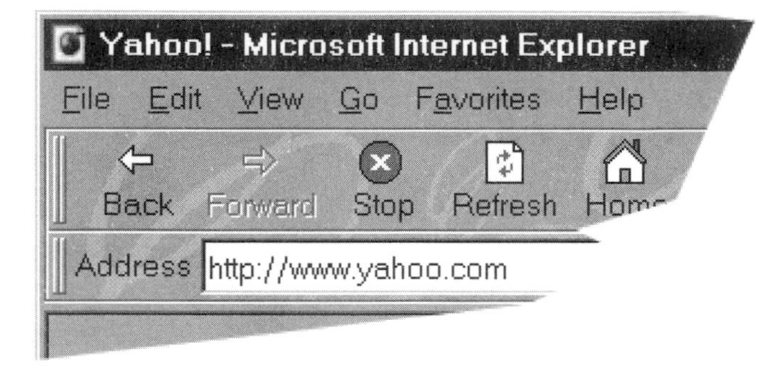

Figure 7: When entering a URL (address), make sure to type it in lower case. The search is case sensitive.

The http:// prefix (HyperText Transfer Protocol) should always be added in front of any web address. The search engines have topical directories of subjects you can access directly, or you can enter your own custom search information. Type in several words to make your search very specific. Otherwise you will receive several thousand "hits" that would take forever to scan through.

Search engines will display about 20 sites at a time. To go to one of the websites listed, click on the underlined text. This is referred to as ***hypertext***, and refers to the website's address. The hypertext link will take you to another website with more information. Each web page will have hypertext links to other related pages, so you can click on them to narrow your search even more. Because of the hypertext links, it doesn't take long to find references to thousands of sites. The biggest challenge on the Internet is sorting through all of them!

When you find a site you like, you can save it as a "bookmark," so that you can go back to it easily at any time. To do this, use your browser's Add Bookmark or Favorites feature to mark the page you are currently on.

Once you get the hang of surfing the web, you will find all types of construction-oriented information. The manufacturers of construction products depend on their product catalogs to sell to the public (customers and builders). A visit to a home show will

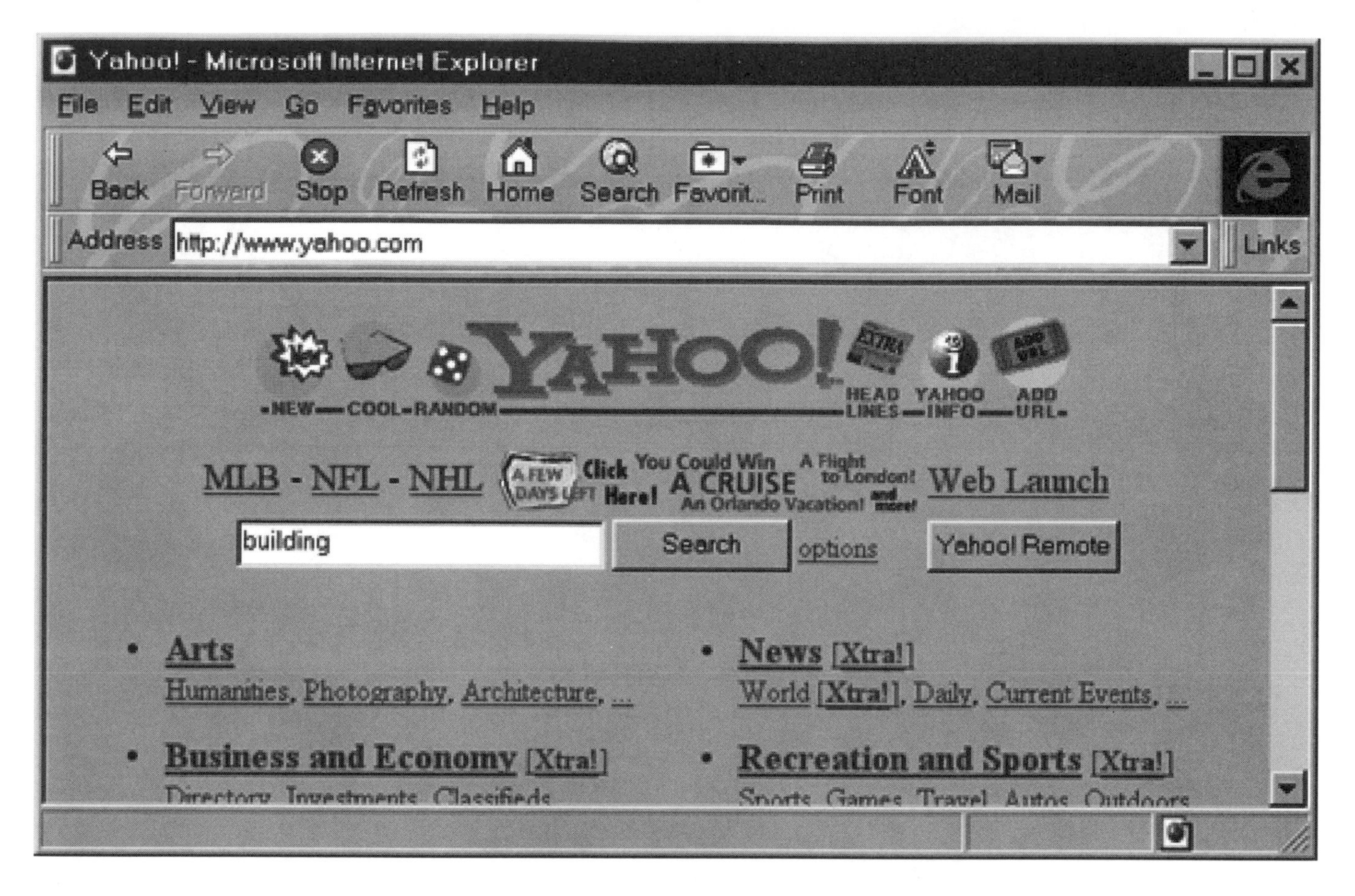

Figure 8: Entering a generic term like *building* will result in several thousand matches. Use specific terms such as *construction windows supplier homebuilding*. The search utility will return only documents that contain all these words. This will narrow your search and return a manageable list of sites. More sophisticated searches are available that will narrow the search even further.

net you an abundance of catalogs from product manufacturers. Now, the Internet provides a perfect arena for these manufacturers to make their catalogs available in electronic form to anyone. These websites also contain a wealth of construction and design hints, all written by the company's building experts. When you surf these sites, you will educate yourself on the most recent information available. Here are a couple of sites with links to thousands of building companies. There are many more sites at the end of this chapter.

BuilderNet (http://www.buildernet.com)
A website designed specifically for architects, engineers and contractors, with links to thousands of manufacturers. The links are organized by material classification code, commonly used in the commercial construction industry, or you can search for a specific product or supplier. Information is free.

Build.Com (http://www.build.com)
Another reference site aimed a little more toward the residential and home improvement market. Companies are listed by construction type, and each hypertext listing has a brief description. Clicking on the

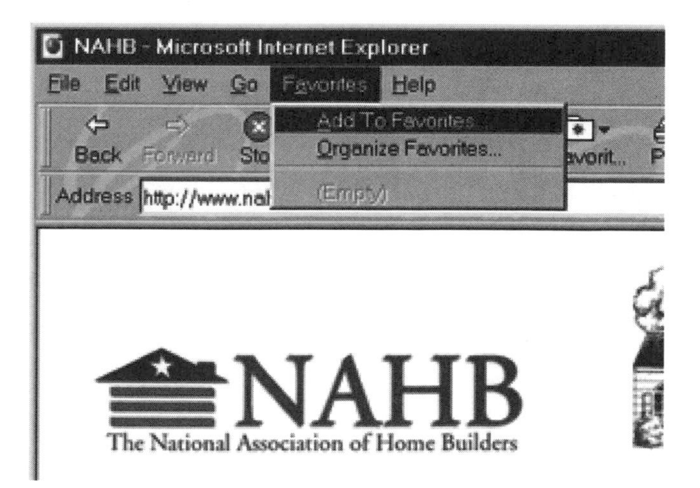

Figure 9: Adding a site to your Favorites list. This will allow you to easily go back to that page at a later date.

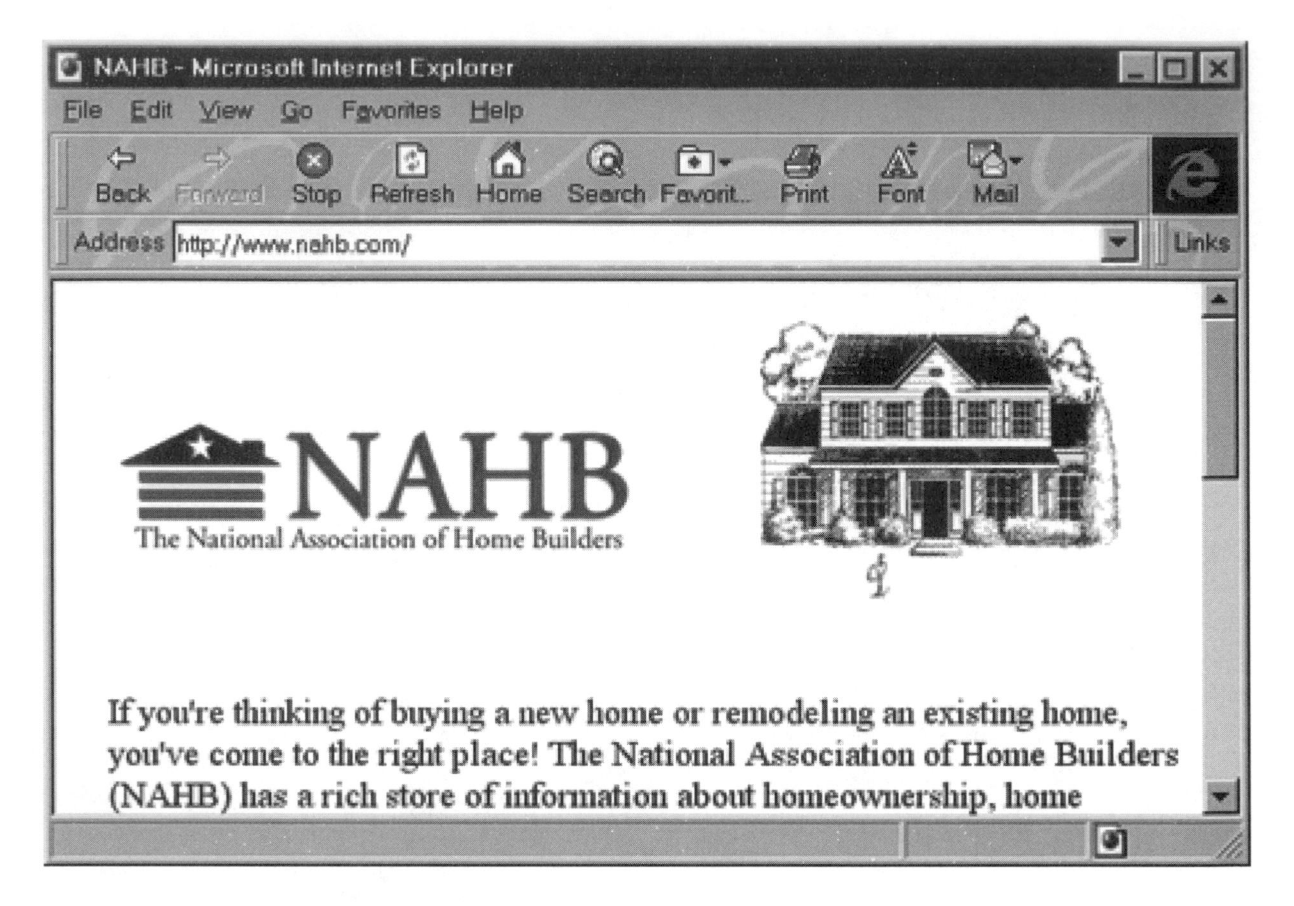

Figure 10: The website for the National Association of Home Builders.

hypertext name will take you to that company's own home page.

Computer Aided Design (CAD) Software

A few years ago, dedicated CAD workstations cost $15,000 and required several days to put in a complete two-dimensional house plan. Now, a $500 CAD package on a home computer can turn out complex floor plans in a fraction of that time. In addition, a few of these programs can do something the old plan programs couldn't: let you view your house from the inside before it is even built. The advances in these programs are nothing short of phenomenal. Even more important, they have become easier to use, as well.

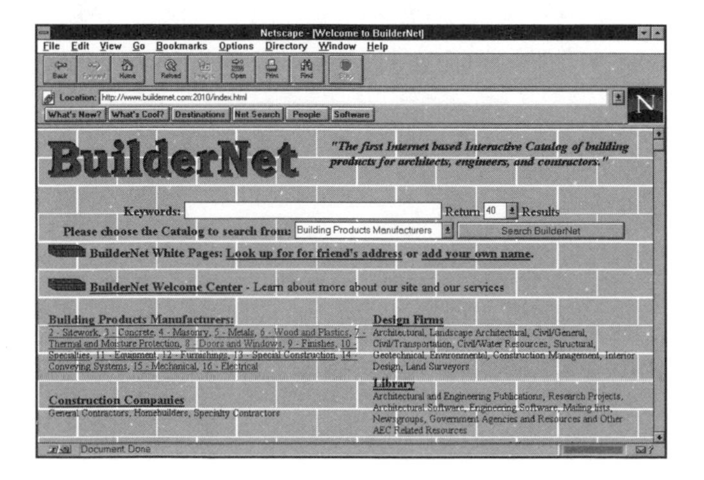

Figure 11: BuilderNet home page.

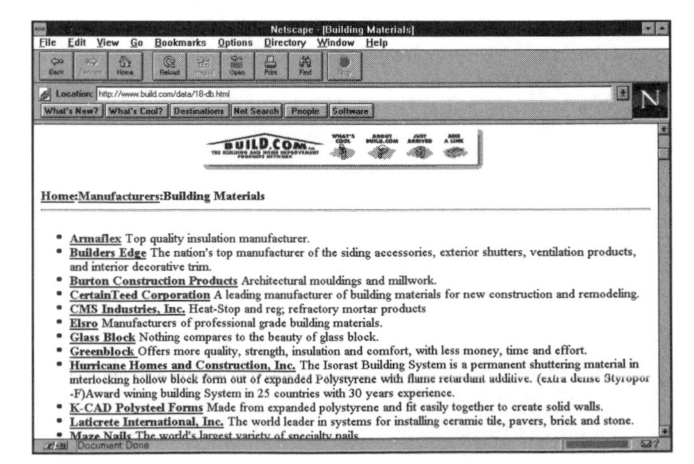

Figure 12: Build.Com list of manufacturers, with brief descriptions.

Even if you don't plan to draw your actual blueprints, the ability to visualize the design in the computer can help you to fix potential design shortcomings. You can place scale versions of your own furniture in the plan to see how they fit. You can use this information to make changes to traffic flow, place electrical outlets and even check out the view from the foyer.

If you choose to draw the entire plan, it is advisable to let a licensed architect or draftsman study the design for structural or stylistic problems. After all, they have studied design for years, and their input can be extremely valuable. CAD packages fall into two basic types:

- Traditional two-dimensional CAD systems
- Newer three-dimensional "virtual reality" systems

Two-Dimensional CAD

These programs fall into the more traditional definition of CAD packages used for construction. They are the ancestors of some older traditional systems that have migrated down from the big CAD workstations. They are the best programs for producing a professional set of blueprints, complete with dimension lines, call outs and accurate scaled images. They can be output to the large electronic pen-drafting machines, which can draw a full-size ¼″-1′ scale blueprint.

Autocad LT
Autodesk Inc.
(415) 507-5000
Price around $500
Baby brother to the famous best-selling Autocad software, one of the most professional and powerful CAD systems available. Autocad LT is a stripped-down version, but still retains the main features necessary to create a full-featured drawing. For advanced users.

DataCAD
Cadkey, Inc.
(203) 298-8888
Price around $250
Another full-featured CAD system capable of developing full-fledged drawings. Excels at drawing double wall lines. For advanced users.

Drafix CAD
Softdesk Retail Products
(816) 891-1040
Price around $500
One of the first Windows-based CAD systems. Easy to use, with ability to store repetitive tasks in macro shortcuts. Medium to advanced.

Visio Home
Visio Corporation
(206) 521-4500
About $50
Very easy to use, productive drafting package. "Smart Shapes" makes altering walls and furniture very easy. Drawings are high quality, but program lacks serious drafting features. Has a big brother called Visio Technical, which includes many traditional drafting tools.

Three-Dimensional CAD

This is one of the major advancements in personal computer drafting software. These programs have borrowed 3D features from their larger, more expensive workstation cousins and integrated them into complete

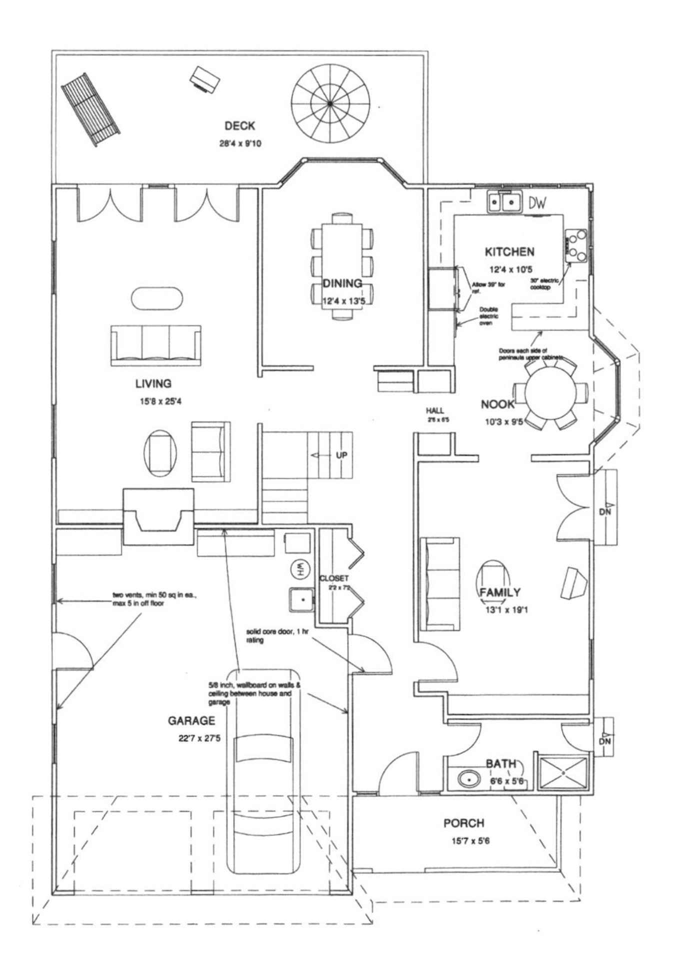

Figure 13: Floor plan drawn in 3D Home Architect, complete with some furniture. The figures on page 62 show the different 3D views that are available after the floor plan is input.

Figure 14: External view of the plan on page 61, complete with roof layout. Note images visible through transparent windows.

plan drawing systems. Paradoxically, instead of making the programs more complex, many of these systems are easier to use than their two-dimensional counterparts. The trade-off is in flexibility. These systems are designed to view the structure, and don't have as many of the drafting features as "professional" 2D packages. Don't plan to use these to create complete blueprints. However, many houses have been built from the backs of paper bags, so blueprints are not always necessary. Three-dimensional CAD can't be beat for planning and viewing your house creation! You input the two-dimensional drawing objects and the programs automatically generate the three-dimensional picture of the house. Many of them allow you to place yourself within the plan and actually view a three-dimensional rendering that looks very much like a real house.

3D Home Architect
Broderbund Software
(415) 382-4745
About $60
Very powerful and easy-to-use plan-viewing software. The basic plan goes in fast and easy. The program handles the complex task of designing stairs extremely well. Once the plan is entered, you can view one or multiple stories of the house at the same time, in virtual reality. This is great when your plan includes two-story or vaulted ceilings in some of the rooms. Profile and whole-house views are also available. Once the plan is complete, you can even print out a whole-house materials estimate. The package also includes a roof design module. Its main function will be for designing and viewing your dream home. It lacks some of the detail-drafting tools of the two-dimensional packages.

Home Design 3D
Expert Software Inc.
(305) 567-9990
About $15 (amazing!)
Low cost and easy to use, this program works well to visualize your dream home. Virtual reality viewing. Features are basic.

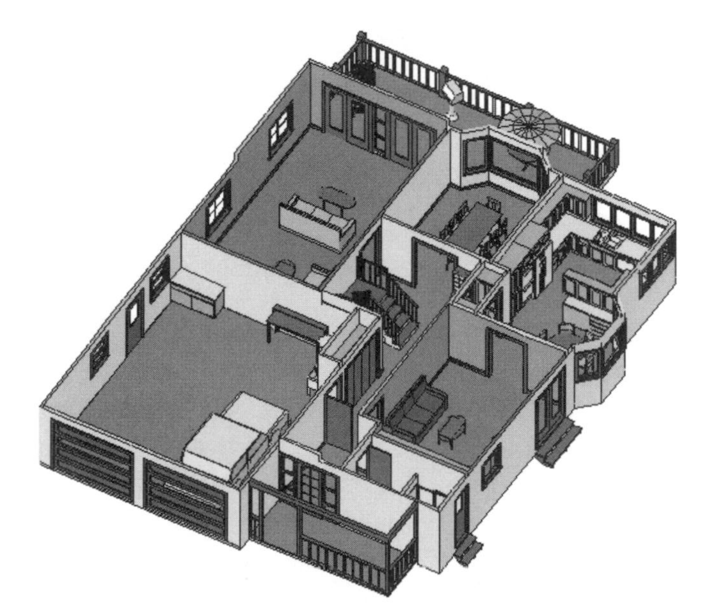

Figure 15: Three-dimensional view of the floor plan on page 61. This image was rendered automatically by the software from the dimensions of the floor plan.

Figure 16: Here's the view of the kitchen from 3D Home Architect, complete with appliances and plumbing fixtures. Virtual reality views are excellent for evaluating clearances and checking out the view from the breakfast table.

myHouse
DesignWare Inc.
(617) 924-6715
About $85
Very easy to use and generates good virtual reality images. Great for viewing. Won't handle more advanced drafting needs.

Estimating Software

The "Achilles' heel" of every construction project is always the cost estimating. The sheer number of items that go into the house is staggering. A computer can help tremendously with this problem. Don't even think of starting a homebuilding project without a complete and accurate estimate. The bank will be impressed with a well-organized printout of costs.

Computers are perfect for calculating costs and making adjustments when plans change, keeping you within your budget. There are two ways to go:

- Dedicated estimating software
- Computer spreadsheets

Dedicated Estimating Software

An in-depth review of estimating software is beyond the scope of this book. Contact the NAHB for its official list of software suppliers and check the Internet for additional information. Many of the construction software companies have their own websites and offer detailed information about their products. Several companies even offer demo versions of their software that you can download directly off the Internet.

There are several dedicated construction estimating programs available. Many of the commercial software packages are geared toward professional residential and commercial builders. They can be quite complex and powerful, allowing the user to estimate costs in several different ways, track multiple suppliers, allow for waste and markup percentages, track retainage, estimate groups of materials together and even enter dimensions using an actual blueprint and digitizing tablet. They range in price from $100 to $1,000. Before you buy one of these packages, make sure you are serious about building and plan to build more than one home. Otherwise, you won't get a good return on investment. Some of the less expensive packages are little more than elaborate spreadsheets, so why not use a generic spreadsheet instead?

Computer Spreadsheets

Another, less intensive method of calculating costs is through the use of a computer spreadsheet such as Lotus 123 or Microsoft Excel. This approach will require more effort on your part, but is totally flexible when it comes to making changes. You need to be pretty familiar with your spreadsheet program before beginning, but this is no different from learning the nuances of a dedicated estimating package. You can follow the same format as this book's estimating forms, providing columns for price and quantity and letting the spreadsheet automatically calculate the formulas as you input the information.

There's an even easier option. Many of the 3D house plan programs will calculate the materials needed from the dimensions of your floor plan. You can then export this list of materials to a special file that can be imported into most popular spreadsheets. This gives you a head start on producing a complete estimate.

Other Construction Software

The list goes on. There are computer software packages available to track job costing, prepare purchase orders, calculate payroll, manage the scheduling of the entire project, manage bids, calculate excavation requirements, engineer trusses and satisfy a host of other special needs. Much of this software is designed for professionals and requires a large investment of time and money to use effectively. If you plan to build only one house, your own, then purchasing more elaborate construction software would be a waste of time. Here are a few other less expensive programs that will work well for managing a one-time construction project. They also have additional advantages: They're inexpensive and can be used for other purposes after the construction is done.

- **Personal information managers (PIM).** These programs track names and addresses and can be purchased at any software store. They include such programs as Act, Microsoft Outlook, Lotus Organizer and many others. Set up a special copy of the PIM to track your suppliers' names. You can use the built-in appointment scheduler to keep up with the subcontractors and to follow the progress of the job. Computer task lists ensure that you keep up with every detail of the project.
- **Word processors** such as Microsoft Word, Ami Pro and WordPerfect. In addition to the obvious

CODE	DIMENSIONS	DESCRIPTION	QTY.	UNIT	PRICE	TOTAL
GN1		heated wall area	1060	sqft	$0.0	$0.0
GN2		heated glass area	257	sqft	$0.0	$0.0
GN3		heated door area	79	sqft	$0.0	$0.0
M1		masonry fireplace	1		$0.0	$0.0
F1	2×4-91½″	fir stud 16″ O.C.	178		$0.0	$0.0
F2	2×4-16ft+	fir plate	1057	ft	$0.0	$0.0
F3	4×12″	door/window header	106	ft	$0.0	$0.0
F4	2×4-91 9/16″	fir stud 16″ O.C.	118		$0.0	$0.0
F5	2×6-91½″	fir stud 16″ O.C.	13		$0.0	$0.0
F6	2×6 16ft+	fir plate	40	ft	$0.0	$0.0
F7	2×4-102⅝″	fir stud 16″ O.C.	29		$0.0	$0.0
F8	2×4-91 11/16″	fir stud 16″ O.C.	10		$0.0	$0.0
F9	2×4-102 11/16″	fir stud 16″ O.C.	4		$0.0	$0.0
S1	4×8′ sheets	ext. wall sheathing	54		$0.0	$0.0
S2		exterior siding	1705	sqft	$0.0	$0.0
IN1	4×16×93″ batts	wall insulation	100		$0.0	$0.0
FL1		landing flooring	13	sqft	$0.0	$0.0
FL2		carpet	1267	sqft	$0.0	$0.0
FL3		vinyl floor	208	sqft	$0.0	$0.0
WB1	14×8′-½″	wall board	210		$0.0	$0.0
D1	30×80×1¾	ext. hinged-glass	6		$0.0	$0.0
D2	26×80×1¾R	ext. hinged-glass	1		$0.0	$0.0
D3	34×80×1¾R	ext. hinged-glass	1		$0.0	$0.0
D4	108×80	garage-glass	2		$0.0	$0.0
D5	36×80×1¾R	ext. hinged	2		$0.0	$0.0
D6	72×80	bifold	1		$0.0	$0.0
D7	30×80×1¾L	hinged	1		$0.0	$0.0
D8	36×80×1¾R	ext. hinged-glass	1		$0.0	$0.0
C1	BD3936×22	bath base cab	1		$0.0	$0.0
C2	BS3636	kit. base cab	1		$0.0	$0.0
C3	BD3336	kit. base cab	1		$0.0	$0.0
C4	CBD3636	kit. crnr base cab	2		$0.0	$0.0
C5	CBD3636	kit. crnr base cab	1		$0.0	$0.0
C6	BDR1536	kit. base cab	1		$0.0	$0.0
C7	BD3636	kit. base cab	1		$0.0	$0.0

Table 2: This is a partial example of an estimate automatically generated by 3D Home Architect. This information was derived from the dimensions of the floor plan and exported to a text file. The file was then imported into a spreadsheet program and manipulated.

function, these programs can create excellent purchase orders. By creating a "mail merge" form, you can automatically generate form letters, bid requests and purchase orders customized with the individual suppliers' names.

- **Personal finance** packages such as Managing Your Money and Quicken. These inexpensive programs will stand in admirably for the more powerful construction accounting software, especially for tracking one job. Make sure to set up a separate checking account just for tracking your construction expenditures. This software can automatically generate checks for payment and categorize the construction expenses.

OTHER USEFUL INTERNET SITES

Associations

Ari CoolNet
http://www.ari.org/
Air-Conditioning & Refrigeration Institute.

ASHRAE
http://www.ashrae.org/
American Society of Heating, Refrigerating and Air-Conditioning Engineers, Inc.

ASID (American Society of Interior Designers)
http://www.wp.com/asid/
Interior design industry news and toll-free numbers to reach ASID World Wide Referral Service.

Fannie Mae
http://www.fanniemae.com/
America's largest source of home mortgage funds.

National Wood Flooring Association
http://www.woodfloors.org/

National Wood Windows and Door Association
http://www.nwwda.org/supply/

OSHA - Safety
http://www.osha.gov/
The government safety organization.

Resilient Floor Covering Institute
http://www.buildernet.com/rfci/index.html/
An industry trade association of resilient floor covering producers.

Southern Pine Council
http://www.southernpine.com/
A lumber industry trade council.

The American Institute of Architects
http://www.aia.org/
The collective voice of America's architects, with more than 58,000 members.

The Better Business Bureau
http://www.bbb.org/bbb/

The Energy Efficient Building Association
http://www.eeba.org/

The National Society of Consulting Soil Scientists, Incorporated
http://www.wolfe.net/~psmall/nscss.html/

Directories

Bricks Resources
http://www.bricknet.com/mfg.html/
The site for manufacturers of masonry products.

Builder Online
http://www.builderonline.com/
This is an excellent directory of product manufacturers created by *Builder* magazine. Manufacturers list is keyword-searchable.

BuildNet
http://abuildnet.com/
Browse more than 15,000 building industry resources with search capability.

Clem Labine's Traditional Building
http://www.traditional-building.com/
Data on historical products and services for restoration and traditionally styled new construction.

Construction
http://www.constructionsite.com/
This site for manufacturers, distributors, contractors, homeowners and architects brings persons needing a service or product in contact with suppliers and manufacturers.

Construction Info
http://www.constructionnet.net/index.htm/
An interactive trade show on the Internet that has teamed up with many of the top national trade associations to provide networking among members

Construction/Remodeling
http://homecentral.com/

A comprehensive directory of other homebuilding sites with reviews.

Cost Estimating
http://www.rsmeans.com/
R.S. Means Company offers cost estimating information, along with advertising for the company's products.

Gardening
http://gardening.com/
A database of over 1,000 plants for the garden; includes search capability.

General Building Info
http://www.cnst.com/net/
Consolidates groups of construction professionals in one user-friendly directory.

HomePoint
http://www.homepoint.com/
A massive database of home product- and service-related sites.

KitcheNet
http://www.kitchenet.com/
Online resource for information about the kitchen and bath industry.

Plumbing
http://www.plumbnet.com/
A site dedicated to plumbers, plumbing engineers, plumbing designers, architects, interior designers and homeowners.

PlumbingSupply.com
http://www.plumbingsupply.com/
Offers advice on common plumbing problems, and lists of manufacturers.

PlumbNet
http://www.plumbnet.com/welcome.html/
Plumbing advice and resources.

Sweet's Group Catalog of Materials
http://database.sweets.com:8080/sweets/
Sweet's is the construction industry's primary product information source and a division of the McGraw-Hill Companies.

WOODnetWORK
http://www.woodweb.com/
Website for woodworking professionals.

Gardening

Cyndi's Catalog of Gardening Catalogs
http://www.cog.brown.edu:80/gardening/
This site contains a listing of more than 1,600 mail-order gardening supply catalogs and their addresses, links to other gardening websites and essays.

LivingHome
http://livinghome.com/
Online periodical guide to home and garden design.

The Ardent Gardener
http://trine.com/GardenNet/ArdentGardener/
Information on every aspect of gardening.

The Garden Gate
http://www.prairienet.org/garden-gate/homepage.htm/
Large site full of gardening information.

Magazines

Electronic House Magazine
http://www.electronichouse.com/
Home automation website with news and products.

Southern Living
http://pathfinder.com/vg/Magazine-Rack/SoLiving/
The online home of the gardening articles from *Southern Living*.

Materials

Appliance Clinic
http://www.phoenix.net/~draplinc/
A veteran appliance repairman imparts, in plain language, his wisdom concerning the buying, maintaining and repairing of your major home appliances—great information for both the consumer and the repair professional.

Aristokraft Kitchen Cabinets
http://www.aristokraft.com/

Builder's Edge - products
http://www.buildersedge.com/
Manufacturer of siding and ventilation equipment.

Certainteed General Building Products
http://www.certainteed.com/
Insulation, PVC pipe, fittings and foundation products, roofing, siding, ventilation, windows.

Delta Faucet Company
http://www.deltafaucet.com/

EcoWater
http://www.ecowater.com/
Manufacturer of a variety of water filtration devices.

Faucet Outlet
http://www.faucet.com/faucet/
Large selection of faucets and information about installing and selecting a faucet.

Flood Family of Products
http://www.floodco.com/
Flood manufactures exterior wood-care products and paint conditioners.

Frigidaire
http://www.frigidaire.com/

Grohe faucets
http://www.grohe.com/
Site featuring Grohe's bath fixtures and faucets.

Kraftmaid Cabinets
http://www.kraftmaid.com/

Lennox Heating and Air
http://www.davelennox.com/

Lite Touch
http://www.hometeam.com/litetouc/
Touch home automation systems.

Lowe's Home Improvement Warehouse
http://www.lowes.com/

Maytag Appliances
http://www.Maytag.com/

Mellco-The 'Decksperts' Home Page
http://www.mellco.com/
Mellco distributes lumber, synthetic products, stains and sealants for decks and other outdoor structures.

Merillat Cabinets
http://www.merillat.com/

Quality Cabinets
http://www.qualitycabinets.com/

The Home Depot
http://www.homedepot.com/

Miscellaneous

Don Vandervort's HomeSource
http://home.earthlink.net/~donvander/
Remodel, renovate, fix up or explore your home with nationally recognized home-improvement authority Don Vandervort.

E design
http://fcn.state.fl.us/fdi/e-design/online/edo.htm
Electronic journal of the Florida Design Initiative.

Home Builders Institute
http://www.hbi.org/
Education and training programs serving the home building industry.

Hometime
http://www.hometime.com/
Dean Johnson's popular PBS television series home site.

The Internet Headquarters for Do-It-Yourselfers
http://doityourself.com/
A vast storehouse of construction and home repair information collected from suppliers and other sources.

Plans

DesignBasics Plans
http://www.designbasics.com
Builders' and homebuyers' plan services.

HomeStyles Interactive—The Home Builders' Connection
http://homestyles.com/
HomeStyles brings together the home plans of 50 of America's leading home designers.

Interactive Home Design Center
http://www.showoff.com/
Home design software.

Safety and Inspections

INSPECTIONS AND INSPECTORS

During the course of your project, you will probably run into two primary types of inspectors:

- County and local officials
- Financial institutions

In general, inspectors are on your side. They are there to help ensure that critical work meets minimum safety and structural standards. Hence, in many cases, an approved inspection does not assure you that a job has been done properly. If certain work fails inspection, you will be required to correct it, reschedule the inspection and probably pay a reinspection fee.

County and Local Officials

A certain number of inspections are normally required by state or county officials during the construction process. When you apply for your building permit, ask for a schedule of required inspections. Although actual inspection points will vary from place to place, your inspections will probably include some of the critical inspection points listed below in order of occurrence:

- Before pouring concrete footings
- Before pouring concrete foundation
- After framing rough-in
- After electrical, HVAC and plumbing rough-in
- After laying sewer or septic tank lines and before filling in
- Final completion

Financial Institutions

If you borrowed money from a financial institution for your project, they have an active interest in seeing that the money is being spent properly, as intended, and at a rate such that the money is used up just as the project is being completed. In addition to requiring approval by local inspectors, bank inspectors will normally want to examine the site each time a loan draw is requested. As your financial institution sees certain phases of work completed, it will release a certain amount of loan money based on the percentage of completion of the project. Hence, it will make sure that you have "earned" the right to use the money as intended. Each lender sets up its own draw schedule, allowing a certain percentage for each construction task.

Preinspection

Nobody enjoys unpleasant surprises. Before you have anyone show up for an inspection, make sure that you are ready to pass the inspection. Check the work yourself. Use checklists and your subs. This will ensure that the formal inspection turns out to be a mere formality. This eliminates rejected inspections, embarrassment, reinspection fees and a bad builder reputation. Inspectors will see you in a favorable light if you are always fully prepared for their inspections. Your major subs—plumbers, HVAC and electricians—normally schedule their own inspections, but you should follow up.

GUIDELINES FOR WORKING WITH INSPECTORS

- MAKE sure the work to be inspected is ready for inspection.
- DETERMINE how much advance notice is required and always give proper advance notice.
- NOTIFY the inspector to cancel or reschedule the inspection if work is not ready.
- BE present for the inspection so that you can determine exactly what is incorrect and what must be done to correct it if necessary. If you don't proceed in this manner, you may have to wait for a form letter in the mail and then follow up.

- STAY on good terms with all your inspectors. Be friendly and courteous. Don't argue when they ask for something to be fixed. They can give you good advice. Cooperate!
- HAVE the site looking clean. This will put your inspector in a good frame of mind and will let him know that you are organized and efficient.

PROJECT SAFETY

Few disasters can affect your project more adversely than the on-site death or severe injury of you, one of your subs, a neighbor's child or a building partner. As they say, "Project safety is no accident." Hence, project safety should play a major part in your overall project planning and routine supervision. The best way to make sure that serious accidents don't happen is to make safety a big issue with your subs. Remain aware of safety and emphasize this to your workers and site visitors. All you have to do is remember to follow a few rules and you'll be OK.

(1) Use common sense. If something seems a little risky, back off and give yourself more time. Most accidents happen when someone is attempting to perform a shortcut to save time.

(2) Wear proper clothing at the site: sturdy leather shoes with sturdy soles—preferably steel-tipped.

(3) Watch out for kids. Construction sites are magnets for children. "Caution Construction" signs should be posted at the site.

(4) You should have builder's risk insurance covering your building site and accidents that may occur there.

SITE SAFETY CHECKLIST

- ☐ Site safety rules communicated to all subs
- ☐ First aid kit available and workers familiar with basic first aid practices
- ☐ Phone numbers for police, ambulance and firehouse available
- ☐ Temporary electrical service grounded
- ☐ All electrical tools grounded
- ☐ All electrical cords kept away from water
- ☐ Warning and danger signs posted in appropriate areas
- ☐ Hard hats and steel-tipped shoes worn where needed
- ☐ Other protective gear available and used: goggles, gloves, etc. Always use protective goggles when flying fragments are possible.
- ☐ Set a good example as a safety-minded individual
- ☐ Adequate slope on edges of all ditches and trenches over 5′ deep
- ☐ Open holes and ditches fenced properly
- ☐ Open holes in subfloor properly covered or protected
- ☐ Scaffolding used and secure where needed
- ☐ Workers on roof with proper equipment
- ☐ Excess and/or flammable scrap not left lying around
- ☐ No nails sticking up out of boards or other materials
- ☐ Gas cans have spark arrest screens
- ☐ Welding tanks shut off tightly when not in use, and stored upright
- ☐ Area where welders working checked for smoldering or slow-burning wood—houses have burned down because of this!
- ☐ Proper clearance from all power lines

Estimating Costs

The cost estimate, or take-off, will be one of the most time-consuming and important phases of your construction project. Jumping into your first construction project without a realistic estimate of your costs could spell financial disaster. Spend a lot of time working on this one before continuing with your project.

"How in the world do I do a take-off with no previous experience?" A fair question indeed. The information in this section should help you along quite a bit. In addition, *Estimating Home Building Costs* (W.P. Jackson, Craftsman Book Company, 6058 Corte del Cedro, P.O. Box 6500, Carlsbad, CA 92008-0992), is an invaluable and time-saving paperback. Also, this is the time to study your construction books in detail to get a feel for the structure of the building and the quantity of materials used in each phase.

The first "materials list" you may come in contact with may be included with your construction plans. These lists are usually offered as an extra at an additional price. As reliable documents, these lists are practically worthless. Many will be inaccurate or based on outdated construction techniques. There is no way of telling whether the list is accurate or not. The best use of this type of take-off is as a checklist while running your own figures.

Another important source of information will be the subcontractors you plan to use. These subs are familiar with the type of construction in your area and will be very knowledgeable in calculating the amounts of materials needed. In fact, most of your subs will offer to calculate the amount of materials needed for you. This can be very valuable, especially for framing, because of the sheer number of different materials used. Don't, however, rely solely on your subs' estimates! Make sure to run your own figures to compare against their estimates. Subs have a tendency to overestimate just to be on the safe side. They hate to run out of material and have to sit around while you run to the supplier for more. Just make sure that you do add in at least a 5 to 10 percent waste factor to each estimate.

Another good source of estimating information is your local materials supplier. In previous years, suppliers used to do quite a bit of estimating for their customers. If you can find a supplier who still does this, he is worth his weight in gold.

When figuring your take-off, use the materials sheet on pages 88-99 as a checklist to make sure that you don't forget any necessary items. As you figure each item, enter it onto your purchase order forms, taking care to separate materials ordered from different suppliers onto different invoices.

EXCAVATION

Excavation is usually charged by the hours of operating time. Larger loaders rent for higher fees. Determine whether clearing of the lot is necessary and whether or not refuse is to be buried on the site or hauled off. Approximate times for loader work are shown below. Remember, these are only approximations. Every lot is different and may pose unique problems. Your grader may supply chain-saw labor to cut up large trees.

One-Half Acre Lot	Small loader	Large loader
Clearing trees and shrubs	2-5 hrs.	1-4 hrs.
Digging trash pit	2-4 hrs.	1-3 hrs.
Digging foundation	2-3 hrs.	1-3 hrs.
Grading building site	1-5 hrs.	1-3 hrs.
Total	**7-17 hrs.**	**4-13 hrs.**

Larger loaders can be an advantage and can be cheaper in the long run, especially if the lot is heavily wooded or has steep topography. If the lot is flat and sparsely wooded, use the smaller loader for the best price.

CONCRETE

When calculating concrete, it is best to create a formula or conversion factor that will simplify calculations and avoid having to calculate everything in cubic inches and cubic yards. For instance, when pouring a 4″ slab, a cubic yard of concrete will cover 81 square feet of area. This is calculated as follows:

1 cubic yard = 27 cubic feet = 46,656 cubic inches

1 sq. ft. of concrete 4″ thick = 576 cubic inches (12″ × 12″ × 4″)

46,656 cu. in. ÷ 576 cu. in. = 81 sq. ft. of coverage

By doing this equation only once, you now have a simple formula for calculating slabs 4″ thick that you can use for driveways, basement floors and slab floors. For instance, to determine the concrete needed for a 1,200 sq. ft. slab, simply divide by 81.

1,200 sq. ft. ÷ 81 = 14.8 cu. yds. of concrete

NOTE: Always make sure to add a 5 to 10 percent waste factor to all calculations.

Use this same principle of creating simple formulas for determining footings, calculating blocks, pouring concrete walls, etc. Just remember to calculate your own set of formulas based on the building codes in your area. For your convenience, refer to the concrete tables to determine the conversion factors for your project.

Footings

Footing contractors will be hired to dig the footings and to supervise pouring of the footings. Footing subs generally charge for labor only, and will charge by the lineal foot of footing poured. Pier holes will be extra. You must provide the concrete. These subs may charge more if they provide the forms. If you use a full-service foundation company, they will charge you for a turnkey job for footing, wall and concrete. This makes estimating easy.

Cu. Yds. concrete needed per lineal foot of footing							
Depth Ins.	Width of footing in inches						
	4″	6″	8″	10″	12″	14″	16″
4″	0.0004	0.006	0.008	0.010	0.012	0.014	0.016
6″	0.006	0.009	0.012	0.015	0.019	0.022	0.025
8″	0.008	0.012	0.016	0.021	0.025	0.029	0.033
10″	0.010	0.015	0.021	0.026	0.031	0.036	0.041
12″	0.012	0.019	0.025	0.031	0.037	0.043	0.049
14″	0.014	0.022	0.029	0.036	0.043	0.050	0.058
16″	0.016	0.025	0.033	0.041	0.049	0.058	0.066

Table 3: Concrete needed for footings.

Footings generally must be twice as wide as the wall they support, and the height of the footing will be the same as the thickness of the wall. The footing contractor will know the code requirements for your area. Ask him for the dimensions of the footing and then figure an amount of concrete per lineal foot of footing. Here is a typical calculation for supporting an 8″ block wall:

Footing dimension: 8″ high × 16″ wide

8″ × 16″ × 12″ (1 ft. of footing) = 1,536 cu. in.

46,656 cu. in. ÷ 1,536 cu. in. = 30.38 lineal ft. of footing per cu. yd.

With this size footing, figure one cubic yard of concrete for every 31 lineal feet of footing. Make sure to include 4 extra feet of footing for every pier hole. Add 10 percent for waste. Refer to the footing table for concrete factors.

Concrete Floors or Slabs

Use the conversion factor of 81 square feet per cubic yard of concrete for 4″ slabs or basement floors. If your slabs must be more or less than 4″, make sure to calculate a new conversion factor or refer to the table.

Monolithic Slabs

Break a one-piece slab into two components—the slab and the footing sections—then figure the items separately using the tables.

Block Foundations and Crawl Spaces

Concrete blocks come in many shapes and sizes, the most common being 8″ × 8″ × 16″. Blocks that are

Square footage of slab that 1 cu. yd. of concrete will fill			
Slab	S.F.	Slab	S.F.
1″	324	7″	46
1½″	216	7″	46
2″	162	8″	41
2½″	130	8½″	38
3″	108	9″	36
3½″	93	9½″	34
4″	81	10″	32
4½″	72	10½″	31
5″	65	11″	29
5½″	59	11½″	28
6″	54	12″	27

Table 4: Cubic yards of concrete in slab.

12″ thick are used for tall block walls with backfill to provide extra stability. Blocks 8″ deep and 12″ deep cover the same wall area: .888 sq. ft.

To calculate the amount of block needed, measure the height of the wall in inches and divide by 8″ to find out the height of the wall in numbers of blocks. The height of the wall will always be in even numbers of blocks plus a 4″ cap block or 8″ half block (for pouring slabs). Cap blocks are solid concrete 4″×8″×16″ blocks used to provide a smooth surface to build on. If you are figuring a basement wall, take the perimeter of the foundation and multiply by .75 (three blocks for every 4 ft.); then multiply this figure by the number of rows of block. To calculate the row of cap block, multiply the perimeter of the foundation by .75. Always figure 5 to 10 percent waste when ordering block.

If your foundation is a crawl space, your footing is likely to have one or more step-downs, or bulkheads, as they are called. Step-downs are areas where the footing is dropped or raised the height of one block. This allows the footing to follow the contour of the land. As a result, sections of the block wall will vary in height, requiring more or fewer rows of block. Take each step-down section separately and figure three blocks for every 4 feet of wall. Multiply the total number of blocks by the total number of rows and then add all the sections together for the total number of blocks needed.

Poured Concrete Walls

Your poured-wall subcontractor will charge by the lineal foot for setting forms. Usually, the price quoted for pouring will include the cost of concrete; if not, you must calculate the amount of concrete needed. Determine the thickness of the wall from your poured-wall sub and find the square foot conversion factor for the number of square feet coverage per cubic yard. Divide this factor into the total square footage of the wall. Example for an 8″ thick wall:

8″ poured concrete wall × 12″ × 12″ = 1,152 cubic inches per square foot of wall area

46,656 cubic inches ÷ 1,152 cubic inches = 40.5 square feet per cubic yard of concrete

Refer to the concrete wall table to quickly calculate the concrete needed.

Brick

To figure the amount of brick needed, figure the square footage of the walls to be bricked and multiply by 6.75. (There are approximately 675 bricks per 100 square feet of wall.) Add 5 to 10 percent for waste.

Cu. Yds. concrete needed per lineal foot of wall							
Height of wall	Width of wall in inches						
	4″	6″	8″	10″	12″	14″	16″
4′	0.049	0.074	0.099	0.123	0.148	0.173	0.198
5′	0.062	0.093	0.123	0.154	0.185	0.216	0.247
6′	0.074	0.111	0.148	0.185	0.222	0.259	0.296
7′	0.086	0.130	0.173	0.216	0.259	0.302	0.346
8′	0.099	0.148	0.198	0.247	0.296	0.346	0.395
9′	0.111	0.167	0.222	0.278	0.333	0.389	0.444
10′	0.123	0.185	0.247	0.309	0.370	0.432	0.494
11′	0.136	0.204	0.272	0.340	0.407	0.475	0.543
12′	0.148	0.222	0.296	0.370	0.444	0.519	0.593
13′	0.160	0.241	0.321	0.401	0.481	0.562	0.642
14′	0.173	0.259	0.346	0.432	0.519	0.605	0.691
15′	0.185	0.278	0.370	0.463	0.556	0.648	0.741
16′	0.198	0.296	0.395	0.494	0.593	0.691	0.790
17′	0.210	0.315	0.420	0.525	0.630	0.7535	0.840
18′	0.222	0.333	0.444	0.556	0.667	0.778	0.889

Table 5: Concrete needed for concrete walls.

Mortar

Mortar comes premixed in bags of masonry cement, which consist of roughly one part portland cement and one part lime. Each bag requires about 20 shovels of sand when mixing. To calculate the amount of mortar needed for brick, figure one bag of cement for every 125 bricks. For block, figure one bag of cement for every 28 blocks. Make sure to use a good grade of washed sand for a good bonding mortar. Most foundations will require at least 10 cubic yards of sand.

FRAMING

Estimating your framing lumber requirements will be the most difficult estimating task and will require studying the layout and construction techniques of your house carefully. Be sure to have a scaled blueprint of your home as well as a scaled ruler for measuring. Consult with a framing contractor before you begin for advice on size and grade of lumber used in your area.

This is the time you will want to pull out your books on construction techniques and study them to familiarize yourself with the components of your particular house. Local building code manuals will include span tables for determining the maximum spans you are allowed without support. Your blueprints will also list the sizes of many framing members and may include construction detail drawings, which can be especially helpful in determining the size and type of lumber needed for framing.

Floor Framing

Determine the size and length of floor joists by noting the positions of piers or beams and by consulting with your framing contractor. Floor joists are usually spaced 16″ on center. Joists spaced 12″ on center are used for extra sturdy floors. If floor trusses are used, figure 24″ on-center spacing. Calculate the perimeter of the foundation walls to determine the amount of sill and box sill framing needed.

Calculate the square footage of the floor and divide by 32 (square footage in a 4 × 8 sheet of plywood) to determine the amount of flooring plywood needed. If the APA Sturdifloor design is used, order ¾″ or ⅝″ tongue and groove exterior plywood.

Bridging between floor joists is used to reduce twisting and warping of floor members and to tie the floor together structurally. Bridging is calculated by taking the total number of floor joists and multiplying by three. This is the lineal feet of bridging needed for one course of bridging. Most floors will require at least two courses of bridging, so double this figure.

If you are using a basement or crawl space, you will probably be using a steel or wood beam to support the floor members. If floor trusses are used, this item may not be needed, since floor trusses can span much greater distances. Consult with your local truss manufacturer.

Wall Framing

When calculating wall framing lumber, add together the lineal feet of all interior and exterior walls. Since there is a plate at the bottom of the wall and a double plate at the top, multiply wall length by three to get the lineal feet of wall plate needed. Add 10 percent for waste. For precut wall studs (make sure to get the proper length if precut studs are used—there are many sizes), allow one stud for every lineal foot of wall and two studs for every corner. Count all door and window openings as solid wall. This will allow enough for waste and bracing.

Headers are placed over all openings in load-bearing walls and are doubled for structural support. Add together the total width of all doors and windows, and multiply by two. Check with your framing sub for the proper size header.

The second floor can be calculated the same way as the first floor, taking into consideration the different wall layouts, of course. The second floor of a two-story house can be figured the same as a crawl space floor with the exception that the joists will be resting on load-bearing walls instead of a beam. This will require drawing a ceiling joist layout.

Roof Framing

If roof trusses are to be used, figure one truss for every 2′ of building length, plus one truss. If the roof is a hip roof or has two roof lines that meet at right angles, extra framing for bridging must be added.

The roof truss manufacturer should calculate the actual size and quantity of roof trusses at the site or from blueprints to ensure the proper fit. Get him to do this, and present a bid and material list, during the estimating process.

"Stick building" a roof is much harder to calculate and requires some knowledge of geometry to figure all the lumber needed. A "stick-built" roof is one that is built completely from scratch. The simplest way of finding the length of ceiling joists and rafters is to measure them from your scaled blueprints. Always round to the nearest greater even length, since all lumber is sold in even numbered lengths.

If you can't measure the length of the rafter from your plans, you can calculate rafter length and roof area using a conversion factor similar to the one used for concrete. Think of a roof as two identical triangles, back to back. If you know the length of the triangle base, you can calculate the length of the long side of the triangle (the rafter) by using trigonometry. We have spared you this headache by providing a conversion factor for each roof slope (rise/run). See the accompanying table for this multiplication factor. A 10″-slope roof means that the roof rises ten inches for every foot of run (distance). This term is sometimes mistakenly referred to as "10-pitch slope."

Pitch (slope)	Conversion Factor
1″	1.01
2″	1.02
3″	1.04
4″	1.06
5″	1.09
3″	1.12
6″	1.16
7″	1.21

Pitch (slope)	Conversion Factor
9″	1.26
10″	1.31
11″	1.36
12″	1.42
13″	1.48
14″	1.54
15″	1.61
16″	1.67

Table 6: Conversion factors for roofs.

To use this table, measure the total width of the house (including roof overhang) from the peak of the roof to the cornice edge. Multiply this by the conversion factor to obtain the rafter length. Don't forget to add in waste for lumber cut from the ends of each rafter (see illustration). This same conversion factor can also be used to calculate roof area for shingles and tar paper. Calculate the total area of the ceiling plus overhang and multiply by the conversion factor to obtain the total roof area. By the way, a 10″-slope roof is very steep!

When calculating ceiling joists, draw a joist layout with opposing joists always meeting and overlapping above a load-bearing wall. Figure one ceiling joist for every 16″, plus 10 percent extra for waste.

Rafters are also spaced 16″ on center. The length of the rafters can be determined by measuring from blueprints, making sure to allow for cornice overhang. Where two roof lines meet, a valley or hip rafter is necessary. Multiply the length of a normal rafter by 1.5 to get the approximate length of this rafter.

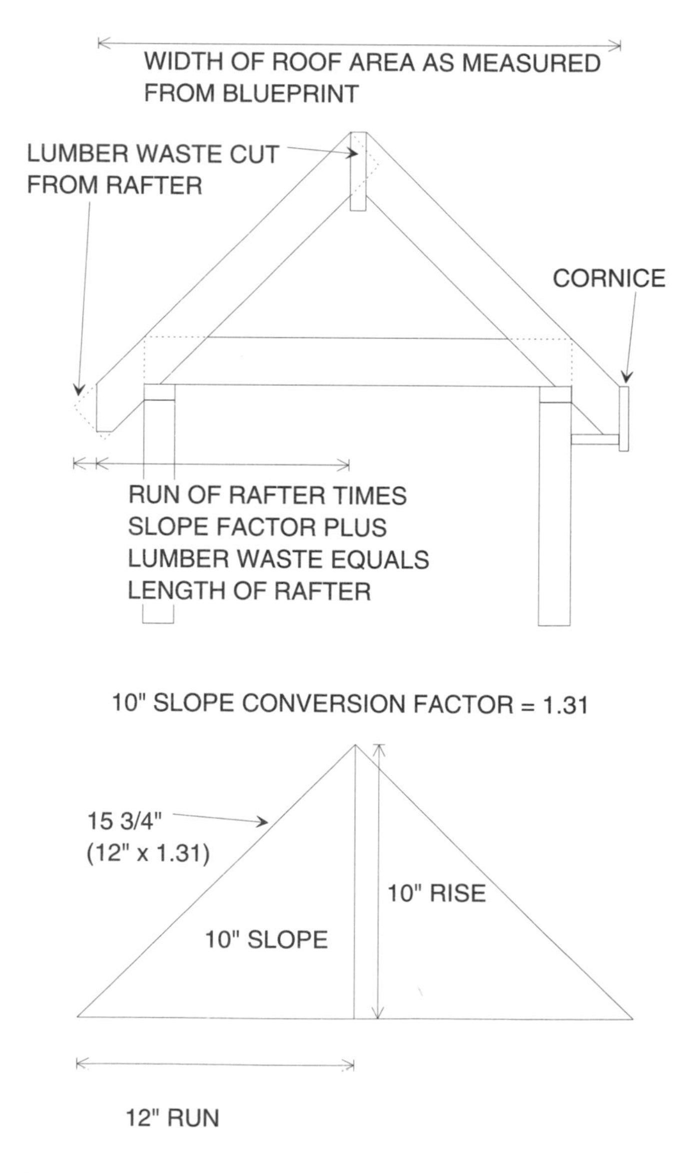

Figure 17: Length of rafter and area of roof calculated by using conversion factor for a 10″-slope roof.

Gable studs will be necessary to frame in the gable ends. The length of the stud should be equal to the height of the roof ridge. Figure one stud per foot of gable width plus 10 percent waste. This will be enough to do two gable ends, since scrap pieces can be used in the short areas of the gable.

Decking for the roof usually consists of ½″ CDX exterior plywood. Multiply the length of the rafter by the length of the roof to determine the square footage of one side of the roof. Double this figure to obtain total square footage area of the roof. Divide this figure by the square footage of a sheet of plywood (32). Add 10 percent waste. This is the number of sheets of plywood needed. Remember the square footage of the roof for figuring shingles.

ROOFING SHINGLES

Roofing shingles are sold in "squares," or the number of shingles necessary to cover 100 sq. ft. of roof. First, find the square footage of the roof, adding 1½ square feet for every lineal foot of eaves, ridge, hip and valley. Divide the total square footage by 100 to find the number of squares. Shingles come packaged in one-third square packages, so multiply the number of squares by three to arrive at the total number of packages needed.

Roofing felt is applied under the roof shingles as an underlayment. It comes in 500 sq. ft. rolls. Divide the total square footage of the roof by 500 and add 20 percent for overlap and waste to determine the number of rolls needed.

Flashing is required around any chimney or area where two roof lines of different height meet. It comes in 50 ft. rolls. Measure the length of the ridge if installing roof ridge vents.

SIDING AND SHEATHING

Multiply the perimeter of the outside walls by the height to obtain the total square footage of outside walls. If gables are to be covered with siding, multiply the width of the gable by the height (this figure is sufficient for both gable ends) and add this to the square footage of the outside walls. This figure is the total square footage of area to be sided.

Different types of siding are sold in different manners. Sheathing and plywood siding are sold in 4×8 sheets (32 sq. ft.). Divide this figure into the total square footage to determine the number of sheets needed. Add 10 percent for waste. Lap siding, on the other hand, is sold by squares or 1,000s. Make sure to ask if siding is sold by actual square footage or by "coverage area" (the amount of area actually covered by the siding when applied). Add 10 percent for waste—or 15 percent if lap siding is applied diagonally.

Corner trim boards are generally cut from 1×2 lumber. Figure two pieces for every inside and outside corner, the length being equal to the height of the wall being sided.

CORNICE MATERIAL

To estimate cornice material, you must first determine the type or style of cornice and the trim materials to be used. Consult blueprints for any construction detail drawings of the cornice.

The following materials are generally used in most cornices:

- Fascia boards—1×6″ or 1×8″
- Drip mold (between fascia and shingles)—1×4″
- Soffit—⅜″ exterior plywood
- Bed mold—1×2″
- Frieze mold—1×8″

Calculate the total lineal feet of cornice, including the gables, to find the total lineal feet of each trim material needed. Calculate the square footage of the soffit by multiplying the lineal feet of cornice by the depth of the cornice. Divide this by the square footage of a plywood sheet to determine the number of sheets needed.

INSULATION

Calculating the amount of wall insulation is easy—batts of insulation are sold by square footage coverage. Simply multiply the perimeter of the exterior walls to be insulated by the wall height to determine the total square footage to be insulated.

To calculate the amount of blown-in insulation needed, first determine the type of insulation to be used. Each type—mineral wool, fiberglass or cellulose—has its own R-factor per inch. This must be known in order to determine the thickness of fill. Then multiply the depth in feet (or fraction thereof) by the square footage of the ceiling area to arrive at the cubic foot volume. Blown-in insulation is sold by the cubic foot. If batt insulation is used in the ceiling, it can be figured in the same manner as the wall insulation.

Most insulation contractors will quote a price including materials and labor when installing insulation. Ask the contractor to itemize the amount of insulation used, for comparison with your figures.

DRYWALL

Gypsum drywall is sold by the sheet (4 × 10, 4 × 12, etc.) but is estimated by square footage. To find the total amount needed for walls, multiply the total lineal feet of inside and outside walls by the wall height. Make sure to count each interior wall twice, since both sides of the wall will be covered. Count all openings as solid wall and add 10 percent for waste. Some subs will charge by the square foot for material and labor—then add extras for tray ceilings and other special work.

For the ceiling, simply take the finished square footage of the house. Add 10 percent for waste and then add this to the wall amount for the total amount of drywall needed. If the house has any vaulted or tray ceilings, extra wallboard and labor must also be figured in.

Joint finishing compound comes premixed in five gallon cans and joint tape in 250 ft. rolls. For every 1,000 sq. ft. of drywall, figure one roll of joint tape and 30 gallons of joint compound.

TRIMWORK

Base molding comes in many styles, with clamshell and Colonial the most common, and will be run along the bottom of the wall in every room. Therefore, the lineal foot figure of walls used to calculate wallboard is equal to the lineal feet of baseboard trim needed. Add 10 percent for waste.

Make sure to measure the perimeter of any room that requires shoe molding or crown molding. Shoe molding (quarter round) is usually used in any room with vinyl or wood flooring to cover the crack between the floor and wall. Crown molding around the ceiling may be used in any room, but is most common in formal areas such as foyers and living and dining areas. Add 10 percent for waste.

When ordering interior doors, you may want to purchase prehung doors with preassembled jambs. These doors are exceptionally easy to install and already have the jamb and trimwork attached. Study the blueprints carefully to be sure that the doors ordered open in the right direction and don't block light switches. (There are right-hand and left-hand doors.) To determine which door to order, imagine standing in front of the door, walking in. If you must use your right hand to open the door, it is a right-hand door. Most blueprints will have the size and type of each door marked in the opening.

FLOORING

Because of the complexity of laying flooring materials—carpet, hardwood floors, vinyl or ceramic tile—it is essential to get a flooring contractor to estimate the flooring quantities. An approximate figure would be equal to the square footage of the area covered plus 10 percent for waste. Carpet and vinyl are sold by the square yard, so divide by nine to determine square yardage. Keep in mind, however, that carpet and vinyl are sold in 12′ wide rolls. This dimension may affect the amount of waste in the estimate.

MISCELLANEOUS COVERINGS

This includes ceramic tile for bathrooms or floors, parquet, slate for foyers or fireplaces and fieldstone for steps or fireplaces. Virtually all these materials are sold by the amount of square footage area covered. Subcontractors who install these items also charge by the square foot, sometimes with material included and sometimes without. Grouts and mortars used in installing these items (with the exception of rock) are premixed and indicate the coverage expected on the container.

PAINT

When hiring a painting contractor, the cost of paint should be included in the estimate unless you specify otherwise. If you should decide to do the painting yourself, you will need to figure the amount of paint needed. The coverage of different paints varies considerably, but most paints state the area covered on the container. Simply calculate the square footage of the area to be covered and add at least 15 percent for waste and touch-up. Also figure the amount of primer needed. If colors are custom-mixed, make sure that you have sufficient paint to finish the job. It is sometimes difficult to get an exact match of a custom color if mixed again at a later date.

CABINETS

Most blueprints include cabinet layouts with the plans. Use this layout if you are purchasing your own cabinets. If the cabinets are installed by a subcontractor, he will take care of the exact measurements. Be sure that painting, staining and installation are included in the price.

WALLPAPER

One roll of wallpaper will safely cover 30 square feet of wall area, including waste and matching of patterns. The longer the repeat pattern, the more waste. Some European wallpapers vary in coverage. Be sure to check coverage area with the wallpaper supplier. When figuring the square footage of a room to be wallpapered, count all openings as solid wall. Also be careful to note if you are buying a single or double roll of wallpaper. Most wallpaper is sold in a roll that actually contains two true rolls of wallpaper, or 60 sq. ft. of coverage. Always buy enough paper to finish the job. Buying additional rolls at a later time can cause matching problems if some of the rolls are from a different dye lot. Look for the same lot numbers (runs) on all rolls purchased to ensure a perfect match. Use only vinyl wallpaper in baths and kitchen areas for water resistance and cleanup.

MILLWORK AND MISCELLANEOUS

The ordering of windows, lights, hardware and other specialty items will not be covered here, as they are fairly straightforward to calculate. Other estimates, such as heating and air, plumbing, electrical and electrical fixtures, must be obtained from the contractors themselves, since they provide the materials and labor.

The following section outlines the way most subcontractors charge for their services. Some subs may or may not include the cost of materials in their estimates. Make sure you know your particular sub's policy.

SUBCONTRACTORS

FRAMING Frame house, apply sheathing, set windows and exterior doors. Charges by the square foot of framed structure (includes any unheated space such as garage). Extras: bay windows, chimney chase, stairs, dormers and anything else unusual.

SIDING Apply exterior siding. Charges by the square of applied siding. Extras: diagonal siding, decks, porches and very high walls requiring scaffolding.

CORNICE Usually the siding subcontractor. Applies soffit and fascia board. Charges by the lineal foot of cornice. Extras: fancy cornice work, dentil molding. Sometimes sets windows and exterior doors.

TRIM Install all interior trim and closet fixtures and set interior doors. Charges a set fee by the opening or by lineal feet of trim. Openings include doors and windows. Extras: stairs, rails, crown molding, mantels, bookcases, chair rail, wainscoting and picture molding.

FOOTINGS Dig footings, pour and level concrete and build bulkheads (for step downs). Charges by the lineal foot of footings. Extras: pier holes.

BLOCK Lay block. Charges by the block. Extras: stucco block.

BRICKWORK Lay brick. Charges by the skid (1,000 bricks).

STONEWORK Lay stone. Charges by the square foot or by bid.

CONCRETE FINISHING Pour concrete, set forms, spread gravel and finish concrete. Charges by the square foot of area poured. Extras: monolithic slab, digging footing.

ROOFING Install shingles and waterproof around vents. Charges a set fee per square foot plus slope of roof. Ex.: $1 per square over slope on a 6/12 slope roof=$7/square. Extras: some flashing, ridge vents and special cutout for skylights.

WALLPAPER Hang wallpaper. Builder provides wallpaper. Charges by the roll. Extras: high ceilings, wallpaper on ceilings and grass cloth.

GRADING Rough grading and clearing. Charges by the hour of bulldozer time. Extras: chain saw operator, hauling away of refuse, travel time to and from site (drag time).

POURED FOUNDATION Dig and pour footings, set forms and pour walls. Charges by the lineal foot of wall. Extras: bulkheads, more than four corners, openings for windows, doors and pipes.

PEST CONTROL Chemically treat the ground around foundation for termite protection. Charges a flat fee.

PLUMBING Install all sewer lines, water lines, drains, tubs, fixtures and water appliances. Charges per fixture installed or by bid. (For instance, a toilet, sink and tub would be three fixtures.) Installs medium-grade fixtures. Extras: any special decorator fixtures.

HVAC Install furnace, air conditioner, all ductwork and gas lines. Charges by the tonnage of A/C or on bid price. Extras: vent fans in bath, roof fans, attic fans, dryer vents, high-efficiency furnaces and compressors.

ELECTRICAL Install all switches and receptacles; hook up A/C compressor. Will install light fixtures. Charges by the receptacle. Extras: connecting dishwasher, disposal, floodlights and doorbells.

CABINETRY Build or install prefab cabinets and vanities and apply Formica tops. Charges by the lineal foot for base cabinets, wall cabinets and vanities. Standard price includes Formica countertops. Extras: tile or marble tops, curved tops, pull out shelves and lazy Susans.

INSULATION Install all fiberglass batts in walls, ceilings and floors. Charges by square foot for batts, by cubic foot for blown-in.

DRYWALL Hang drywall, tape, finish and stipple. Charges by the square foot of drywall. Materials are extra. Extras: smooth ceilings, curved walls, tray and vaulted ceilings and open foyers.

SEPTIC TANK Install septic tank. Charges fixed fee plus extra for field lines.

LANDSCAPING Level with tractor, put down seeds, fertilizer and straw. Charges fixed fee. Extras: trees, transplanted shrubs, pine straw and bark chips.

GUTTERS Install gutters and downspouts. Charges by lineal foot plus extra for fittings. Extras: half-round gutters, collectors, special water channeling and gutters that cannot be installed from the roof.

GARAGE DOOR Install garage doors. Fixed fee. Extras: garage door openers.

FIREPLACE Supply and install prefab fireplace and flue liner. Extras: gas log lighter, fresh air vent and ash dump.

PAINTING AND STAIN Paint and stain interior and exterior. Charges by square foot of finished house. Extras: high ceilings, stained ceilings and painted ceilings.

CERAMIC TILE Install all ceramic tile. Charges by the square foot. Extras: fancy bathtub surrounds and tile countertops.

HARDWOOD FLOOR Install and finish real hardwood floors. Charges by square foot. Extras: beveled plank, random plank and herringbone.

FLOORING Install all carpet, vinyl, linoleum and prefinished flooring. Charges by the square yard. Extras: contrast borders and thicker underlayments.

Bids Beat Estimates Every Time

Now that you know how detailed an estimate can be, look at the other side of the coin. You can spend a day calculating how many cubic yards of concrete you will need and approximately how many hours of labor it will take, or you can call up a full-service foundation company and nail down a bid. The moral here is to use bids whenever possible in completing your take-off. The bottom line is what counts; if your estimate comes out to be $600 to tile a floor and your lowest bid is $750, you've obviously missed the boat—and wasted a little time.

CODE NO.	DESCRIPTION	QTY.	MATERIAL		LABOR		SUBCONTR.		TOTAL
			UNIT PRICE	TOTAL MATL	UNIT PRICE	TOTAL LABOR	UNIT PRICE	TOTAL SUB	
1	LAYOUT								
2	STAKES								
3	RIBBON								
4	ROUGH GRADE								
5	PERK TEST								
6	WELL AND PUMP								
7	Well								
8	Pump								
9	BATTER BOARDS								
10	2×4								
11	1×6								
12	Cord								
13	FOOTINGS								
14	Re-Rods								
15	2×8 Bulkheads								
16	Concrete								
17	Footings								
18	Piers								
19	FOUNDATION								
20	Block								
21	8″								
22	12″								
23	4″								
24	Caps								
25	Headers								
26	Solid 8s								
27	Half 8s								
28	Mortar Mix								
29	Sand								
30	Foundation Vents								
31	Lintels								
32	Portland Cement								

Form 1: Cost Estimate Summary.

CODE NO.	DESCRIPTION	QTY.	MATERIAL		LABOR		SUBCONTR.		TOTAL
			UNIT PRICE	TOTAL MATL	UNIT PRICE	TOTAL LABOR	UNIT PRICE	TOTAL SUB	
33	WATERPROOFING								
34	Asphalt Coat								
35	Portland Cement								
36	6 Mil Poly								
37	Drain Pipe								
38	Gravel								
39	Stucco								
40	FOUNDATION								
41	(Poured)								
42	Portland Cement								
43	Wales								
44	TERMITE TREAT								
45	SLABS								
46	Gravel								
47	6 Mil Poly								
48	Re-Wire								
49	Re-Rods								
50	Corroform								
51	Form Boards								
52	2×8								
53	1×4								
54	Concrete								
55	Basement								
56	Garage								
57	Porch								
58	Patio								
59	AIR CONDITIONING								
60	House								
61	FRAMING SUBFLOOR								
62	I-Beam								
63	Steel Post								
64	Girders								

Cost Estimate Summary (continued).

CODE NO.	DESCRIPTION	QTY.	MATERIAL		LABOR		SUBCONTR.		TOTAL
			UNIT PRICE	TOTAL MATL	UNIT PRICE	TOTAL LABOR	UNIT PRICE	TOTAL SUB	
65	Scab								
66	Floor Joist								
67	Bridging								
68	Glue								
69	Plywood Subfloor								
70	5/4″ T&G								
71	3/4″ T&G								
72	1/2″ Exterior								
73	Joist Hangers								
74	Lag Bolts								
75	Subfloor								
76	FRAMING—WALLS								
77	Treated Plate								
78	Plate								
79	8′ Studs								
80	10″ Studs								
81	2×10 Headers								
82	Interior Beams								
83	Bracing								
84	1×4								
85	1/2″ Plywood								
86	Sheathing								
87	4×8								
88	4×9								
89	Laminated Beam								
90	Flitch Plate								
91	Bolts								
92	Dead Wood								
93	FRAMING—CEILING/ROOF								
94	Ceiling Joist								
95	Rafters								
96	Barge Rafters								

Cost Estimate Summary (continued).

CODE NO.	DESCRIPTION	QTY.	MATERIAL		LABOR		SUBCONTR.		TOTAL
			UNIT PRICE	TOTAL MATL	UNIT PRICE	TOTAL LABOR	UNIT PRICE	TOTAL SUB	
97	Beams								
98	Ridge Beam								
99	Wind Beam								
100	Roof Bracing Material								
101	Ceiling Bracing								
102	Gable Studs								
103	Storm Anchors								
104	Decking								
105	⅜″ CDX Plywood								
106	½″ CDX Plywood								
107	2×6 T&G								
108	Plyclips								
109	Rigid Insulation								
110	Felt								
111	15 #								
112	40 #								
113	60 #								
114	FRAMING—MISCELLANEOUS								
115	Stair Stringers								
116	Firing-in								
117	Chase Material								
118	Purlin								
119	Nails								
120	16DCC								
121	8DCC								
122	8″ Galv. Roof								
123	1½″ Galv. Roof								
124	Concrete								
125	Cut								
126	ROOFING								
127	Shingles								
128	Ridge Vent								

Cost Estimate Summary (continued).

CODE NO.	DESCRIPTION	QTY.	MATERIAL		LABOR		SUBCONTR.		TOTAL
			UNIT PRICE	TOTAL MATL	UNIT PRICE	TOTAL LABOR	UNIT PRICE	TOTAL SUB	
129	End Plugs								
130	Connectors								
131	Flashing								
132	Roof to Wall								
133	Roll								
134	Window								
135	Ventilators								
136	TRIM—EXTERIOR								
137	DOORS								
138	Main Entrance								
139	Rear Door								
140	Side Door								
141	Sliding Glass								
142	Windows								
143	Fixed Window								
144	Glass								
145	Window Frame								
146	Material								
147	Corner Trim								
148	Window Trim								
149	Door Trim								
150	Flashing								
151	Roof to Wall								
152	Metal Drip Cap								
153	Roll								
154	Beams								
155	Columns								
156	Rail								
157	Ornamental Iron								
158	NAILS								
159	16d Galv. Casing								
160	8d Galv. Casing								

Cost Estimate Summary (continued).

CODE NO.	DESCRIPTION	QTY.	MATERIAL		LABOR		SUBCONTR.		TOTAL
			UNIT PRICE	TOTAL MATL	UNIT PRICE	TOTAL LABOR	UNIT PRICE	TOTAL SUB	
161	4d Galv. Box								
162	LOUVERS								
163	Triangular								
164	Rectangular								
165	SIDING								
166	CORNICE								
167	Lookout Material								
168	Fascia Material								
169	Rake Molding								
170	Dentil Molding								
171	Continuous Eave Vent								
172	Screen								
173	⅜″ Plywood								
174	DECKS AND PORCHES								
175	MASONRY—BRICK								
176	Face								
177	Fire								
178	MORTAR								
179	Light								
180	Dark								
181	Sand								
182	Portland Cement								
183	Lintels								
184	Flue								
185	Ash Dump								
186	Clean Out								
187	Log Lighter								
188	Wall Ties								
189	Decorative Brick								
190	Lime Putty								
191	Muriatic Acid								
192	Sealer								

Cost Estimate Summary (continued).

CODE NO.	DESCRIPTION	QTY.	MATERIAL		LABOR		SUBCONTR.		TOTAL
			UNIT PRICE	TOTAL MATL	UNIT PRICE	TOTAL LABOR	UNIT PRICE	TOTAL SUB	
193	Under-Hearth								
194	Block								
195	Air Vent								
196	PLUMBING								
197	Fixtures								
198	Miscellaneous								
199	Water Line								
200	Sewer								
201	HVAC								
202	Dryer Vent								
203	Hood Vent								
204	Exhaust Vent								
205	ELECTRICAL								
206	Lights								
207	Receptacles								
208	Switches—Single								
209	Switches—Three-Way								
210	Switches—Four-Way								
211	Range—One Unit								
212	Range—Surface								
213	Oven								
214	Dishwasher								
215	Trash Compactor								
216	Dryer								
217	Washer								
218	Freezer								
219	Furnace—Gas								
220	Furnace—Electric								
221	A/C								
222	Heat Pump								
223	Attic Fan								
224	Bath Fan								

Cost Estimate Summary (continued).

CODE NO.	DESCRIPTION	QTY.	MATERIAL		LABOR		SUBCONTR.		TOTAL
			UNIT PRICE	TOTAL MATL	UNIT PRICE	TOTAL LABOR	UNIT PRICE	TOTAL SUB	
225	Smoke Detector								
226	Water Heater/Electric								
227	Floodlights								
228	Doorbell								
229	Vent-a-Hood								
230	Switched Receptacle								
231	150 Amp Service								
232	200 Amp Service								
233	Disposal								
234	Refrigerator								
235	Permit								
236	Ground Fault Interrupter								
237	Fixtures								
238	FIREPLACE								
239	Log Lighter								
240	INSULATION								
241	Walls								
242	Ceiling								
243	Floor								
244	WALLBOARD								
245	Finish								
246	Stipple								
247	STONE								
248	Interior								
249	Exterior								
250	TRIM—INTERIOR								
251	Doors								
252	Bifold								
253	Base								
254	Casing								
255	Shoe								
256	Window Stool								

Cost Estimate Summary (continued).

CODE NO.	DESCRIPTION	QTY.	MATERIAL		LABOR		SUBCONTR.		TOTAL
			UNIT PRICE	TOTAL MATL	UNIT PRICE	TOTAL LABOR	UNIT PRICE	TOTAL SUB	
257	Window Stop								
258	Window Mull								
259	1×5 Jamb Matl.								
260	Crown Molding								
261	Bed Molding								
262	Picture Molding								
263	Paneling								
264	OS Corner Molding								
265	IS Corner Molding								
266	Cap Molding								
267	False Beam Matl.								
268	STAIRS								
269	Cap								
270	Rail								
271	Baluster								
272	Post								
273	Riser								
274	Tread								
275	Skirtboard								
276	Bracing								
277	LOCKS								
278	Exterior								
279	Passage								
280	Closet								
281	Bedroom								
282	Bath								
283	Dummy								
284	Pocket Door								
285	Bolt Single								
286	Bolt Double								
287	Flush Bolts								
288	Sash Locks								

Cost Estimate Summary (continued).

CODE NO.	DESCRIPTION	QTY.	MATERIAL		LABOR		SUBCONTR.		TOTAL
			UNIT PRICE	TOTAL MATL	UNIT PRICE	TOTAL LABOR	UNIT PRICE	TOTAL SUB	
289	TRIM—MISCELLANEOUS								
290	Door Bumpers								
291	Wedge Shingles								
292	Attic Stairs								
293	Scuttle Hole								
294	1 × 12 Shelving								
295	Shelf and Rod Brackets								
296	Rod Socket Sets								
297	FINISH NAILS								
298	3d								
299	4d								
300	6d								
301	8d								
302	Closet Rods								
303	Sash Handles								
304	SEPTIC TANK								
305	Dry Well								
306	BACKFILL								
307	DRIVES AND WALKS								
308	Form Boards								
309	Re-Wire								
310	Re-Rods								
311	Expansion Joints								
312	Concrete								
313	Asphalt								
314	Gravel								
315	LANDSCAPE								
316	PAINT AND STAIN								
317	House								
318	Garage								
319	Cabinets								
320	WALLPAPER								

Cost Estimate Summary (continued).

CODE NO.	DESCRIPTION	QTY.	MATERIAL		LABOR		SUBCONTR.		TOTAL
			UNIT PRICE	TOTAL MATL	UNIT PRICE	TOTAL LABOR	UNIT PRICE	TOTAL SUB	
321	TILE								
322	Wall								
323	Floor								
324	Miscellaneous								
325	GARAGE DOORS								
326	GUTTERS								
327	Gutters								
328	Downspouts								
329	Elbows								
330	Splash Blocks								
331	CABINETS								
332	Wall								
333	Base								
334	Vanities								
335	Laundry								
336	GLASS								
337	SHOWER DOORS								
338	MIRRORS								
339	ACCESSORIES								
340	Tissue Holder								
341	Towel Racks								
342	Soap Dish								
343	Toothbrush Holder								
344	Medicine Cabinet								
345	Shower Rods								
346	FLOOR COVERING—BATH								
347	Vinyl								
348	Carpet								
349	Hardwood								
350	Slate or Stone								
351	MISCELLANEOUS								
352	CLEANUP								

Cost Estimate Summary (continued).

CODE NO.	DESCRIPTION	QTY.	MATERIAL		LABOR		SUBCONTR.		TOTAL
			UNIT PRICE	TOTAL MATL	UNIT PRICE	TOTAL LABOR	UNIT PRICE	TOTAL SUB	
353	HANG ACCESSORIES								
354	FLOOR COVERING—OTHER								
355	Vinyl								
356	Carpet								
357	Hardwood								
358	Slate or Stone								
359									
360									
361									
362	MISCELLANEOUS								

Cost Estimate Summary (continued).

SECTION II:
PROJECT MANAGEMENT

Preconstruction Preparation

The Project Management section covers each of the trades normally encountered in residential construction. They will be covered in the order in which they normally occur in the construction process. The stages covered are:

- Preconstruction
- Excavation
- Foundation, waterproofing and pest control
- Framing
- Roofing and gutters
- Plumbing
- HVAC (heating, ventilation and air conditioning)
- Electrical
- Masonry and stucco
- Siding and cornice
- Insulation and soundproofing
- Drywall
- Trim and stain
- Painting and wallcovering
- Cabinetry and countertops
- Flooring, tile and glazing
- Landscaping

NOTE: The multitude of combinations and options in a home are infinite, as are the ways in which to get them done. The steps and checklists outlined on the following pages attempt to cover *most* of the possible steps needed to build *most* homes.

For this reason, there are bound to be a number of steps in this manual relating to features you may not have planned. For example, you may not have a basement or require stucco work.

Skip steps that do not apply to your project. Likewise, there may be special efforts required in your project that are not covered in this manual. Also, there may be a number of steps you will undertake in special or unconventional circumstances that will not be covered in this book. In many instances, the exact order of steps is not critical; use your common sense.

PREPARATION

Before you run out and start building, there are a number of preparation steps that you should complete. These are somewhat frustrating because they show no visible signs of progress and are often time-consuming. There are a lot of forms to fill out and chores to do here. But you will be better off to get them all out of the way. For most of the steps listed in this section, order of completion is not critical: Just get them done.

Give yourself ample time to complete these steps. Rome wasn't built in a day. If you are a first-time builder, it is typical to spend at least as much time planning your project as you spend executing it. That is, if you plan to have your home completed six months after breaking ground, you should spend at least six months in planning and preparation prior to breaking ground.

Once you break ground, there's no turning back and no time to stop, so you'll be relieved that everything has been thought out ahead of time. This section will cover the items that should be completed prior to breaking ground.

PRECONSTRUCTION STEPS

PC1 BEGIN construction project scrapbook. This is where you will begin gathering all sorts of information and ideas you may need later.

Start with a large binder or a file cabinet. Break into sections such as:

- Kitchen ideas
- Bath ideas
- Floor plans
- Landscaping ideas
- Subcontractor names and addresses
- Material catalogs and prices

Your scrapbook will probably evolve into a small library, a wealth of information used to familiarize yourself with the task ahead, and serve as a constant and valuable source of reference for ideas to use in the future.

PC2 DETERMINE size of home. Calculate the size of home you can afford to build. Refer to the House Plans section in this book for details on this step.

PC3 EVALUATE alternate house plans. Consider traffic patterns, your lifestyle and personal tastes. Refer to the House Plans section for details on evaluating house plans.

PC4 SELECT house plan or have house plan drawn. If you use an architect, he will probably charge you a certain fee per square foot. Make sure you really like it before you go out and buy a dozen copies of the blueprints!

PC5 MAKE design changes. Unless you have an architect custom design your home, there will always be a few things you might want to change—kitchen and kitchen cabinets, master bath, a partition, a door, etc. This is the time to make changes. Refer to the material on making changes in the How to Be a Homebuilder and House Plans sections.

PC6 PERFORM material and labor take-off. This will be a time-consuming but important step. Refer to the Cost Estimating section for details on this step. Get bids to home in on true costs.

PC7 CUT costs. After you have determined estimated cost, you may need to pull the cost down. Refer to the Saving Money section for details on reducing costs without reducing quality.

PC8 UPDATE material and labor take-off. Based on changes you may have made, update and recalculate your estimated costs based on bids.

PC9 ARRANGE temporary housing. If you own a home and must sell it in order to build, or your current home is too far from your future construction site, consider renting an apartment or home close to your site. If possible, select a place between your site and your office (if you have one). This will greatly facilitate site visits. Since it will take approximately six months to build even the smallest house (unless you are using a prefabricated kit), you should get at least a six-month lease with a month-to-month option thereafter, or a one-year lease. For first-time, part-time builders, count on ten months to a year before move-in. Delays caused by weather, sub absenteeism, material delays and other problems always seem to add months to the effort. Keep in mind that during the course of the project, you will be paying rent, construction interest and miscellaneous expenses, as well as additional travel costs. Plan extra money to handle this.

PC10 FORM a separate corporation if you want to set your enterprise off from your personal affairs. This may give you a better standing among members of the building trade if they feel that they're dealing with a company instead of an individual. Be advised that your lender may not take kindly to this.

PC11 APPLY for and obtain a business license if required. This may be a legal requirement to conduct business as a professional builder in the county where your site is located. Check with your local authorities for advice.

PC12 ORDER business cards. These will be helpful in opening doors with banks, suppliers and subs. List yourself as president for extra clout.

PC13 OBTAIN free and clear title to land. This must be done before applying for a construction loan. Bankers want to make sure that you have something at stake in this project, too. It is strongly recommended that you only build on land that you own outright.

PC14 APPLY for construction loan. In proper terms, you should be applying for a "construction and permanent loan." This saves you one loan closing, since the construction loan turns into a permanent loan when construction is completed.

PC15 ESTABLISH project checking account. This will be used as the single source of project funds to minimize reconciliation problems. Write all checks from this account and keep an accurate balance.

PC16 CLOSE construction loan. This allows you access to funds up to your loan amount. You will pay interest only on money drawn from your reserve. You will not be paying interest on money you have not yet used. Remember, though, that money is paid out as work is completed, based on the judgment of bank officials. Hence, you will need a certain amount of money to begin work (excavation and perhaps some concrete work). Do not begin any physical site work until this is done. Many financial institutions will not loan you money if you have already started construction prior to contacting them.

PC17 OPEN builder accounts with material suppliers. This will involve filling out forms and waiting for mail replies notifying you of your account numbers. Read and be familiar with the payment and return policies of each supplier.

PC18 OBTAIN purchase order forms. Look for bound, prenumbered ones with a carbon copy. You will use these forms for all purchases. Pay only those invoices that reference one of your purchase order numbers. (Make sure to advise suppliers that you are using a purchase order system.)

PC19 OBTAIN building permit. This involves filling out building permit applications with your county planning and zoning department. Build a friendly relationship with all county officials. You may need to count on them later.

PC20 OBTAIN builder's risk policy. Your construction site is a liability. You, laborers, subcontractors or neighborhood children could get hurt. Materials may get damaged or stolen. An inexpensive insurance policy is a must. Your financial institution and/or your county officials can point you in the right direction. You should also check with your home and/or auto insurance company for a quote.

PC21 OBTAIN worker's compensation policy. This is normally charged out as a percentage of the cost of your project. There may be a minimum-size policy you can purchase.

PC22 ACQUIRE minimum tools. Certain items are a must.

- Assortment of coated, galvanized nails, framing and finish nails
- Builder's level (6″ line level and 4′ recommended)
- Can of WD-40 (everything is always getting rusty)
- Chisels (a few flat-bladed wood chisels and perhaps a concrete chisel)
- Circuit tester
- Circular saw (Skil Saw)
- Crowbar and/or nail claw (for tearing out framing mistakes)
- Fiberglass hard hat—preferably yellow or orange; scratch your name inside
- 50′ or 100′ steel tape measure
- First aid kit
- Garden hose
- Good all-purpose 14 oz. framing hammer; hammers now feature fiberglass construction, providing strength and rustproofing
- Good ankle-height leather construction boots, preferably with steel-tipped toes
- Heavy-duty leather gloves
- Heavy-duty utility markers (for making site markings on framework and leaving notes)
- Large plastic trash can and metal trash can

- Large push broom or shop vacuum
- Orange spray paint (for marking things)
- Pick
- Pliers (rubber-coated handles are recommended)
- Plumb bob (for attaching to string to check for true vertical)
- Pocketknife
- Power drill and a set of wood drill bits
- Roll of heavy-duty cotton twine and a blue chalk marker
- Set of screwdrivers, including flathead and Phillips styles (rubber handles recommended)
- Set of wrenches, crescent wrench or other adjustable wrench (rubber handles recommended)
- Shovels (flat for cleanup, and pointed for digging)
- Small sledgehammer (for pounding stakes and adjusting framework)
- Steel square
- Strong flashlight
- UL-approved 50′ to 100′ grounded extension cord and work lamp
- Utility knife and extra blades

If you do not own a pickup truck and can't borrow one, try to buy an inexpensive trailer that can be attached to the back of your car. You can sell it later, after the project is over. As an alternative, you can probably borrow a pickup truck from your subs from time to time.

PC23 VISIT numerous residential construction sites in various stages of completion. Observe the progress, storage of materials, drainage after heavy rainfall and other aspects of those sites.

PC24 DETERMINE need for an on-site storage shack. If you do a lot of the work, this could be helpful.

PC25 NOTIFY all subs/labor. Call up all subs and labor to tell them of the date you plan to break ground. This should be done approximately a week before you plan to break ground. Confirm that they still to plan to do the work and at the agreed-upon price. If you must get a backup sub, this is the time to do it—look on your sub list for number two.

PC26 OBTAIN compliance bond, if necessary. This is primarily for your protection if you have a builder assist you during construction or build for you.

PC27 DETERMINE minimum requirements for certificate of occupancy. As soon as you meet these requirements, you can move into your home while you finish it, avoiding unnecessary rent. During final construction stages, concentrate on meeting these requirements. For example, you can do landscaping and certain other tasks later. Don't move in unless you have finished the rooms you plan to live in immediately.

PC28 CONDUCT spot survey of lot. Locate a good, reliable licensed surveyor. Have him furnish a signed survey report. This confirms that the planned project and excavation are within allowable boundaries. Ask your surveyor to help evaluate water drainage if you are uncertain.

PC29 MARK all lot boundaries and corners with 2′ stakes and bright red tape.

PC30 POST construction signs. You should have at least one "DANGER" or "WARNING" sign. Have one sign for each street exposure if you have a corner lot. You may also want another sign that has your name, "company" and phone number on it. Your lender may have its own sign also.

PC31 NAIL building permit to a tree (that will not be torn down) or a post in the front of the lot visible from the street. You may want to provide a little rain protection for it. You cannot do any work on-site until it is posted.

PC32 ARRANGE for rental toilet. This is often viewed as a luxury among builders. They rent by the month at about $40 and up, including maintenance.

PC33 ARRANGE with the phone company for a site telephone. If the site is located far away from anything, this becomes more of a necessity, especially in cases of emergency. If you get one, have the phone company install it with a strong metal case with a lock. You will have to provide a sturdy 4×4 post or tree to hang it on. Make sure you don't have to pay if it gets stolen. It should be in a place where you can hear the phone ring if you are inside. Specify a loud bell.

Site Location and Excavation

DIGGING IN

Now it's time to really get mud on your new construction shoes! You are entering that phase of construction where your decisions will be much harder to change. Take great care in locating and excavating your homesite. A mistake here will be almost impossible to repair later. Site location consists of several phases:

- Locating the house on the lot
- Excavating the area where the house is located
- Marking the foundation lines with batter boards

Site Location

Before locating the house on the lot, make sure to obtain a site plan from your surveyor. The site plan should indicate the dimensions of the lot, public and private easements and the setback requirements of your county's building code. With this information in hand, you can ensure that you place the house properly, with no encroachment on another person's property or on public easements. You can also request that the surveyor place the house on the site plan and install corner stakes when surveying.

If you decide to locate the corner stakes, check the drainage of the property and look for any subsoil rocks or springs. After the house is located and the grading is properly done, the lot should drain away from the house, but not onto any other property. If you change the natural drainage of the lot so that water now flows onto someone else's property, you can be held liable for any damage that results. You should know before excavation starts where the top of your foundation will be located. You may want to run a string at this height between the corner stakes to aid in excavating to the proper depth. Use a string level, leveling hoses or a builder's transit to ensure that the string is plumb and level.

Leveling hoses are simple but effective tools to find the absolute level of two distant objects. You can use simple garden hoses for this, but a much better version is available at most do-it-yourself stores for a very reasonable price (much cheaper than renting a builder's transit). Fill the hose with water and mount one end at the highest corner of the foundation. Raise or lower the other end of the hose (while adding more water) until both ends of the hose are completely full of water. Since water seeks its own level, the two hose ends will then be exactly level. Mark this elevation for string placement.

Once excavation commences, make sure that the depth of excavation will match your intended foundation height. You can measure from the string to ensure proper depth.

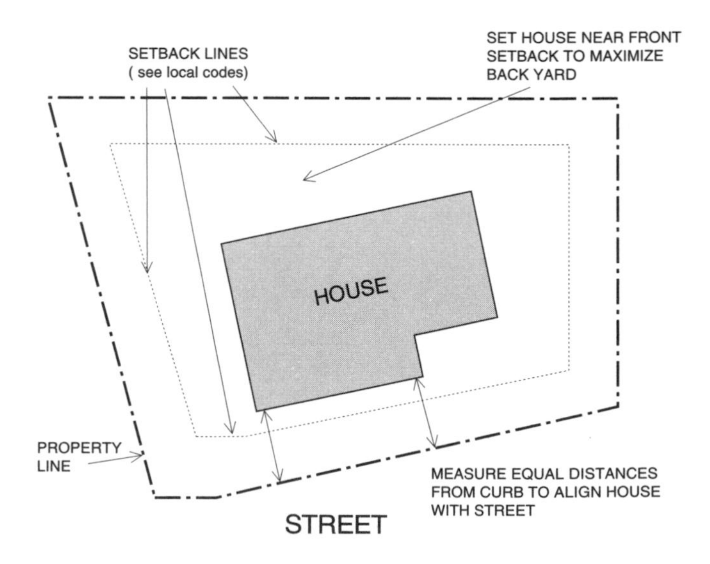

Figure 18: Reference lines on a site plan.

To lay out the corner stakes, start from an established reference line, such as the curb of the street. Most houses look best when located parallel to the street. However, you may violate this rule if you are concerned about other factors, such as solar orientation. Just make sure to place the house well within the front, side and back setback limits. Establish the front line of the house first by measuring from your reference lines. Then locate the back corners of the house. Make sure that your stakes are reasonably square by measuring the diagonals of the square (see illustration). The length of each diagonal should be the same. To obtain a 90° angle, use the old geometric principle of the "golden triangle." The golden triangle has three sides in multiples of 3 ft., 4 ft. and 5 ft., and will form a 90° angle (see illustration next page).

Breaking Ground

This is perhaps one of the most exciting moments during your homebuilding project! This phase actually involves several tasks. You may use your grader four times during the project:

- Clear site and excavate.
- Backfill and smooth out basement floor area.
- Cut the driveway, rough grade and haul away trash.
- Finish grade

The excavation subcontractor is one sub you will pay by the hour. This is usually tracked by the 'hour' meter on the bulldozer. Don't let the machine idle while talking with the operator—shut it off. The rates can vary, so shop around. You may be able to get a better rate by hiring excavators to work on off-hours, such as weekends.

Avoid clearing or excavating when the ground is wet or muddy: It will take more time, be less accurate and create a mess. Know your lot. Do you have large, heavy trees that need heavy-duty clearing equipment? Or could you use lighter equipment that rents for less?

Remember: The more horsepower, the higher the hourly rate, but the more it can accomplish in less time. But never use overpowered equipment for the job, or you'll end up paying too much. Negotiate hours with the sub. Does he include travel time to and from the site (drag time)?

If you want to save money by cutting the trees down yourself, go ahead. But cut them 4′ from the ground: The bulldozer needs a good piece of the tree to pull the roots out of the ground. If you don't have a fireplace, consider selling the wood.

Batter Boards

Once the site is properly excavated, you need to lay out the batter boards. These boards provide a precise reference for the foundation subs to install foundation forms. If you have excavated for a full basement, you may be able to get your poured-foundation sub to install forms without the need for batter boards. This is one of the advantages of using a poured-concrete foundation.

If you are using a block wall foundation or a slab foundation, batter boards will be necessary for the subs. Since the subs will be laying out the foundation to the dimensions of the batter boards, you must make sure that the layout is completely plumb and square! A mistake here can cause a great deal of frustration and expense to correct later. Crooked slabs are almost impossible to correct without repouring. So use a pound of prevention here.

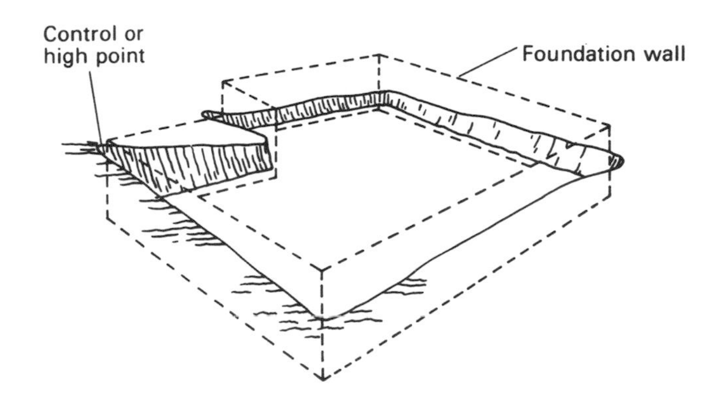

Figure 19: Establishing depth of excavation.

EXCAVATION—Steps

EX1 CONDUCT excavation bidding process on all site preparation work.

EX2 LOCATE underground utilities (gas main, water line, electrical or telephone cables) if they cross your property. You may want to call a locater service or your local utilities for help in locating utility lines. Mark their location with color-coded flags, and mark them on a copy of the plat. Don't place flags down more than two or three days in advance. Kids love to pull them up. Your excavator will need to know where the utility lines run. Damaging utilities can be time-consuming and expensive.

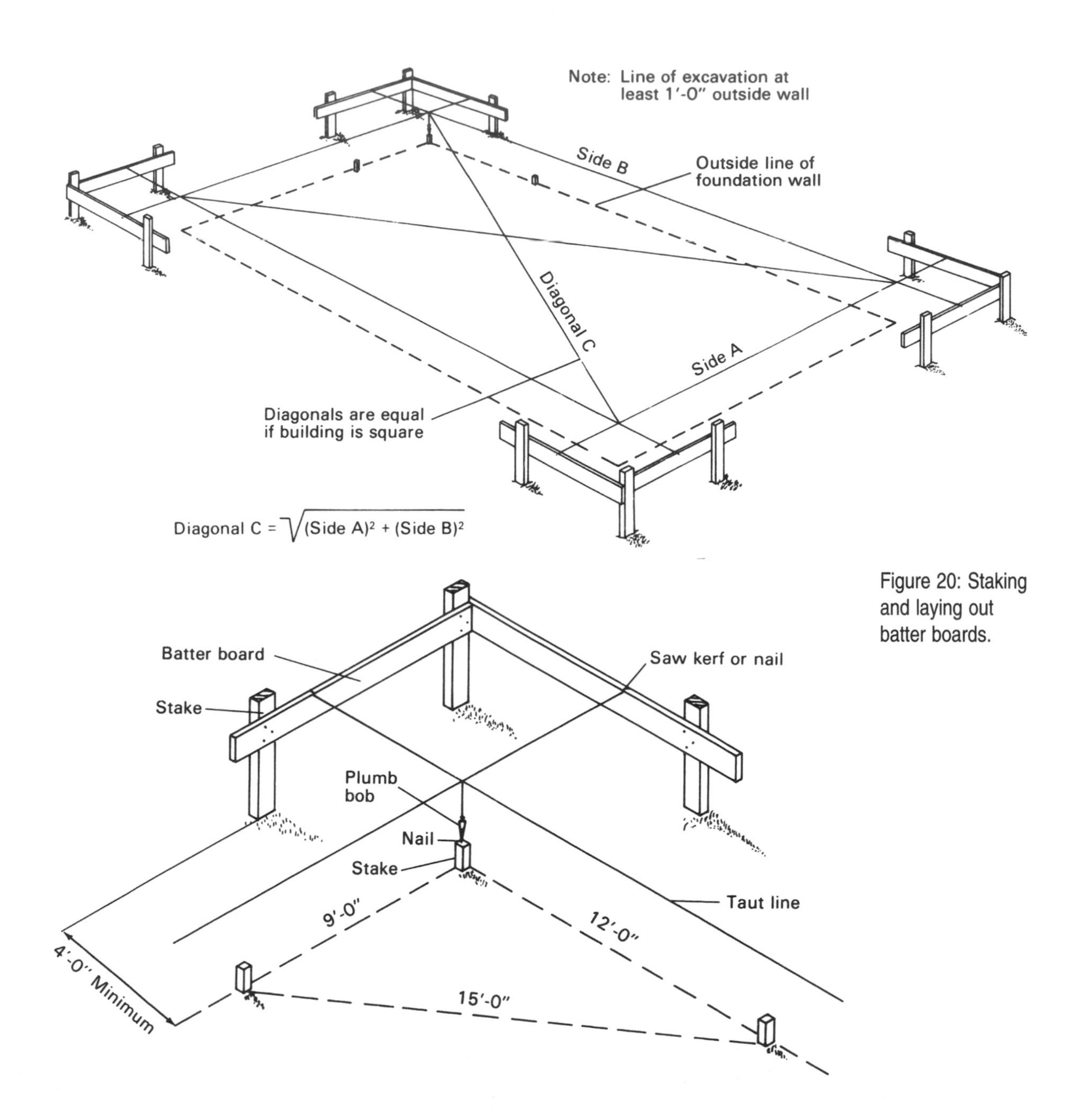

Figure 20: Staking and laying out batter boards.

EX3 DETERMINE where dwelling is to be situated exactly. Your surveyor can help you with this. Make sure to allow both for zoning setbacks and for any utility or highway easements. A landscape architect can help you determine exactly where the home should face from a passive solar and appearance aspect. Mark all corners of the house with stakes so the grader will know where to clear. The grader should dig an extra 3′ beyond the corner stakes of the house. For consistency, you should have the same setback as the adjacent homes, if there are any.

EX4 MARK area to be cleared. Trees to be saved should be marked with red tape or ribbon. Allow clearance for necessary deliveries and several parked vehicles. Remind your excavator to be careful about knocking the bark off of trees to be saved. This could kill the tree or invite wood-hungry pests. Now is the time to clear an area for the driveway, patio and septic tank (if you plan to have one).

EX5 MARK where curb is to be cut (if it is to be cut). The standard curb cut is 14′ wide. This allows for a slight flare on each side of the driveway area. In some cases, if you do not cut the curb, you will have a tough bump to deal with getting into your driveway. If it is to be cut later, you can remedy the problem by piling dirt or gravel up at the curb. You may need a variance if you plan to cut more than one curb area—as in the case of a circular driveway on a corner lot.

EX6 TAKE last picture of site area. This is the last chance you have to see it untouched.

EX7 CLEAR site area. Be present for this step. This is normally not something that is worthwhile for you to do because the grader can clear in an hour what would take you a week or more by hand. Cut down necessary trees and brush. Leave at least 4′ to 6′ feet of the tree trunk. Tractors need the trunk to act as a lever when pulling the roots out of the ground. Pull stumps out of the ground and remove as necessary. Make sure to go over the plan with the grader. If your plan is reversed, make sure the grader is aware of this fact. At this point you may wish to mark the foundation area with powdered lime or corner stakes. Do not try to save time by cutting down trees yourself; the grader needs the weight of the tree to pull the stump out of the ground.

EX8 EXCAVATE trash pit. Trash pit location should be a minimum of 20′ from the building or driveway. Make sure not to locate the pit near where you plan to add a swimming pool at a later date. A trash pit can save money normally spent on hauling a large amount of scrap off-site, as well as a dumping fee at a landfill. Some builders don't like them because they attract termites that might eventually be attracted to your home. The area may settle over time as materials decompose. The pit is also a hazard for young children and adults during construction. In most areas it is unlawful to burn or bury trash.

EX9 INSPECT cleared site. Check for proper building dimensions, scarred trees and thoroughness. Have trash, roots, stumps and other debris hauled away.

EX10 STAKE out foundation area. This is the process of locating all corners of the structure to tell the excavator where to dig. It is customary to excavate about 2′ to 3′ out in all directions from the house footprint. This will allow room for workers to set up poured-foundation walls and to perform waterproofing.

EX11 EXCAVATE foundation/basement area. You must be present when this step is being performed. To execute this step properly, you should have a transit set up. Take depth measurements continuously. This will help you to keep the foundation grade level and at the proper depth. You'll need to shoot the basement depth as you get close to final grade. Cutting too low can cause your sewer line to be lower than the one at the street, and can also cause water problems in your basement. Cutting too high can cause you to have a steeper driveway and extra steps at the front porch. Strike a nice balance. But overall, it's better to be 6″ too high than 6″ too low. This will help prevent water problems later. You are normally paying this person by the hour, unless, of course, you have decided on a firm dollar bid. Make sure he works hard doing exactly what you have in mind. It would be handy to have a copy of the written contract with you. Have the excavator also clear the topsoil off the driveway and foundation area. This must be done before the crushed stone is put down on it or the stone will just sink into the loose topsoil. Have the topsoil stacked in its own pile so that you can redistribute it on the lawn area after construction. Your yard will be much healthier with an extra layer of topsoil.

Make sure the operator digs down to firm soil and below the frost line, no matter how deep. Don't scrimp here! A poor footing can ruin your house. If you are not sure of the firmness of the soil, use a wider footing to provide more stability.

EX12 INSPECT site and excavated area. Refer to attached inspection sheet and contract specifications.

BATTER BOARDS—Steps

EX13 LAY OUT outer four corners of house with stakes and square up using a transit or the "golden triangle." Make sure tops of stakes are level with the proposed top of foundation. Check this with a builder's transit or leveling hoses. Remeasure distances and drive a small finishing nail or mark top of stake at exact corner of the building line. This will act as a reference for the batter boards.

EX14 DRIVE three 2×4 stakes of suitable length in a triangular pattern 4′ outside the corner stake with the tops of the stakes 6″ to 12″ higher than the corner stake. Make sure to sharpen the ends of the stakes before using them!

EX15 NAIL 1×4 or 1×6 boards horizontally between the three stakes so that the tops are all level with the tops of the corner stakes. You may want to clamp the horizontal boards initially until final height adjustments are complete.

EX16 PULL a string between opposite batter boards and adjust until the string is exactly over the nails in the corner stakes. Attach string with a finishing nail. Repeat with each batter board until the four outside corners are marked properly.

EX17 RECHECK the distance of each string—measuring from the intersections of the strings. Measure the diagonal distance of the corners of the strings—if the distance is the same, the corners are square. Check the height and level of each string. Make minor adjustments. Once the strings are positioned accurately, mark the top of the batter board or saw a kerf in the board where the lines touch the board so that they can be replaced if broken or moved.

EX18 LAY OUT all inside corners and remaining batter boards using the existing batter board strings as a reference. Recheck the measurements, height and diagonals of the strings as you add batter boards.

DRAINAGE—Steps

EX19 APPLY crusher run to driveway area. This will provide an excellent base for a paved drive and allows for easy site access for your subs and supply trucks when the ground is muddy.

EX20 SET up a silt fence. Local ordinances will probably not allow mud runoff from your site to drain into the street or adjacent proper-

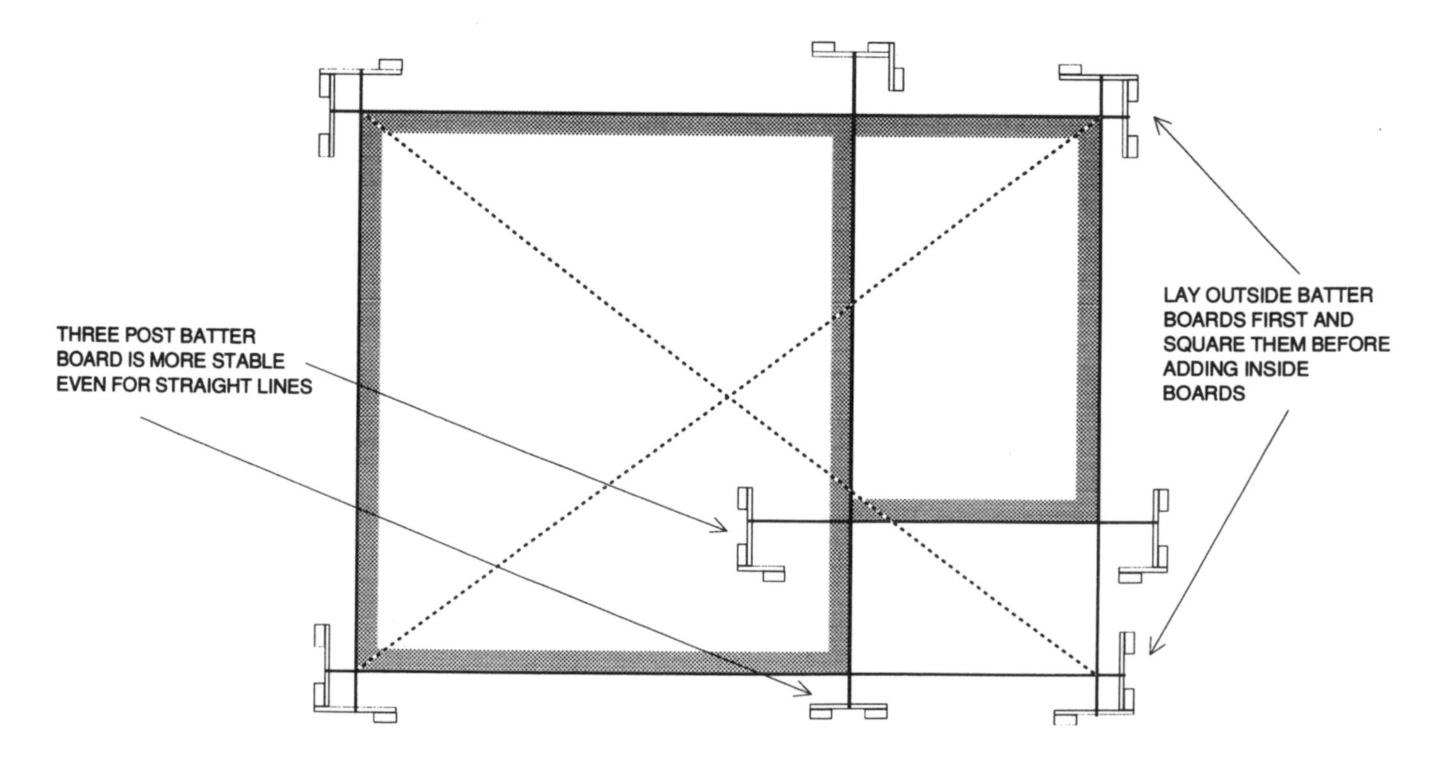

Figure 21: Lay batter board lines to the outside of the foundation, leaving room for foundation subcontractor to dig around batter boards and lines.

ties. A silt fence prevents mud from flowing into the street and neighboring yards. Two-foot stakes will support the fence. Staple the covering to the stakes with a stapler. Rolls of plastic, mesh or burlap are available at material suppliers for this purpose, or you may choose to use bales of hay. These bales can be reused later when landscaping.

EX21 PAY excavator. Have him sign a receipt. Pay only for work done up to this point. Give him an idea of when you will be needing him for the backfill work.

Backfill

EX22 PLACE supports on interior side of foundation wall to support wall during backfill. Don't backfill until the house is framed in to provide additional support for the foundation walls.

EX23 INSPECT backfill area. Refer to related checklist in this section.

EX24 BACKFILL foundation. Remind excavator not to puncture waterproofing poly with large roots or branches in dirt.

EX25 CUT driveway area.

EX26 COVER septic tank and septic tank line.

EX27 COVER trash pit. Area should be compacted with heavy equipment and covered with at least two feet of dirt.

EX28 PERFORM final grade. Topsoil should be spread over top surface. If you have truck-loads of topsoil hauled in, you may want to get a free soil pH test done by the county extension service. This could save you money in buying expensive fertilizers to compensate for inferior soil.

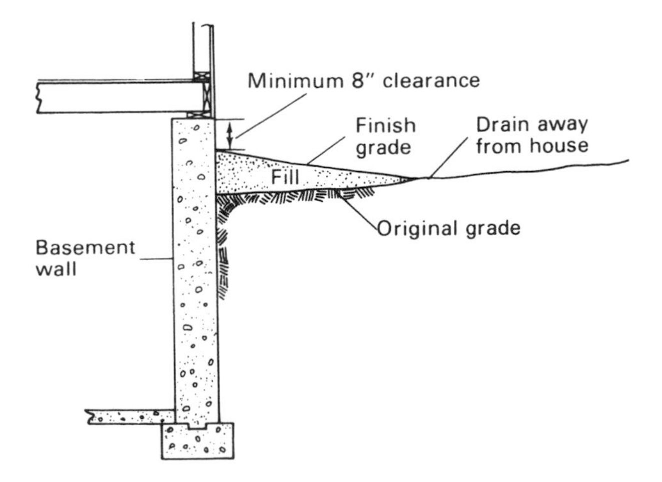

Figure 22: Finished grade sloped for drainage.

EX29 CONDUCT final inspection. Check all excavation and grading work.

EX30 PAY excavator (final). Have sub sign an affidavit.

EXCAVATION—Sample Specifications

- Excavator to perform all necessary excavation and grading as indicated below and on the attached drawing(s), and to provide all necessary equipment to complete the job.
- Bid to include all equipment and equipment drag time for two trips.
- Bid to include necessary chain saw work.
- Trees to be saved will be marked with red ribbon. Trees to be saved are to remain unscarred and otherwise undamaged.
- Foundation to be dug to depth indicated at foundation corners. Foundation hole is to be 3′ wider than foundation for ease of access. Foundation base not to be dug too deeply in order to pour footings and basement floor on stable base.
- Foundation floor to be dug smooth within 2″ of level.
- Excavate trash pit (10′ × 12′) at specified location, as depicted on attached drawing.
- Driveway area to be cleared of topsoil and cut to proper level. All topsoil to be piled where specified.
- All excess dirt to be hauled away by excavator.
- All stumps and trees to be dug up and hauled away. Hardwoods to be cut up to fit fireplace and piled on-site where directed.
- Grade level shall be established by the builder, who will also furnish a survey of the lot showing the location of the dwelling and all underground utilities.

- Finish grade to slope away from home where possible.
- Backfill foundation walls after waterproofing, gravel and drain tile have been installed and inspected.
- Builder to be notified prior to clearing and excavation. Builder must be on-site at start of operation.

EXCAVATION—Inspection

Clearing

- ☐ Underground utilities left undisturbed.
- ☐ All area to be cleared is thoroughly cleared and other areas left as they were originally.
- ☐ All felled trees removed or buried. All remaining trees are standing and have no scars from excavation. Firewood cut and stacked if requested.

Excavation

- ☐ All area to be excavated is excavated, including necessary work space:
  - Foundation area (extra 3′ at perimeter)
  - Porch and stoop areas
  - Fireplace slab
  - Crawl space cut and graded with proper slope to ensure dry crawl space.
- ☐ Trash pit dug according to plan if specified.
- ☐ All excavations done to proper depth with bottoms relatively smooth. Should be within 2″ of level.

Checklist Prior to Digging and Clearing

- ☐ All underground utilities marked.
- ☐ All trees and natural areas properly marked off.
- ☐ Planned clearing area will allow access to site by cement and other large supply trucks. Room to park several vehicles.
- ☐ All survey stakes in ground.

Checklist Prior to Backfill

- ☐ Any needed repairs to foundation wall complete.
- ☐ Form ties broken off and tie holes covered with tar (poured wall only).
- ☐ All foundation waterproofing completed and correct. This includes parging, sprayed tar, poly, etc.
- ☐ Joint between footing and foundation sealed properly and watertight.
- ☐ Footing drain tile and gravel installed according to plan.
- ☐ All necessary runoff drain tile in place and marked with stakes to prevent burial during backfill.
- ☐ All garbage and scrap wood out of trench and fill area.
- ☐ Backfill supports in place and secure.
- ☐ Backfill dirt contains no sharp edges to puncture waterproofing.

Checklist After Backfill

- ☐ All backfill area completely filled in.
- ☐ Lot smoothed out to a rough grade.
- ☐ Necessary dirt hauled in and spread.

Checklist After Final Grade Completed

- ☐ Two to three percent slope or better away from dwelling.
- ☐ Berms in proper place, of proper height and form.
- ☐ Water meter elevation correct.
- ☐ Final grade smooth. Topsoil replaced on top if requested.
- ☐ No damage to curb, walkways, driveway, base of home, water meter, gutter downspouts, HVAC, splash blocks and/or trees.

Foundation

The foundation will be one of the most critical stages of construction. Foundation work has one chance to be right—the consequences of poor craftsmanship are permanent. Additionally, this effort will involve individuals from more than one company, which can complicate the job. Hire the best.

This phase of construction requires coordination between a small army of subcontractors, and all of them must work with you over a relatively short period of time. The following list briefly describes the sequence of events:

(1) Footing subs dig footings.

(2) Exterminator treats soil in and around footings to prevent termites.

(3) County inspector checks footings before pouring.

(4) Footings are poured.

(5) Foundation walls are laid or poured (basement construction).

(6) Plumber and HVAC subs lay water, sewer and gas pipes before slab floor is poured (slab construction only).

(7) Slab is inspected.

(8) Slab is poured by concrete finishers.

(9) Basement wall is waterproofed.

(10) Basement wall drainage system is installed.

As you can see, foundation work can be an exercise in scheduling and diplomacy. Make sure to contact all involved parties well in advance of the pour and plan around rain delays. If possible, cover the foundation area with poly well in advance to prevent rain from ruining your well-laid plans. This poly can be used during the pour as the vapor barrier under the slab, if left undamaged.

PEST CONTROL

Below ground, all slab and footing areas must be permanently poisoned prior to laying any gravel, poly or concrete. Poison must be allowed to soak in after application without rain. Rain will reduce or nullify the effectiveness of the pest control treatment. Your pest control treatment can be taken care of in one or two visits, depending upon how thorough you want to be. Some builders have a second treatment applied around the base of the home after all work is done. This is just more insurance, especially if you are building for yourself.

THE CONCRETE SUBS

You will likely deal with several different parties when working with concrete for foundations, slabs and driveways:

- A full-service foundation company
- A subcontractor to dig and pour footings
- Labor to lay concrete block walls
- Concrete finishers to finish concrete slab, patios, walkways and driveways
- Concrete supplier

Full-Service Foundation Company

This subcontractor can be one of the best bargains you will encounter in your project. This sub replaces most of the other foundation subs listed above and may not cost a penny more. A full-service foundation company can:

- Lay out the entire foundation for you, including installing batter boards. Only the corners of the foundation need to be marked with stakes. The foundation sub will do the rest.
- Lay out, dig and pour the footings.
- Pour all basement walls.

- Handle all scheduling of inspectors and concrete deliveries.
- Waterproof (tar, gravel and drain pipe).

Generally, poured foundations are slightly more expensive than other types, but this cost can be offset by savings in time, labor, hassle and interest paid on your construction loan. However, this sub is only cost-effective to use if you plan a full basement.

The Footings Sub

The footings sub will normally charge you by the lineal foot of footings required. He will provide either steel or wood forms, but may ask you to supply some of the form material. Form work can cost approximately one-third of the total footing job. In many areas of the country with stable soil, the footings subs can pour directly into the footings trench, with no forms required.

Concrete Block Masons

If you do plan to have a basement made of concrete block, you'll need this sub. Concrete block masons will likely be a different group than those laying your brick veneer (if you have any). They will charge for labor by the unit (block) laid.

Concrete Finishers

The concrete finisher may be the footing sub, but not necessarily. He will charge by the square footage finished. Costs will also vary depending on the types of finishes you want. Slick, rough, smooth, trowel finish or pea gravel are a few of the surfaces that this sub can produce. Don't use a crew that has fewer than three men—this is a tough job. Finishing must be done fast, and even faster in hot weather. Concrete can set in half an hour. The finishers will finish the slab and then come back to do drives and walkways later.

Concrete Supplier

The concrete supplier will charge by the cubic yard of concrete used. A standard cement truck can hold 8 to 10 cubic yards. The supplier may charge you a flat rate just to make the trip, so pour and finish as much of the concrete work as you can at one time to minimize his trips.

CONCRETE STRUCTURES

The concrete structures that you will use will vary somewhat, depending on whether or not you have a basement. If you plan to have a basement, you will have footings, foundation walls and a foundation floor. If you don't plan to have a basement, you will have a slab with a footing built into the edge. These structures are described briefly below.

FOOTINGS are what your home rests on. A footing is a solid, subgrade rectangle of concrete around the perimeter of the dwelling. Its width and thickness will vary depending on the local building codes, soil and weight of materials supported. Footings are normally twice as wide as the wall they support. Additional footings are provided for fireplaces and column supports.

CONCRETE BLOCKS will be used if you have a basement and you decide not to pour the walls. Concrete blocks are normally cast hollow units measuring 7⅝″ wide, 7⅝″ high and 15⅝″ long. These are often referred to as 8″ × 8″ × 16″ units because, when laid with ⅜″ mortar joints, they will occupy a space 16″ long and 8″ high. Concrete masonry units come in a wide variety of shapes and sizes. Your local building code may have certain guidelines in the use of concrete units.

STEEL REINFORCEMENT is used to strengthen concrete slabs, walls and footings and help them resist the effects of shrinkage and temperature change. Reinforcing comes in two major forms: bars and mesh. Reinforcing bars used for residential construction are normally about ⅜″ in diameter. Reinforcing wire, normally with a 6″ grid and ¼″ wire, is adequate to do the job. For monolithic slabs, mesh is used more than bar steel. Steel reinforcement should be laid before cement is poured, and secured so that no movement will occur during the pouring process. Normally the steel is located two-thirds of the way up from the bottom of a slab and in the center of poured concrete walls. Your concrete sub and concrete supplier will have primary responsibility for proper reinforcement.

Foundation Walls

Foundation walls are made of either concrete blocks or poured concrete.

Concrete block: This method is normally a little less expensive than a poured wall. However, concrete block walls must be leveled by hand. The

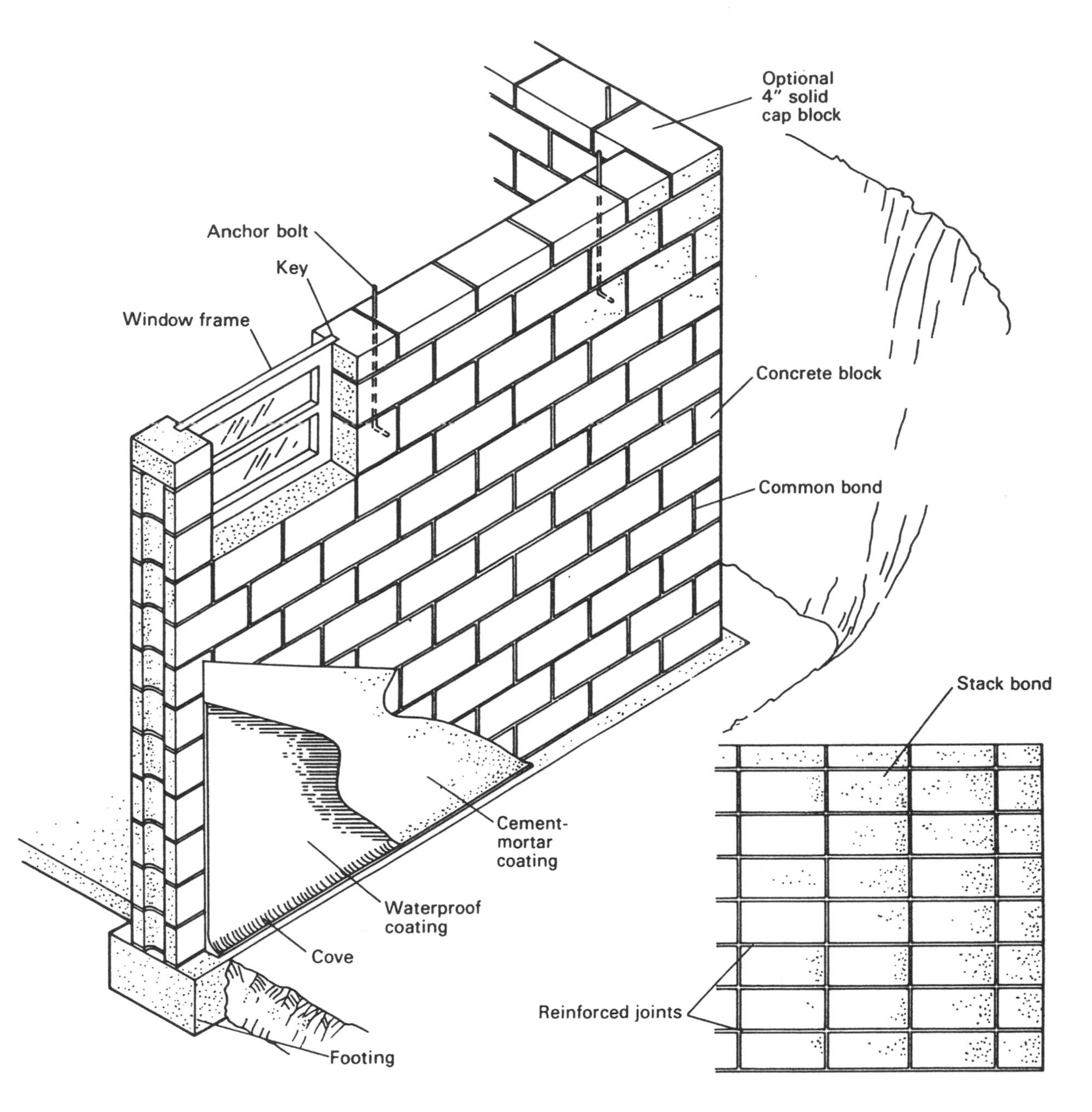

Figure 23: Concrete block foundation walls.

concrete block wall must be coated with two ¼″ layers of portland cement below grade level.

Poured wall: This method, which is growing in popularity, costs a bit more but yields a wall three times as strong as a block wall and more impervious to water. If you are planning to have a brick veneer exterior, you must tell the form sub to provide for a brick ledge in the forms. This is what your brick wall will rest on.

Foundation Floors

Foundation floors are poured after the footings and walls have been properly cured. Coarse, washed gravel is normally used as an underlayment. The reinforcing is raised to the proper level with rocks or other items. The concrete is normally poured to a depth of 4″. The floor should be finished as close to level as possible, with a slight slope toward the drains.

Monolithic slab: This is a combination of footings and foundation floor poured all at once. If fill dirt is used below slab, extra reinforcing bars should be used. The dirt should be well compacted with machinery prior to pouring.

Driveways: Driveway slabs are generally 4″ thick

and should be allowed to cure for two to three days before any vehicles are allowed to drive on it. Driveways are not usually installed until trimwork starts.

Sidewalks and patios: These are simple structures that should only be poured after the surface ground has had ample time to settle. This is usually done when the driveway is poured.

Avoiding Cracks in Concrete

To avoid cracks in slabs and foundation walls for pennies a yard, use concrete with special fiberglass admixtures. This adds to the tensile strength of the concrete and reduces the potential for cracking. You may still want to use wire mesh for reinforcement. The fibers are everywhere. If you don't need to install wire mesh, the concrete contractors may even charge less than for a conventional pour because they have less work.

Concrete Joints

One of the most common reasons for concrete cracking, aside from unstable soil and tree roots, is due to expansion and contraction. Expansion and contraction are best controlled by using two types of joints:

(1) **Isolation (expansion control) joints:** These joints are placed between concrete structures in order to keep one structure's expansion from affecting adjacent structures. Common places used are between:
   - Driveway and garage slab
   - Driveway and roadway
   - Driveway and sidewalk
   - Slabs and foundation wall

(2) **Control (contraction) joints:** These joints are placed within concrete units, such as driveways (normally every 15′), sidewalks (normally every 5′ or closer) and patios (normally every 12′).

Construction joints are temporary joints marking the end of a concrete pour due to delays in concrete supply. If you are lucky, you won't need any of these. Your concrete forms sub and/or finisher should be placed responsible for setting all expansion joints, and finishers are responsible for all control joints.

WEATHER CONSIDERATIONS

RAINY WEATHER can adversely affect almost any concrete work. Heavy rain can affect poured concrete surfaces and slow the drying process. Schedule your concrete pouring for periods of forecast dry weather. Have rolls of heavy plastic available if the weather forecast is wrong.

COLD WEATHER is a challenge. The main problem to prevent here is freezing. Concrete work in temperatures below 40° Fahrenheit becomes difficult. Several methods to protect work include heating water and heating aggregate, covered with tarpaulins, over hot pipes. Frozen subgrade must be heated to ensure that it will not freeze during the curing process. Mix the heated aggregate and water before adding the concrete. Consider using high-early-strength concrete to reduce the time required to protect the concrete from freezing. Once concrete has been poured, if freezing weather is likely, the concrete should be protected either by special heated enclosures or insulation such as batts or blankets.

HOT WEATHER is also a challenge. The main problem to prevent here is rapid drying, which induces cracking and reduction in concrete strength. Regardless of the external temperature, concrete should not be allowed to exceed 85° Fahrenheit. This can be achieved by using a combination of the following methods:

- Cool the aggregate with water before mixing.
- Mix concrete with cold water. Ice can be used in extreme cases.
- Provide sources of shade for materials.
- Work at night or very early in the morning.
- Protect drying concrete from sun and winds, which accelerate the drying process. Wet straw or tarpaulins can be used.

Concrete trucks normally hold about 9 cubic yards. Make sure to schedule trucks to provide a constant supply of concrete.

WATERPROOFING

Waterproofing normally applies to foundation wall treatments and drainage tile used to maintain proper foundation water drainage. This is critical to ensure that you will have a dry basement in the many years to come.

Wall Treatment
Wall treatments are applied below the finished grade (below ground level). Normally, this treatment consists of hot tar, asphalt or bituminous material applied to the exterior surface. A layer of 6-mil (millimeter) poly (polyethylene) is then applied to the tar while still tacky. This is an added protection against water seepage. If masonry block walls are used, two ¼″ coats of portland cement are applied to the masonry blocks, extending 6″ above the unfinished grade level. The first coating is roughened up before drying, allowing the second coat to adhere properly. Masonry blocks and the first coating of portland cement are dampened before applying coats of portland cement. The second coating should be kept damp for at least forty-eight hours. The joint between the footing and wall should be filled with portland cement before applying the tar coating. This is called *parging* and further protects against underground water and moisture.

A new waterproofing compound is available on the market that is superior to asphalt and carries a 10-year warranty if applied correctly. This compound has the ability to bridge cracks in masonry and block and will not degrade with age. It is considerably more expensive than asphalt; however, since wet basements top the list of major homeowner complaints, the extra expense seems well worth it. Ask your waterproofing contractor if he is trained to apply this compound.

Drain Pipe
Four-inch perforated drain pipe (plastic is preferred over tile) is placed around the perimeter of the foundation with #57 gravel below and 8″ to 12″ of #57 gravel above. The pipe should be sloped at least 1″ every 20′. Pipe sections should be spaced ¼″ apart, with a strip of tar paper (roofing paper) over each joint. The drain pipe will only serve its purpose if it drains off somewhere. You should make it so that the drain slopes to one corner. If you like, have a runoff pipe installed to carry the water farther away.

PEST CONTROL—Steps
PS1 CONDUCT standard bidding process and select pest-control sub. Use a company that has been around for at least ten years, since you may want to get annual pest control and a warranty. In some cases, builders are not charged for the pre-treating service. The pest control sub looks for a contract with the homebuyer.

PS2 NOTIFY pest control sub of when to show up for the pretreatment.

PS3 APPLY poison to all footing, slab ground and surrounding area. Since it is hard to test the effectiveness and coverage of treatment, it is advised that someone be present to oversee this critical step. This treatment should only be done if rain is unlikely for forty-eight hours.

After Construction
PS4 APPLY poison to ground-level perimeter of home. Again, it is advised to have someone oversee this step.

PS5 OBTAIN signed pest control warranty and file it away. Many lenders require a pest warranty at closing to insure proper protection.

PS6 Pay pest control sub and have him sign an affidavit.

CONCRETE—Steps
CN1 CONDUCT standard bidding process (concrete supplier). Bid will be measured in cubic feet. If you use a full-service foundation sub, this will probably be unnecessary.

CN2 CONDUCT standard bidding process (formwork and finishing). If you use a full-service foundation company, subs will lay out the foundation, pour footings and walls and arrange for the concrete. Pricing can get pretty complicated, depending on the height and size of the walls.

CN3 CONDUCT standard bidding process (concrete block masonry).

Footings
CN4 INSPECT batter boards. This is quite critical, since your entire project is based on the position of your batter boards and string. This is done with a transit.

CN5 DIG footings for foundation, A/C compressor, fireplaces and all support columns.

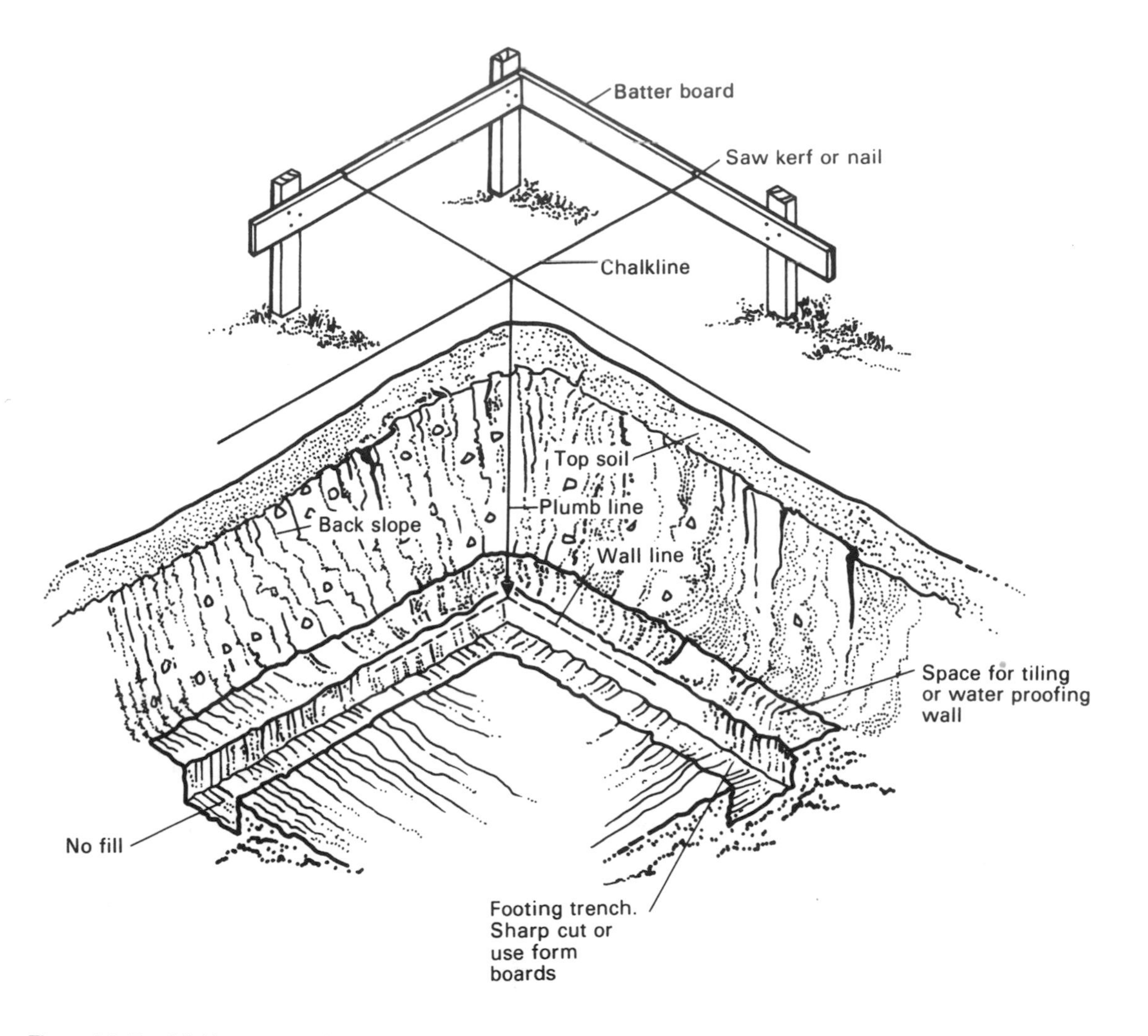

Figure 24: Establishing corners for excavation and footings.

CN6 SET footing forms as necessary. Be very careful that the footing sub places footings so that walls will sit in the center of the footing. Subs can either use their own forms or use lumber at the site. If the sub asks you to provide forms material, have three times the foundation perimeter in 16′ 2×4s.

CN7 INSPECT footing forms. Refer to related checklist in this section. Make sure that the side forms are sturdy, as they will have to support lots of wet concrete. Make sure that the tops of the forms are level, as they will be used as a guide in finishing the footing surface. Also make sure that the forms are parallel with each other. Footings are normally coated on the inside with some form of oil so that the concrete doesn't stick to them. Check for this. If you plan to have poured walls on your footings, have the forms sub supply the "key" forms, which will be pressed into the footings after they are poured. In many areas, steel rods are inserted vertically in footings prior to pouring.

CN8 SCHEDULE and complete footing inspection. Obtain signed, written approval from local building inspector.

CN9 CALL your concrete supplier to schedule delivery of concrete if you are not using a full-service finisher. This should be done at least twenty-four hours in advance. Make sure good weather is forecast for the next few days. Have rolls of 6-mil poly available in case of rain. Make sure to add calcium carbonate if freezing is a potential problem.

CN10 POUR footings. This should be done as quickly as possible once the footing trenches

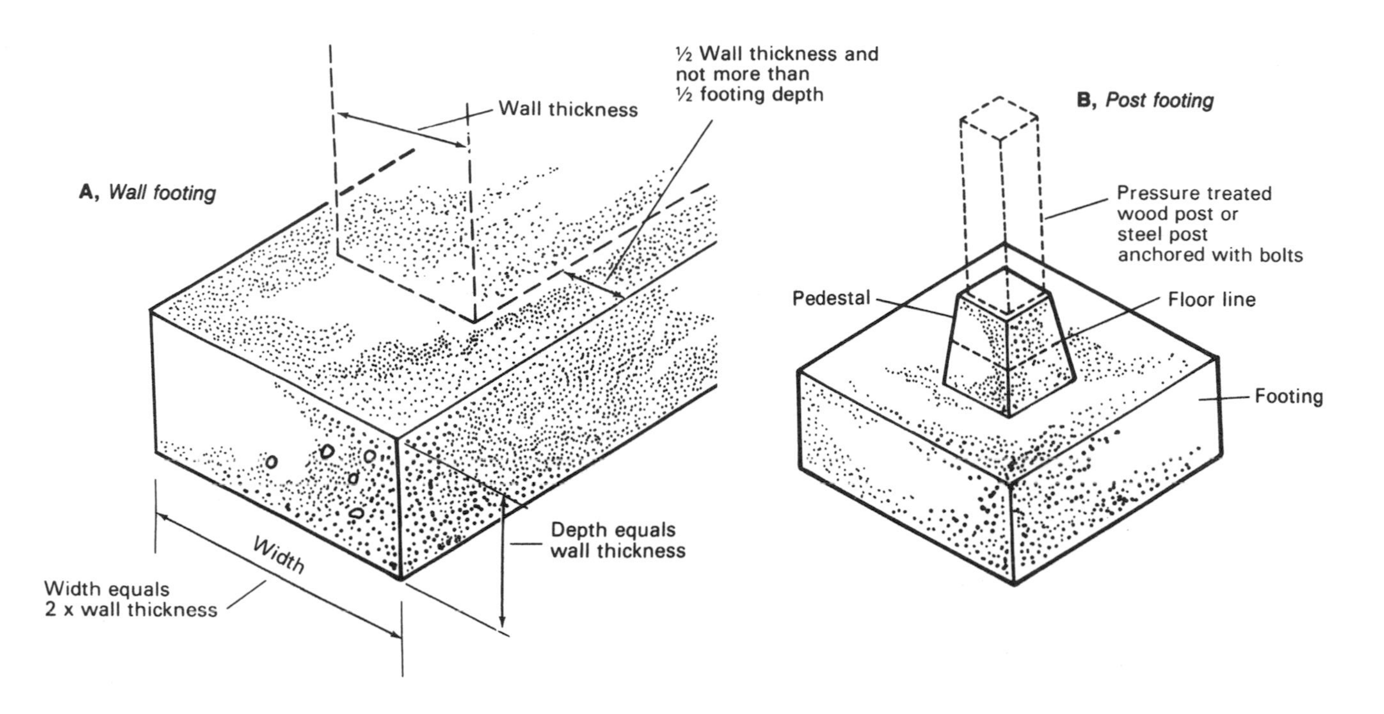

Figure 25: Concrete footing: A, Wall footing; B, Post footing.

have been dug so that the soil does not get soft from rain and exposure.

CN11 INSTALL key forms in footings before setting has taken place. Your forms sub should do this. Budget finishers will just cut a groove in the footing with a 2×4.

CN12 FINISH footings as needed. This normally involves simply screeding the top and smoothing out the surface a bit.

CN13 REMOVE footing forms. If subs used your lumber, have them clean and stack it to be used later for bracing. Framers won't use this lumber because it will dull their saws.

CN14 INSPECT footings. Make sure that the footings are level and that there are no visible cracks.

CN15 PAY concrete supplier for footing concrete, unless you are using a full-service finisher—in which case you will make one payment after the walls have been poured and inspected.

Poured Wall

CN16 SET poured-wall foundation and stoop arm forms. These are normally special units made out of steel held together with metal ties. Make stoop arms extra long to ensure that the front porch will not settle or crack.

CN17 CALL concrete supplier in advance to schedule concrete. This step will usually be done by a full-service foundation contractor.

CN18 INSPECT poured-foundation wall and stoop arm forms. Again, make sure they are level, parallel, sturdy and oiled.

CN19 POUR foundation walls and stoop arms. If more than one truckload is required, make sure to schedule deliveries so that a constant supply of concrete is available for pouring. It is unsatisfactory to have cement setting when wet cement will be poured over it. This can result in a fault crack in the foundation. Make sure that additional loads of concrete are on-site or on the way to ensure a continuous pour. If the weather is rainy, have a grader available in case you have to push the concrete truck up a muddy slope.

CN20 FINISH poured-foundation walls. Again, the tops of the forms serve as screed guides. The surface is smoothed, and lag bolts should be set into the surface. Make sure the bolts are at least 4″ in the concrete and are standing straight up. Also make sure that the bolts are

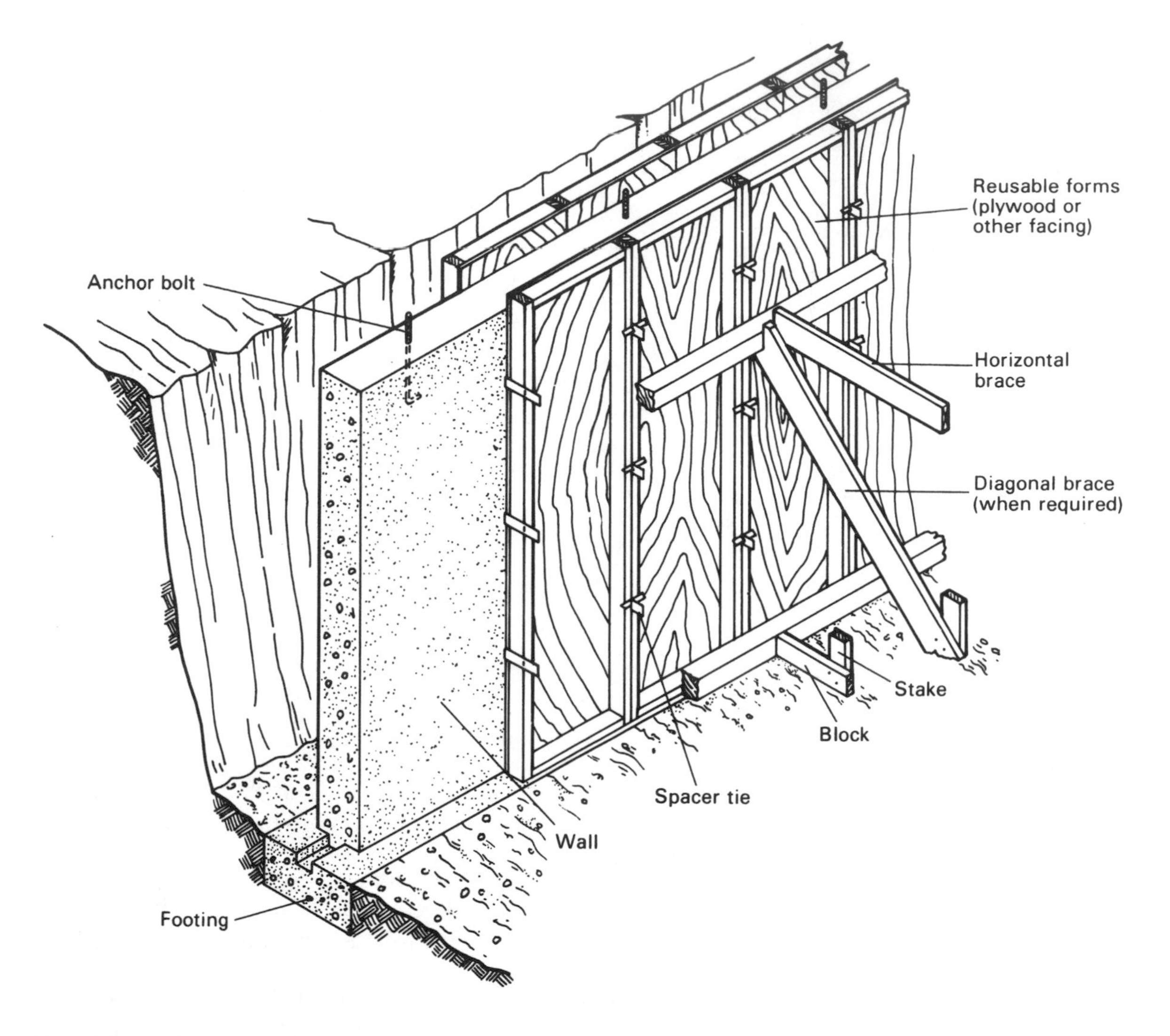

Figure 26: Forming for poured concrete walls.

not in places that will interfere with any studs or door openings. Put wide washers on the sunken end of the lag bolts for greater holding strength. Many builders are omitting lag bolts and just using concrete nails to anchor sill plate to the foundation.

CN21 REMOVE poured-foundation wall forms. This should only be done several days after the concrete has had time to set.

CN22 INSPECT poured-foundation walls. Check for square, level and visible surface cracks.

CN23 BREAK off tie ends if metal footing forms are used.

CN24 PAY concrete supplier for poured-foundation wall concrete.

Block Wall

CN25 ORDER concrete blocks and schedule block masons. Make sure the blocks are placed as near to the work area as possible. Block masons will charge extra if they have to move excessive amounts of block very far.

CN26 LAY concrete blocks. You may want to watch some of this in progress since it will be very hard to correct mistakes when the blocks have set. If you see globs of cement on blocks, wait till it dries to clean it off—it will come off in one piece. Compare against blueprints as the work proceeds.

CN27 FINISH concrete blocks. While mortar is still setting, the masons will tool the joints concave if you like. However, for subgrade work, it is better to have the mortar flush with the blockwork. If any of the block wall is to be exposed, have the block mason stucco the block to provide a smooth, more attractive surface.

CN28 INSPECT concrete block work. Compare work with blueprints. Refer to your specifications and inspection guidelines.

CN29 PAY concrete block masons. Have them sign an affidavit.

CN30 Clean up after block masons. Stack or return extra block. Spread leftover sand in driveway area. Bags of mortar mix should be protected from moisture.

Concrete Slab

CN31 SET slab forms for basement, front stoop, fireplace pad, garage, etc. Check forms against blueprints as they are being erected. If the slab is the floor of the basement, the walls will act as the forms.

CN32 PACK slab subsoil. Soil should be moistened slightly, then packed with a power tamper. This has to be a solid surface.

CN33 INSTALL stub plumbing. If the slab is to be the floor of the house, schedule the plumber to set all water and sewer lines under slab area. Ask the plumber to install a foam collar around all sewer lines that protrude out of the slab. This makes final adjustments to the pipe easier during finish plumbing. If the pipe needs to be moved slightly, the foam can be chipped away instead of the concrete. Make sure any resulting gap is resealed with spray foam insulation.

CN34 SCHEDULE HVAC sub to run any gas lines that may run under the slab, such as gas fireplace starters.

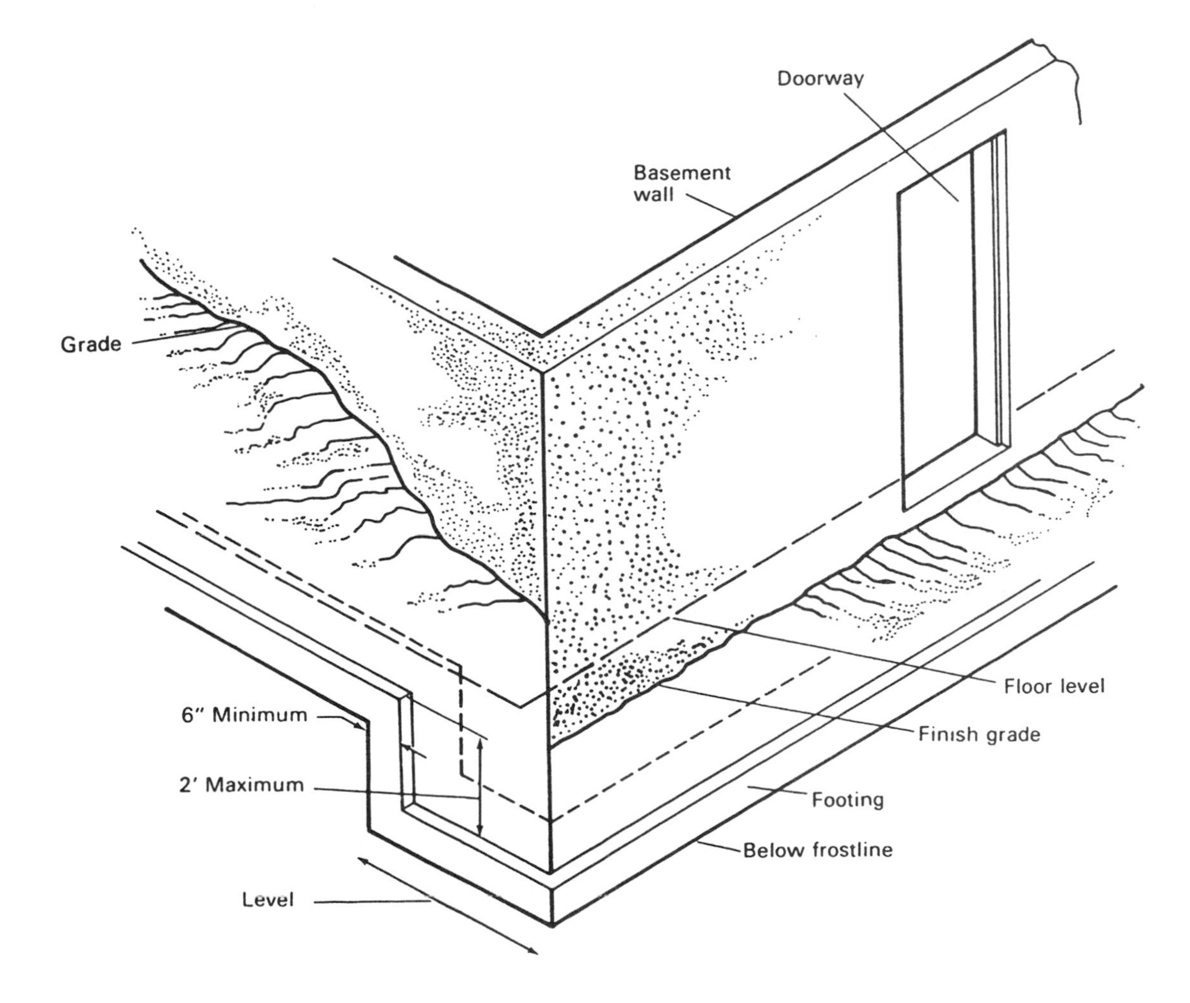

Figure 27: Stepped footing.

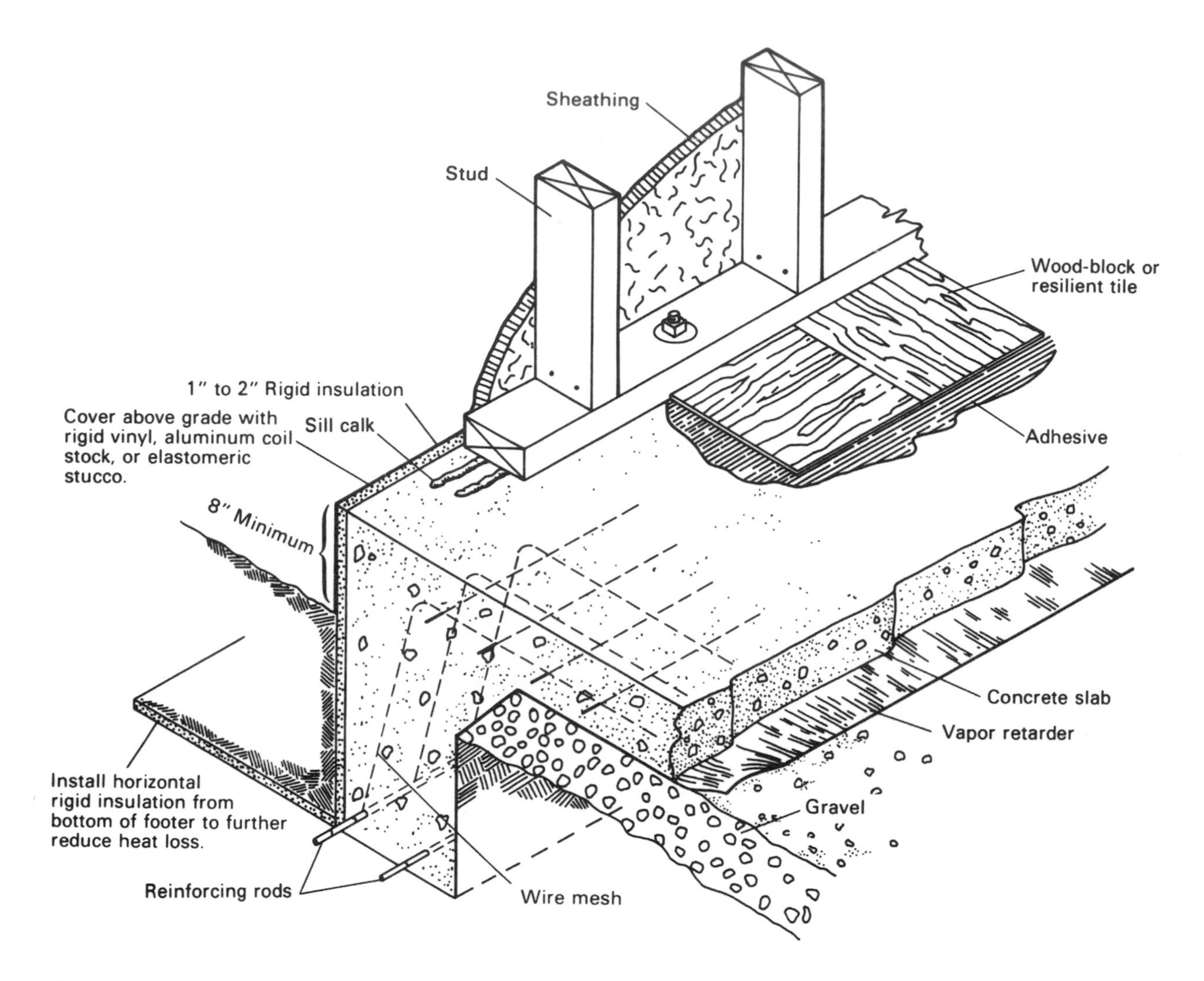

Figure 28: Combined floor slab and foundation (thickened-edge slab).

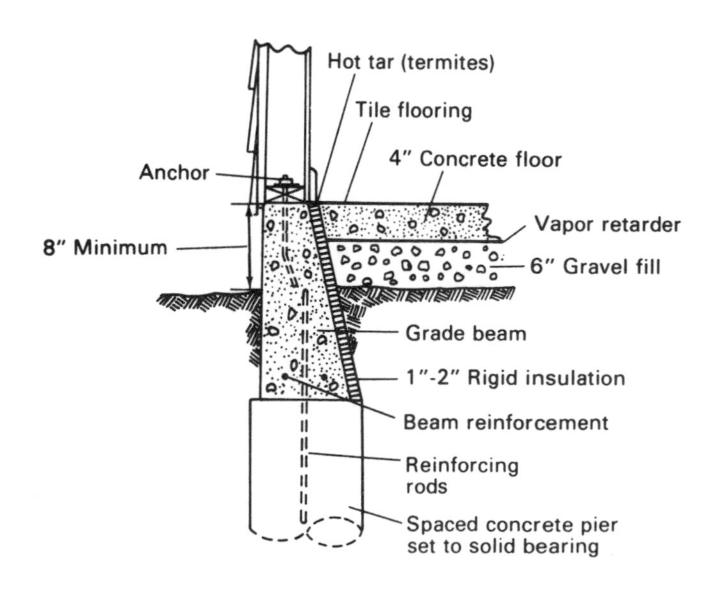

Figure 29: Reinforced grade beam for concrete slab. Beam spans between concrete piers located below frostline.

CN35 INSPECT plumbing and gas lines against blueprint for proper location. Make sure that plumbing is not dislodged, moved or damaged in subsequent steps.

CN36 POUR and spread crushed stone (crusher run) evenly on slab or foundation floor and garage slab area.

CN37 INSTALL 6-mil poly vapor barrier. Can be omitted in garage slab.

CN38 LAY reinforcing wire as required. The rewire should lie either in the middle of the slab and garage slab or at a point one-third from the bottom. It can be supported by the gravel or rocks.

CN39 INSTALL rigid insulation around the perimeter of the slab to cut down on heat loss through the slab.

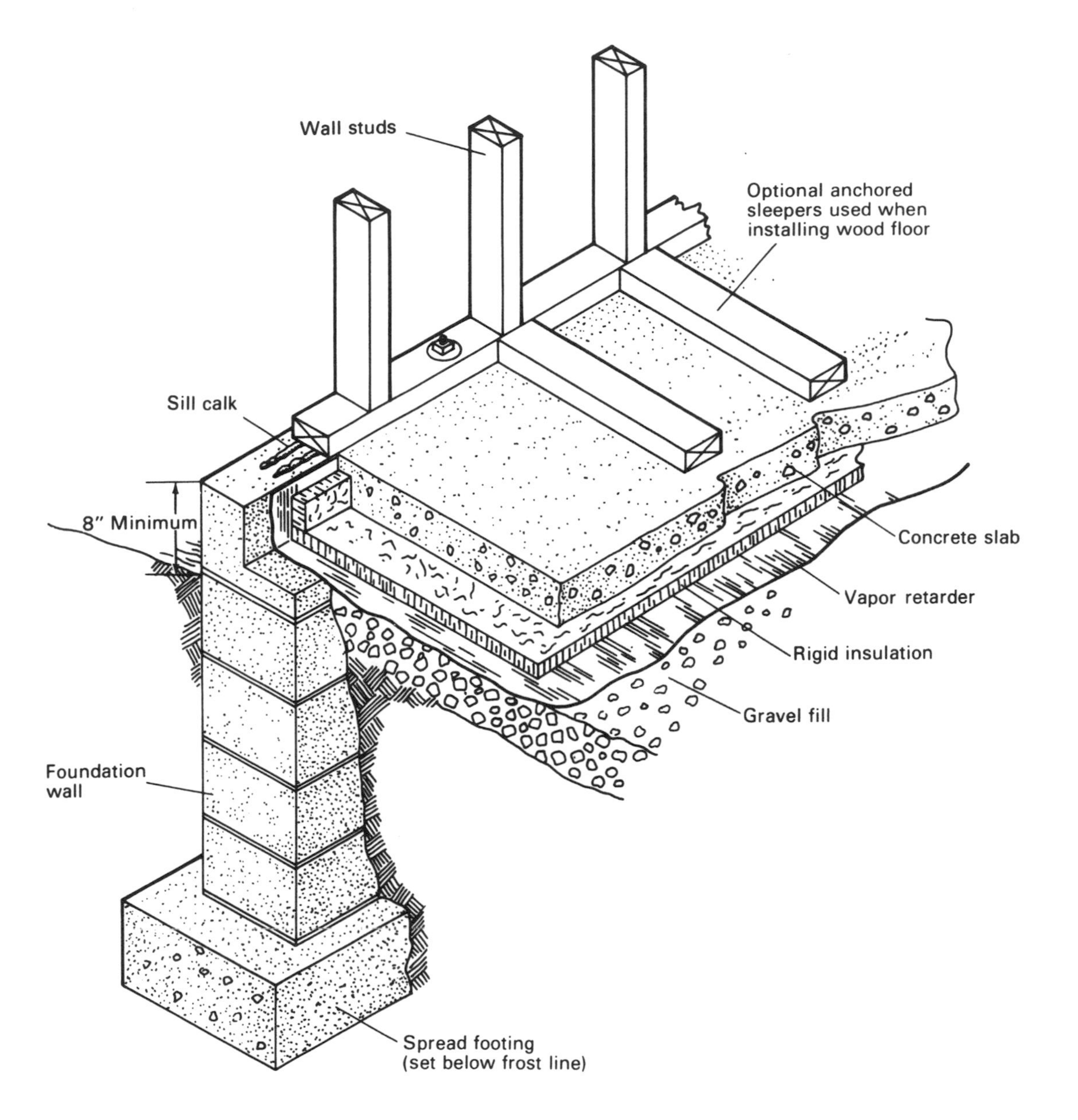

Figure 30: Independent concrete slab and foundation wall system for climates with deep frostline.

CN40 CALL concrete supplier to schedule delivery of concrete for slab. Confirm concrete type, quantity, time of pour and site location. Finishers should also be contacted.

CN41 INSPECT slab forms. Compare them against the blueprints. Make sure that all necessary plumbing is in place! Check that forms are parallel, perpendicular and plumb as required. Forms should be sturdy and oiled to keep from sticking to concrete. See checklist.

CN42 POUR slab and garage slab. This should be done as quickly as possible. Call it off if rain looks likely. Have 6-mil poly protection available.

CN43 FINISH slab and garage slab. Finishers should put their smoothest finish on all slab work. This normally requires four steps:

- trough screed with a floater
- rough trowel
- power trowel
- finish trowel

CN44 INSPECT slab. Check for rough spots, cracks and high or low spots. High and low spots can be checked with a hose. See where the water collects and drains. Water should never completely cover a nickel laid flat. Make sure any backfill does not cover the brick ledge if one is present.

CN45 PAY concrete supplier for slab and garage slab concrete.

Driveways and Patios

CN46 COMPACT driveway, walkway, patio and mailbox pad subsoil. This must be a solid surface.

CN47 SET driveway, walkway, patio and mailbox forms. You should not pour less than a 3″-thick driveway. If you did not lay down crushed stone previously, you may want to do it now. If your electrical or gas service crosses the driveway area, it must be installed before this step. Use scrap lumber for forms.

CN48 CALL your concrete supplier to schedule delivery of concrete. Confirm type and quantity of concrete, delivery time and site location. Schedule finishers.

CN49 INSPECT driveway, walkway, patio and mailbox forms. Pull a line level across forms to determine slope. Forms should be installed so that driveway will slope slightly away from house and toward an area with good drainage.

CN50 POUR driveway, walkway and patio areas. Make sure finishers do not thin concrete for easier working.

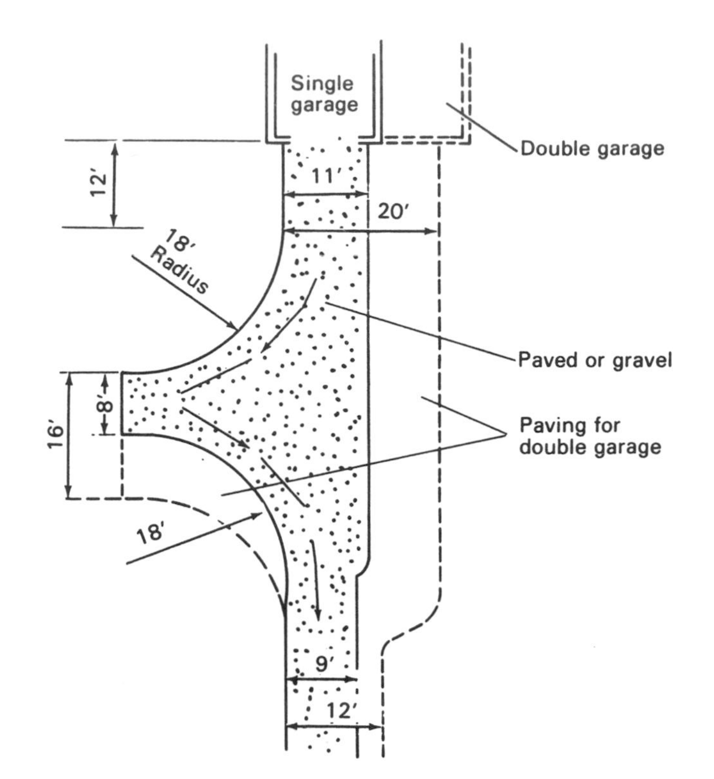

Figure 31: Driveway turnaround.

CN51 FINISH driveway, walkway, patio and mailbox areas. Make sure that the finishers put isolation joints between the slab, driveway, patio and walkways. Expansion joints should be put about every 12′ in the driveway and every 5′ or less in the walkways. Driveway should be formed where it meets the street.

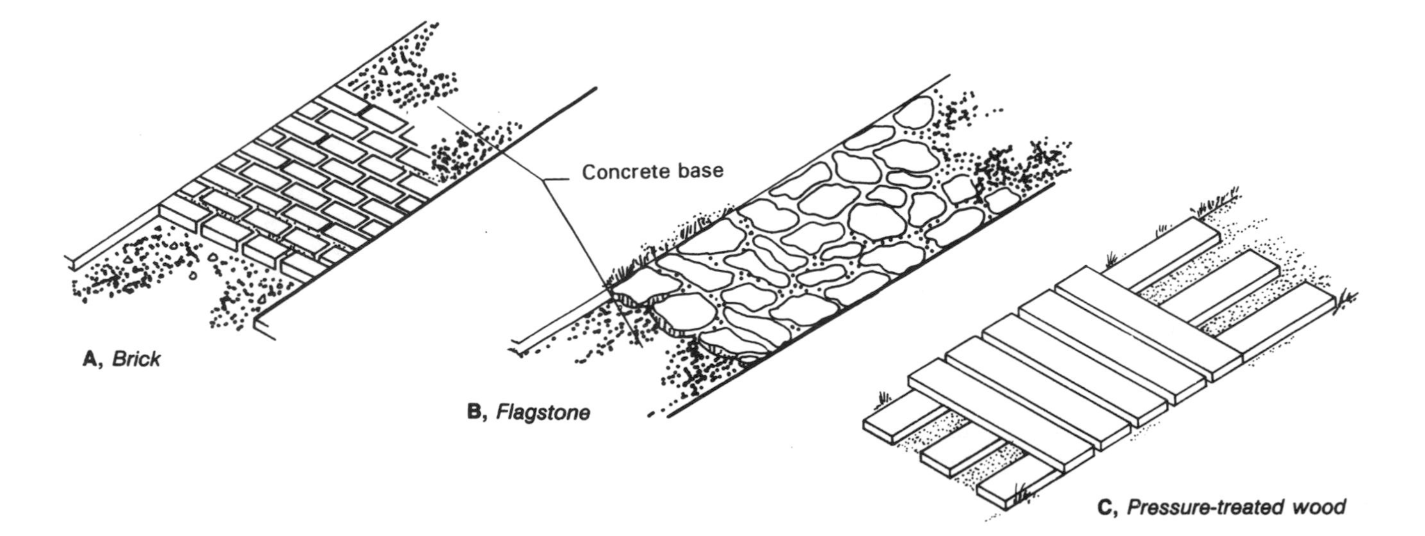

Figure 32: Other sidewalk types.

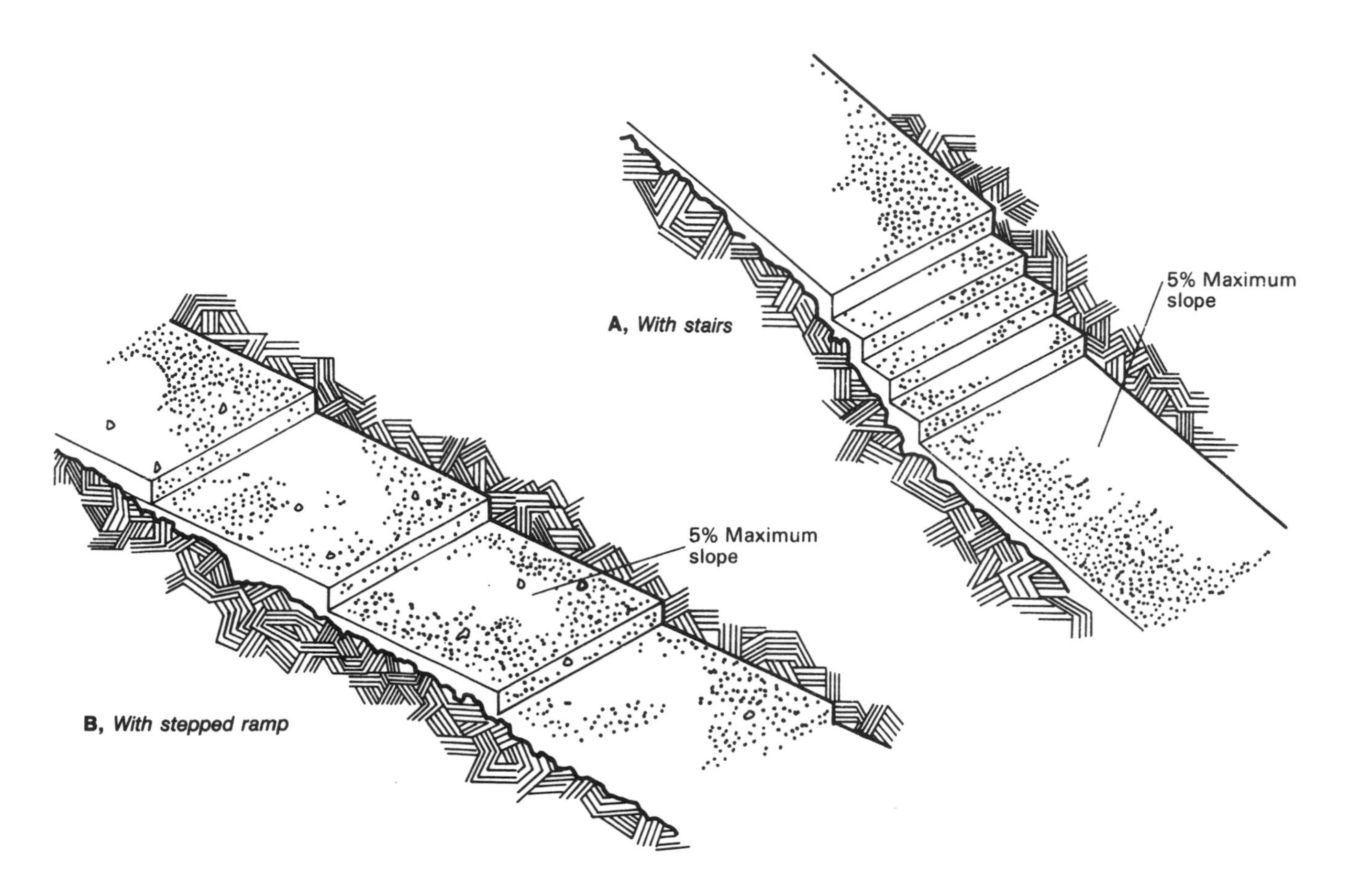

Figure 33: Sidewalks on slopes.

CN52 INSPECT driveway, walkway and patio areas.

CN53 ROPE off drive, walk, patio and mailbox areas so that people and pets don't make footprints. Put some red tape on the strings so that they can be seen.

CN54 PAY concrete supplier for driveway, walkway, patio and mailbox pad concrete.

CN55 PAY concrete finishers and have them sign an affidavit.

CN56 PAY concrete finishers' retainage.

WATERPROOFING—Steps

WP1 CONDUCT standard bidding process. Invite your full-service company to bid.

WP2 SEAL footing and wall intersection with portland cement. Skip if you have no basement.

WP3 APPLY ¼″ portland cement to moistened masonry wall. Roughen surface and allow to dry twenty-four hours. Perform only if you have a masonry foundation.

WP4 APPLY second ¼″ portland cement to moistened wall. Keep moist for forty-eight hours and allow to set. Perform only if you have a masonry foundation.

WP5 APPLY waterproofing compound to all below-grade walls.

WP6 APPLY black 6-mil poly (4-mil is a bit too thin) to tarred surface while still tacky. Roll it out horizontally and overlap seams.

WP7 INSPECT waterproofing (tar/poly). No holes should be visible.

WP8 INSTALL 1″ layer of gravel as bed for drain tile. Optional for crawl spaces; may use even if you have no basement.

WP9 INSTALL 4″ perforated plastic drain tile

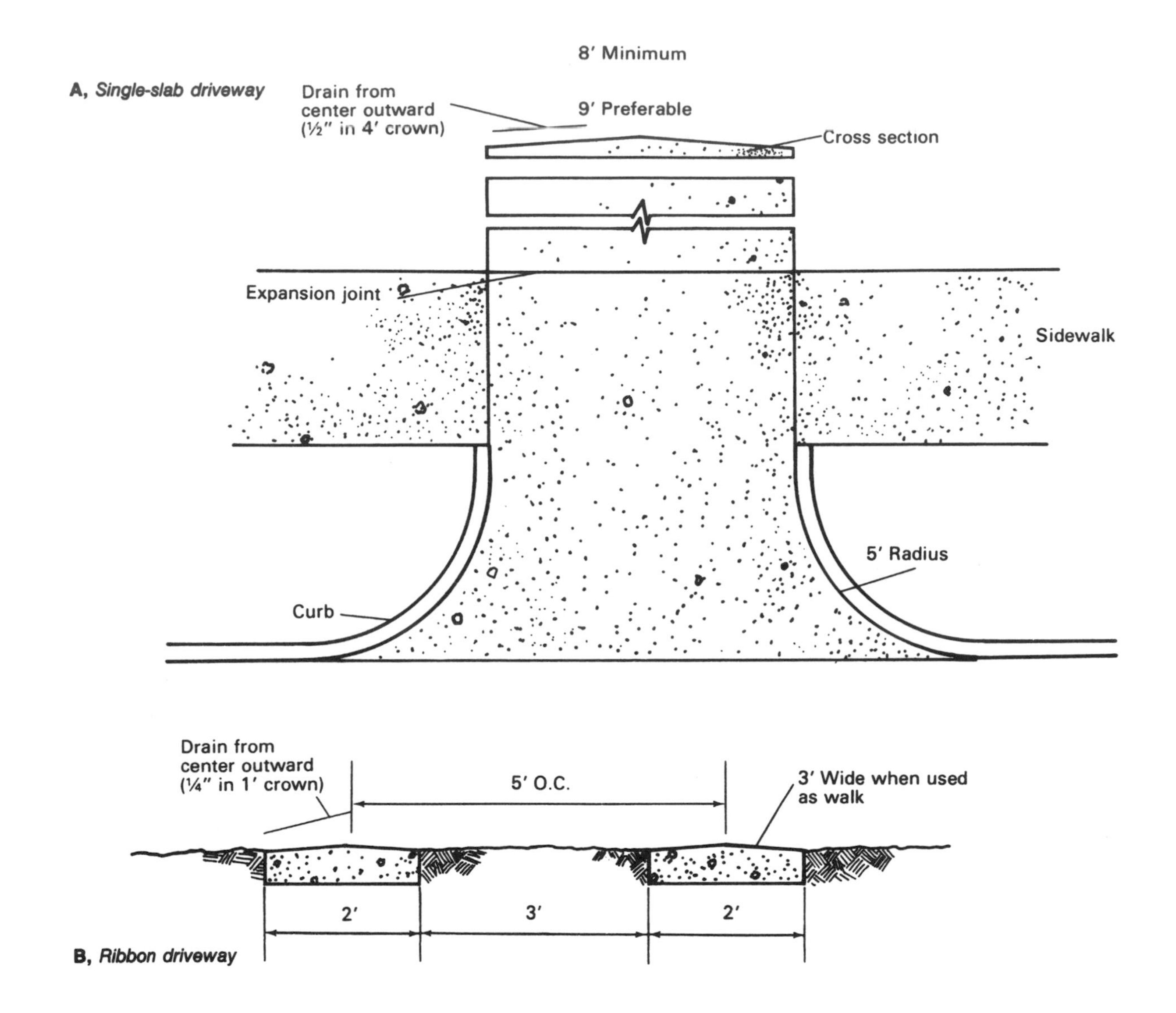

Figure 34: Driveway details.

around entire foundation perimeter. Also install PVC runoff drainpipe.

WP10 INSPECT drain tile for proper drainage. Mark the ends with a stake to prevent them from being buried.

WP11 APPLY 8″ to 12″ of top gravel (#57) above drain tile. *Hint:* Surround water spigot area with any excess gravel to prevent muddy mess during construction. Spigot will be used quite often.

WP12 PAY waterproofing sub and have him sign an affidavit.

WP13 PAY waterproofing sub retainage.

CONCRETE—Sample Specifications

General

- All concrete, form and finish work is to conform to the local building code.
- Concrete will not be poured if precipitation is likely or unless otherwise instructed.
- All form, finishing and concrete work MUST be within ¼″ level.
- All payments to be made five working days after satisfactory completion of each major structure as seen fit by builder.

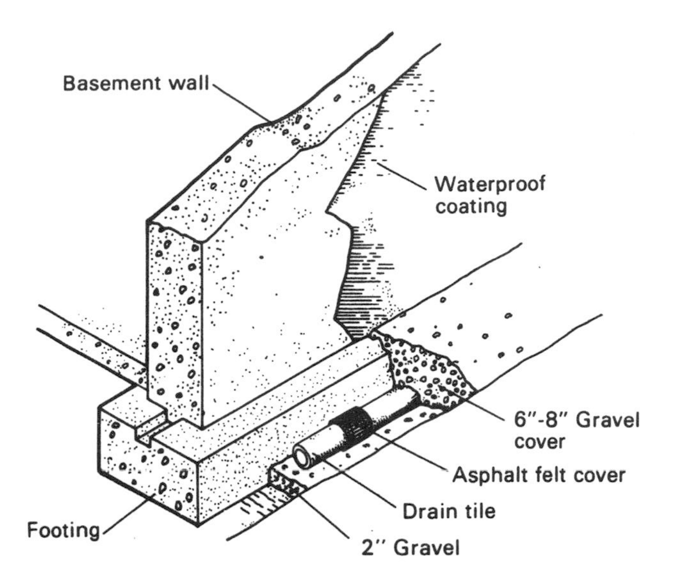

Figure 35: Drain tile for soil drainage at outer foundation walls.

Concrete Supplier

- Concrete is to be air-entrained ASTM Type I (General Purpose), 3000 psi after 28 days.
- Concrete is to be delivered to site and poured into forms in accordance with generally accepted standards.
- Washed gravel and concrete silica sand are to be used.
- Each concrete pour is to be done without interruption—no more than one-half hour between loads of concrete to prevent poor bonding and seams.
- Concrete is to be poured near to final location to avoid excessive working. Concrete will not be thinned at the site for easier working.

Formwork

Bid is to perform all formwork per attached drawings, including footings for:

- Exterior walls
- Monolithic slab
- Poured walls
- Bulkheads
- Garage
- A/C compressor slabs
- Patio
- Basement piers

According to specifications:

- All footing forms to be of 2″ or thicker wood or steel
- All forms to be properly oiled or otherwise lubricated before being placed into service
- All forms, washed gravel and reinforcing bars to be supplied by form subcontractor unless otherwise specified in writing; all of these materials to be included in bid price
- All forms will be sufficiently strong so as to resist bowing under weight of poured concrete.
- Form keys to be used at base of footings and for brick ledge
- Expansion joints and expansion joint placement are to be included in bid. Expansion joints: Driveway every 15′, sidewalks every 5′
- Pier footings and perimeter footings to be poured to exactly the same level
- All concrete to be cured at the proper rate and kept moist for at least three days
- Foundation and garage floor to be steel-troweled smooth with no high or low points
- Finished basement floor will slope toward drains
- Garage floor, driveway and patios will slope away from dwelling for proper drainage
- Concrete to be poured and finished in sections on hot days in order to avoid premature setting

Concrete Block

- All courses to be running bond within ¼″ level
- Top course to be within ¼″ level
- Standard 12″ block to be used on backfilled basement area
- 4″ cap block to finish all walls
- Horizontal reinforcing to be used every three courses
- Trowel joints to be flush with block surface

WATERPROOFING—Specifications

- Bid is to include all material and labor required to waterproof dwelling based on specifications described below.
- Two ¼″ coats of Portland cement are to be applied smoothly and evenly. First coat is to be applied to moistened masonry blocks and roughened before drying. Second coat is to be kept moist for forty-eight hours to set (block wall only).
- Portland cement is to be applied smoothly to intersection of footing and wall.
- Sharp points in Portland cement coating are to be removed prior to application of tar coating, as they will puncture 6-mil poly.
- Asphalt coat is to be applied hot, covering entire subgrade area. Coat is to be a minimum of 1/64″.
- 4″ diameter perforated PVC drainpipe is to be installed around entire perimeter of foundation. Tile sections are to be separated ¼″ with tar paper to cover all joints. Drain tile should have a minimum slope of 1″ in 20′. Drain holes every 6″ in pipe.

PEST CONTROL—Specifications

- Bid is to provide all materials, chemicals and labor to treat all footing and slab areas.
- Base of home is to be sprayed after all construction is complete.
- Termite treatment warranty is to be provided.

CONCRETE—Inspection

Batter Boards

- ☐ All batter boards installed:
  - Exterior walls
  - Piers and support columns
  - Garage or carport
  - Fireplace slabs
  - Porches and entryway
  - Other required batter boards
- ☐ All batter boards installed properly:
  - Level
  - Square
  - Proper dimensions according to blueprints
  - All strings tight and secure
  - All string nails to be marked to prevent movement
- ☐ All bulkheads well formed and according to plan. Step down depth dimension correct.

Checklist Prior to Pouring Any Concrete

- ☐ All planned concrete and form work has been inspected by state and/or local officials if necessary prior to pouring
- ☐ Forms are laid out in proper dimensions according to plans—length, height, depth, etc.
- ☐ Correct relationship to property lines, setback lines, easements and site plan (front, back and all sides)
- ☐ No rain predicted for six to eight hours minimum
- ☐ Forms secure with proper bracing
- ☐ Forms on proper elevation
- ☐ No presence of groundwater or mud
- ☐ No soft spots in ground area affected
- ☐ Ground adequately packed
- ☐ Forms laid out according to specifications: parallel, perpendicular and plumb where indicated
- ☐ All corners square
- ☐ Necessary portions are straight; no bowing. Wood forms should be nominal 2″ thick.
- ☐ Spacers used where necessary to keep parallel forms proper distance apart
- ☐ Expansion joints in place and level with top of surface; approximately ½″ wide
- ☐ All plumbing underneath or in concrete is installed, complete, inspected and approved
- ☐ Reinforcing bars bent around corners—no bars just intersecting at corners
- ☐ Concrete supplier scheduled
- ☐ Adequate finishers available to do finish work
- ☐ Holes under forms filled with crushed stone
- ☐ Proper concrete ordered

- ☐ Forms tested for strength prior to pouring
- ☐ Backfill stable and not in the way, will not roll onto freshly finished cement or get in finisher's way
- ☐ Garage, carport, patio, entrance and porch slabs angled to slope away from home
- ☐ Any necessary plumbing has been inspected.
- ☐ Gravel, re-wire and 6-mil poly have been placed properly when and where required
- ☐ All water and sewer lines installed and protected from abrasions during pouring of concrete
- ☐ Finishers have been scheduled or are present.
- ☐ Crawl space, if any, has been cleared with a rake
- ☐ Foam collars have been placed over all water and sewer pipes where they protrude out of the concrete slab

General Checklist After Concrete and Blockwork

- ☐ Surface checked for level with string level, builder's level or transit
- ☐ Corners squared using a builder's square
- ☐ No cracks, rough spots or other visible irregularities on surface
- ☐ Sill bolts vertical and properly spaced (if used)

Foundation Wall (Concrete Block)

- ☐ Masonry joints properly tooled and even
- ☐ Blocks have no major cracks or irregularities
- ☐ Cap blocks installed level as top course
- ☐ Room allowed for brick veneer above, if any is to be used
- ☐ Joint between wall and footing properly parged (sloped away from wall) with mortar

WATERPROOFING—Inspection

- ☐ Portland cement is relatively smooth and even, with no sharp edges protruding.
- ☐ Portland cement completely covers intersection of wall and footings.
- ☐ Asphalt coating completely covers entire subgrade area, including intersection of wall and footings.
- ☐ Black 6-mil poly completely covers entire tar coating and adheres firmly to it. Look for any tears or punctures in poly which need to be patched before backfill. Remove large rocks and roots from backfill area so that poly will not be torn during backfill process later.
- ☐ Tarred area does not go above grade level or where stucco is to be placed.
- ☐ Poly is secured to stay in place during backfill.
- ☐ No rain is in basement.
- ☐ No wet spots are on interior basement walls.

Framing

Framing is one of the most visible signs of progress on the building project. Become close friends with your framing contractor. He will be involved with the project longer than any other contractor. His skills will have a large impact on the quality of your home and your ability to stay on schedule.

Take adequate time and choose your framer carefully. A good framer is worth his weight in 2×4s! A good framing crew should consist of at least three people in order to move lumber around efficiently. Check your carpenter's ability to read blueprints. He will be making many interpretations of these plans when framing. Your carpenter should have experience as the master carpenter on several jobs, and not just as a helper.

When you feel comfortable with your carpenter's experience, proceed by informing him of the quality of work you demand. Ask him to be sure to "crown" all material during construction. What is "crowning"? All lumber has a natural curve to it. Observe the concentric arcs when looking at the end of a board. When framing, the carpenter should be sure to turn all the lumber so that the natural curves of the lumber are in the same direction. When framing horizontal joists, this curve should bow up. The top of the curve is the "crown." Ask your carpenter to lay aside any badly warped or curved lumber to send back to the materials supplier or to be used for short framing pieces, such as cripples or dead wood.

A, *Northern Hardwood and Pine Manufacturers Association, Inc.;*

B, *Pacific Lumber Inspection Bureau, Inc.;*

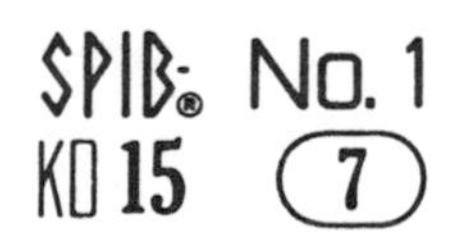

C, *Southern Pine Inspection Bureau;*

D, *Canadian Lumberman's Association.*

Figure 36: Examples of symbols or lumber logos of quality control agencies.

LUMBER

Lumber (that is, boards) is used to build a home. Wood is used to build a fire. Lumber comes in lengths starting at about 8′ and increasing in length by even numbers. The most common lengths are 8′, 10′, 12′, 14′, 16′, 18′ and 20′.

Use surfaced, kiln-dried lumber to reduce warping due to shrinkage. Warpage in framing pieces can also be reduced by using a higher grade of lumber, and will result in a neater framing job and a happier framing sub. The additional cost is worth it unless you have an unusually careful carpenter willing to spend hours picking out bad lumber. Specify Douglas fir for the best job. Avoid mixing fir and pine lumber on the same job. Pine can be $\frac{1}{4}$″ to $\frac{1}{2}$″ different in size because of different shrinkage rates. Any lumber in contact with the foundation or concrete must be pressure-treated pine, or "celcure" as it is commonly called. This lumber provides moisture and termite protection between the foundation and the framing of the rest of the house.

Studs

A 2×4 is different from a stud. A stud is a high-quality 2×4 cut to a specific length, such as 92$\frac{5}{8}$″. A stud is designed to make it easy to frame a wall that will be exactly 8 feet tall. Therefore the stud is cut less than the 8′ length to allow room for the top and bottom wall plates. The term "2×4" is generic and refers to the lumber's nominal dimensions. Actually, 2×4 lumber is about 1$\frac{5}{8}$″ × 3$\frac{5}{8}$″. It was 2″×4″ when it was cut, but shrinkage and planing of the rough lumber results in a smaller dimension. So when you order "2×4s", you are ordering all that "extra" lineal framing material used for plates, temporary bracing and other components that are cut to varying lengths. For your job site, this 2×4 lumber will normally be sold to you by the lineal foot. For example, when you frame a home, you will order a certain number of studs for vertical wall components, and perhaps 1,000 feet of 2×4 lumber for everything else. You may want to request longer lengths, such as 14′ or 16′, just so you have plenty of long pieces when you need them.

PAYMENT OF FRAMING CONTRACTOR

The framing contractor does not usually provide any of his own materials for the job, except perhaps nails. It is your responsibility to make sure that all necessary materials are on the site when needed. Carpenters usually charge by the square footage of the area framed, plus extras, or will appraise the job and supply a total bid price. This includes all space covered by a roof, not just heated space: the garage area, enclosed porches, tool sheds, garden houses, etc. The framing charge usually includes setting the exterior doors and windows and installing the sheathing. Special items, such as bay windows, stairs, curved stairs, chimney chases and recessed or tray ceilings, usually require additional charges.

TYPES OF FRAMING

The two main types of framing used for residential construction are balloon framing and platform framing. For a more complete description of these two

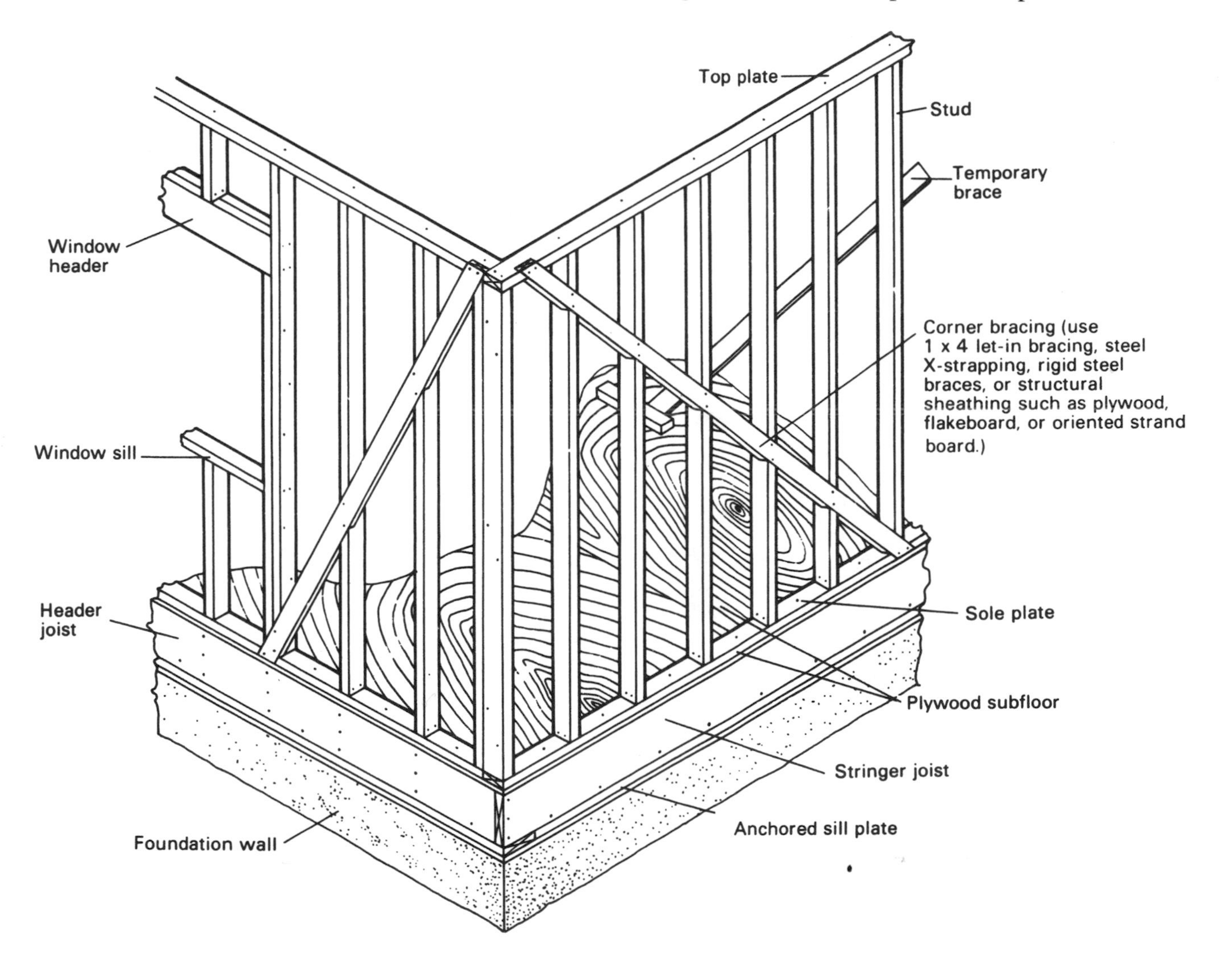

Figure 37: Wall framing used in platform construction.

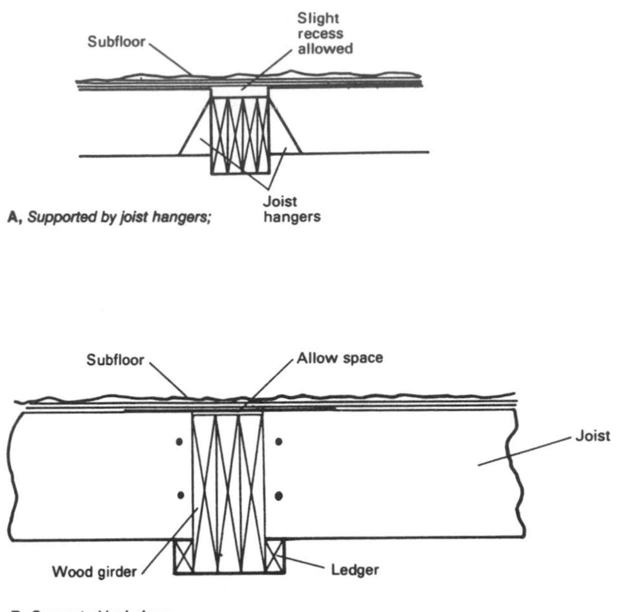

Figure 38: Joints butted to side of wood beam.

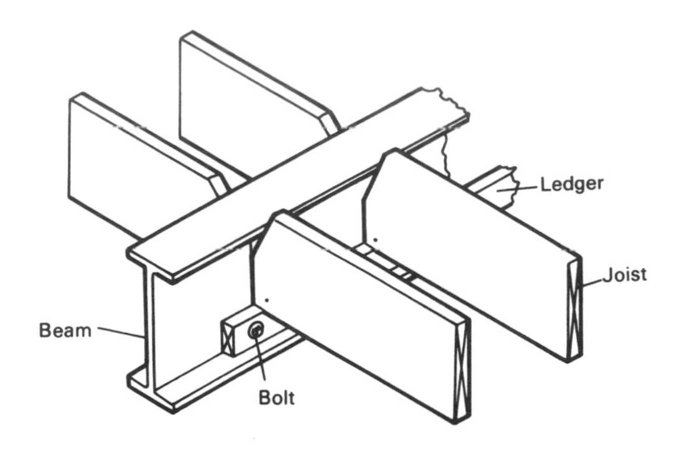

Figure 40: Steel beam with joists bearing on ledger.

methods, refer to a good carpentry manual. Platform framing is the most widely used type and will be the type your carpenter is probably familiar with. In this book, only platform framing will be covered, as it is by far the most popular framing method.

SUBFLOOR FRAMING

There are two types of floor framing: joist floors and floor trusses. Joist floors usually consist of 2×8s or 2×10s spaced 12″, 16″ or 24″ on center with bridging in between. Bridging consists of diagonal bracing between joists to improve stability and to help transfer the load to adjacent joists. Joist floors are by far the most common type in use today, but floor trusses are becoming more popular due to several advantages over the conventional method. A floor

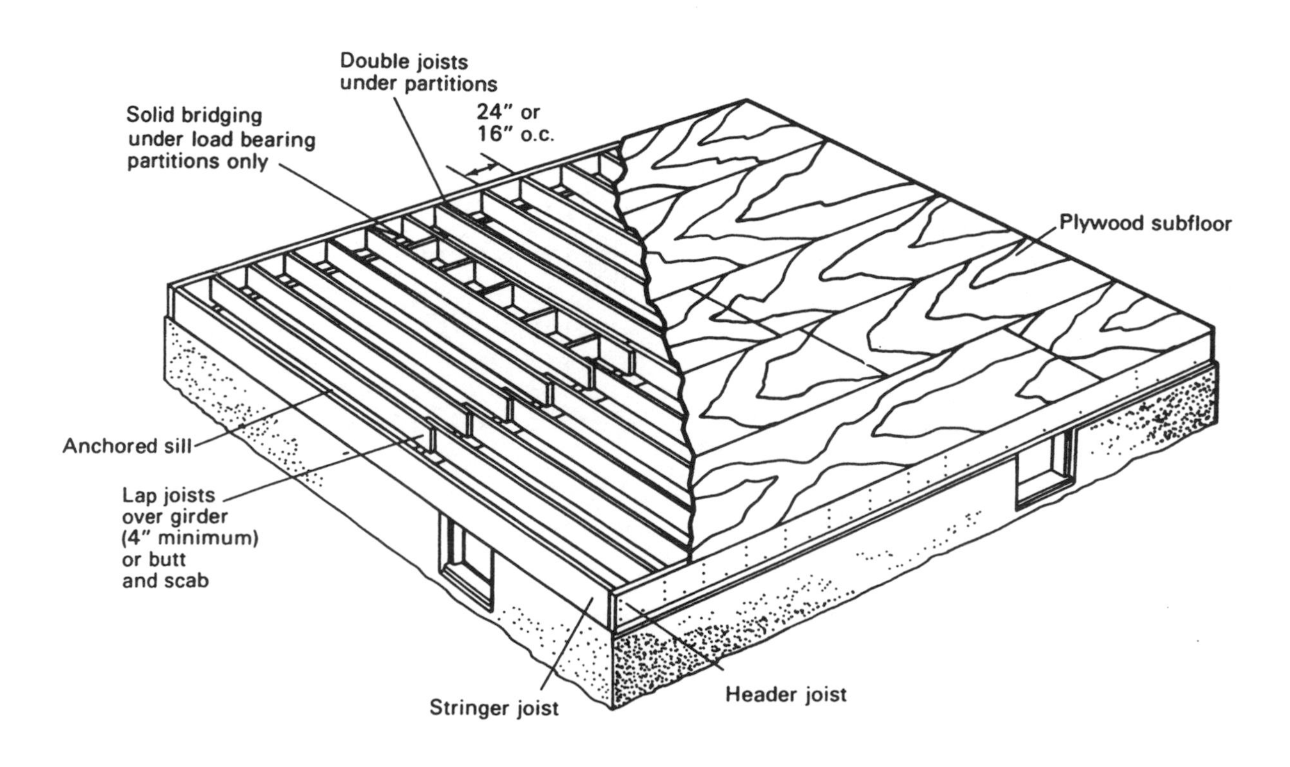

Figure 39: Typical floor framing.

truss, like a roof truss, consists of 2×4s with diagonal bracing in between, which makes the entire unit very strong for its weight.

Floor trusses have the following advantages over standard joists:

- Because of their added rigidity, floor trusses can span longer distances without the need for beam support.
- Wiring and heating ducts may be run within the floor truss, eliminating the need to box in heating ducts below ceiling level. This makes the jobs of the plumber, electrician, HVAC and telephone sub much easier. Make sure to point this out to these contractors when bargaining for rates.
- Floor trusses provide better sound insulation between floors because there is less solid wood for sound to travel through.
- Floor trusses warp less, squeak less and are generally stronger than comparable joist floors.
- Even though floor trusses are a bit more expensive, their advantages can simplify a construction project and save money in the long run. Your supplier for these units will be the same as for roof trusses.

FLOORING MATERIALS

The flooring design is another area where recent innovations have produced a simpler and superior product. Most standard flooring consists of a subfloor, usually of ½″ or ⅝″ CDX plywood, and a finish

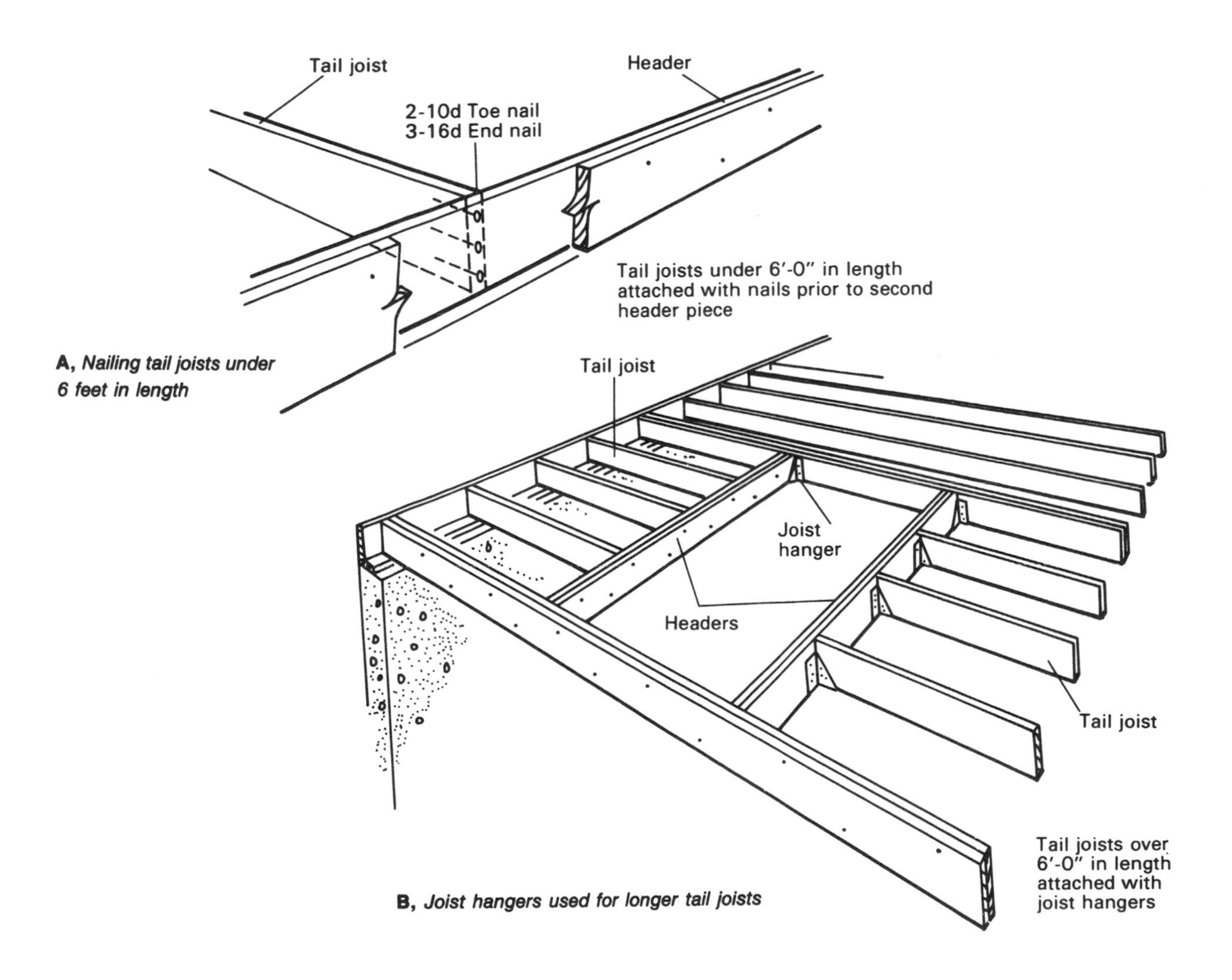

Figure 41: Floor opening framed with double header and double trimmer joist.

	Length of maximum clear span										
	1.0 × 10^6 psi	1.1 × 10^6 psi	1.2 × 10^6 psi	1.3 × 10^6 psi	1.4 × 10^6 psi	1.5 × 10^6 psi	1.6 × 10^6 psi	1.7 × 10^6 psi	1.8 × 10^6 psi	1.9 × 10^6 psi	2.0 × 10^6 psi
Living areas (40 lb/ft² live load) Minimum required bending stress (lb/in²)	1,050	1,120	1,190	1,250	1,310	1,380	1,440	1,500	1,550	1,610	1,670
Joist size											
2×6	7' 3"	7' 6"	7' 9"	7'11"	8'2"	8'4"	8'6"	8'8"	8'10"	9' 0"	9'2"
2×8	9' 7"	9'11"	10' 2"	10' 6"	10'9"	11'0"	11'3"	11'5"	11' 8"	11'11"	12'1"
2×10	12' 3"	12' 8"	13' 0"	13' 4"	13'8"	14'0"	14'4"	14'7"	14'11"	15' 2"	15'5"
2×12	14'11"	15' 4"	15'10"	16' 3"	16'8"	17'0"	17'5"	17'9"	18' 1"	18' 5"	18'9"
Sleeping areas (30 lb/ft² live load) Minimum required bending stress (lb/in²)	1,020	1,080	1,150	1,210	1,270	1,330	1,390	1,450	1,510	1560	1,620
Joist size											
2×6	8'0"	8' 3"	8'6"	8' 9"	8'11"	9'2"	9'4"	9'7"	9' 9"	9'11"	10'1"
2×8	10'7"	10'11"	11'3"	11' 6"	11'10"	12'1"	12'4"	12'7"	12'10"	13' 1"	13'4"
2×10	13'6"	13'11"	14'4"	14' 8"	15' 1"	15'5"	15'9"	16'1"	16' 5"	16' 8"	17'0"
2×12	16'5"	16'11"	17'5"	17'11"	18' 4"	18'9"	19'2"	19'7"	19'11"	20' 3"	20'8"

Source: National Forest Products Association (1977). Span Tables for Joists & Rafters.
Note: Use table 8 for joists spaced 16 inches on center.
The modulus of elasticity (E) measures stiffness and varies with the species and grade of lumber as shown in the technical note on design values. The bending stress (F) measures strength and varies with the species and grade of lumber as shown in the technical note on design values.

Table 5: Allowable spans for simple floor joists spaced 24 inches on center for wood with modulus of elasticity values of 1.0 to 2.0 × 10^6 pounds per square inch.

floor that can be ⅝" particleboard or another layer of plywood. The second layer is installed after drying-in the structure. If this method is used, be sure to install a layer of asphalt felt (the type used in roofing) or heavy craft construction paper between the two layers. This acts as a sound buffer and reduces floor squeaking. Often, a two-story house will use ½" flooring on the first floor and ⅝" on the second floor.

A potentially superior flooring method has been introduced by the American Plywood Association. Called the APA Sturdifloor, it is a single-layer floor that saves labor and materials. This floor is constructed with one layer of Sturdifloor-approved CDX plywood (usually tongue and groove) that is nailed and glued to the floor framing members. Tongue-and-groove (T & G) plywood has edges that interlock along the two longest edges. This reduces the tendency for boards to flex separately when a load is applied directly to the joint. The act of gluing the plywood to the joists produces an extremely rigid floor and virtually eliminates floor squeaking. This method works extremely well in conjunction with floor trusses, and produces a sturdy floor with a minimum of time, hassle and materials.

NOTE: Whatever method you use, make sure you purchase only approved plywood for this purpose. Plywood used for single-layer construction will have an approved APA stamp on it. Ask your supplier about this type of plywood. All flooring plywood must be CDX grade. The *C* refers to the grade of plywood on one side, the *D* refers to the grade on the other side and the *X* refers to exterior grade. This grade of plywood is made with an exterior type glue that is waterproof and will stand up to prolonged exposure to the elements during construction. If you use a two-layer floor with particleboard as your second layer, do not leave this material exposed to the weather. It is not waterproof and will swell up like a sponge.

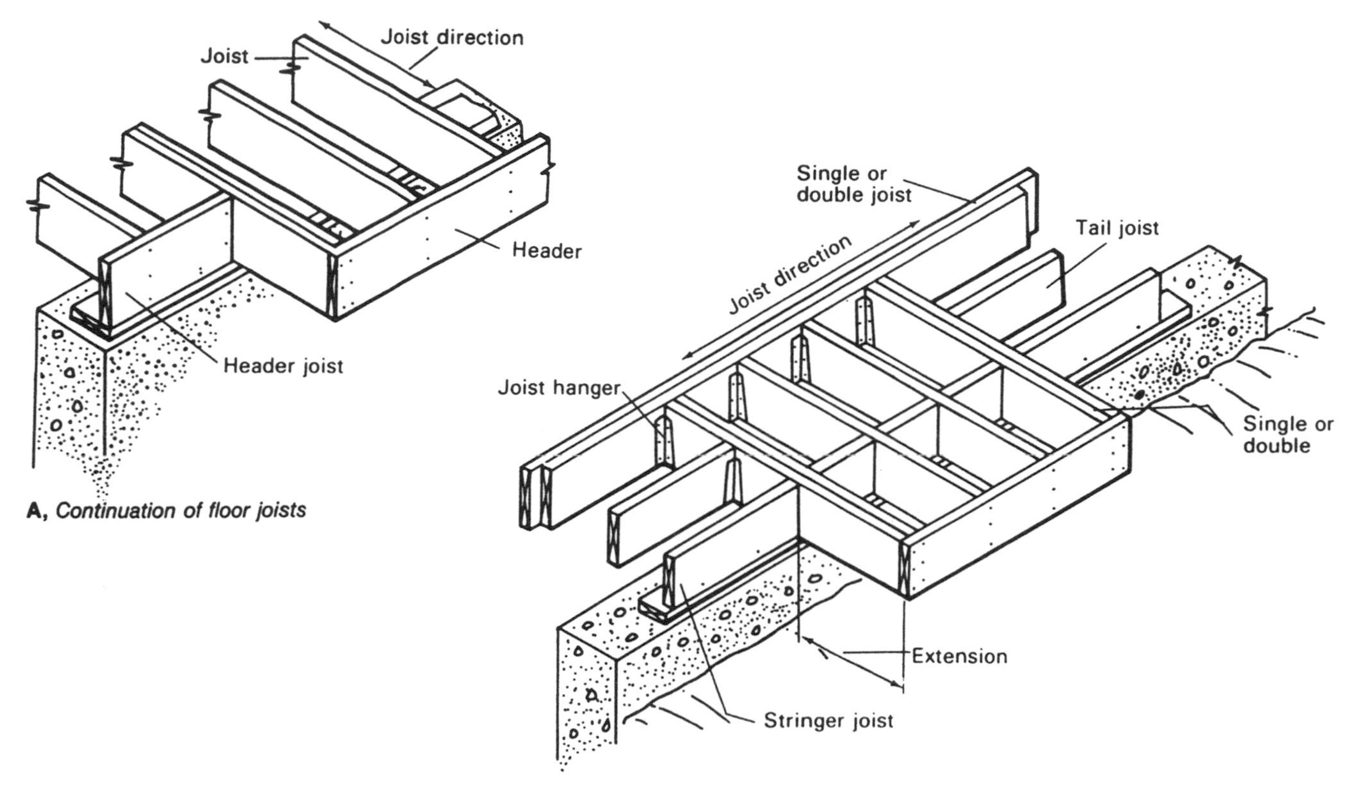

Figure 42: Floor framing at wall projections.

WALL FRAMING

Wall framing is a critical stage in construction. You must be sure that everything is in its right place so that surprises do not occur later. If you make any last minute changes to the plans, consult with your framer or architect to ensure that you don't create new problems. Even a simple change of framing height from 8′ to 9′ can cause many design problems. Windows may not appear proportional. The length of the stairs' run will change, because more steps will be needed to reach the second floor. Your plan may not have the room for additional steps.

Supervision

During construction, make sure that the walls are spaced properly in relation to plumbing that may be cast in a concrete slab. By the time the plumber discovers a problem, you might have a major reconstruction project ahead. Don't forget: If you plan to install one-piece fiberglass shower stalls, you must set them in place before putting up the walls! Some fiberglass stalls will not fit through bathroom door openings.

Make sure to check periodically to be sure that all the carpenter's walls are square. It is easier to get this wrong than you think. Stairs and fireplaces are other sources of potential problems. Check them carefully. Many older carpenters have the habit of installing bridging between studs to prevent warpage. Discourage this practice, as it is an unnecessary waste of lumber and can make insulation more difficult to install—and less effective. Also, energy-minded individuals should consider using 2×6 instead of 2×4 stud walls, which allows space for more insulation. This applies to the exterior walls only, of course.

Your carpenter will be required by building code to brace the corners of the structure for added rigidity. This can be accomplished by using a sheet of plywood in the corners, a diagonal wood brace cut into the studs, or a metal strap brace made for that purpose. Although more tedious, diagonal bracing or metal strapping is advantageous because it is cheaper and does not interfere with the insulating ability of sheathing. Plywood is not a good insulator.

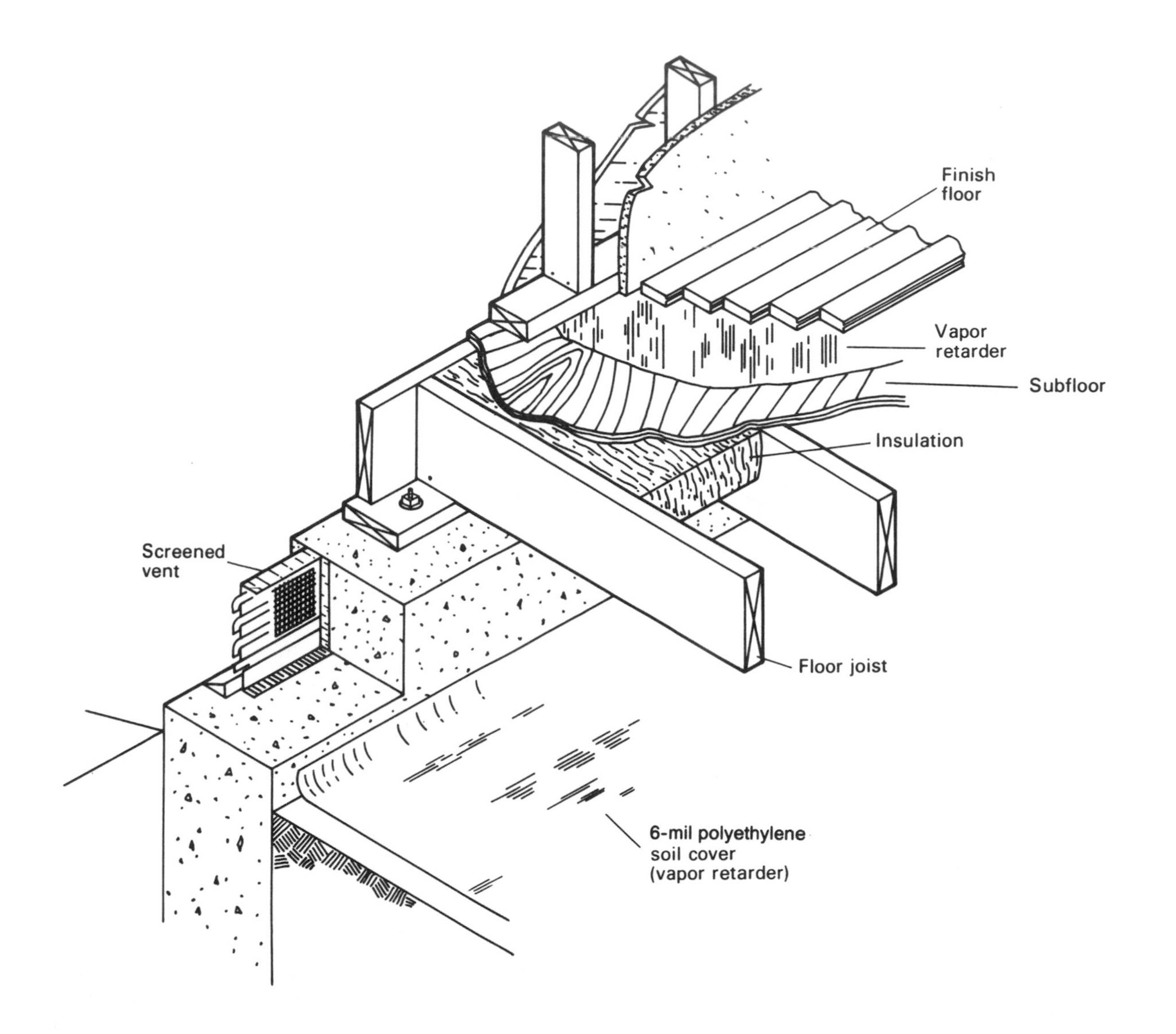

Figure 43: Crawl space ventilator and soil cover.

As the house is dried-in, make sure to cover openings in the walls and ceiling as soon as possible to protect the interior from rain. Place poly over all openings until the windows and doors arrive. Once the windows and doors are installed, lock them with the locksets or by nailing them shut with finishing nails.

Stud Spacing

The vast majority of houses in the United States are built with 2×4s spaced 16″ on center (apart, or O.C.). Framing studs are a genuine classification of framing lumber. Stud grade is more expensive than standard framing lumber. Studs are more uniform and straight, with fewer cracks. You should not frame walls with utility grade 2×4s. They will be more uneven, making drywall application a nightmare. Utility grade lumber also may not be capable of supporting the proper load. To save costs, however, you can use standard #3 fir for the top and bottom plates and jacks, since they do not support any load. The plates only serve to hold the studs together.

A recent trend is to deviate from the normal 16″ framing for cost and energy reasons. Many houses are now constructed with 2×6s to allow more room for insulation. This construction change can increase a wall's insulation factor from R-13 to R-19, a significant energy-conservation measure. When this method is used, studs are spaced 24″ apart. This simplifies construction and reduces the number of studs used by 20 to 30 percent. This method is not without its problems, however. The biggest problem is that most doors and windows are designed to be set in

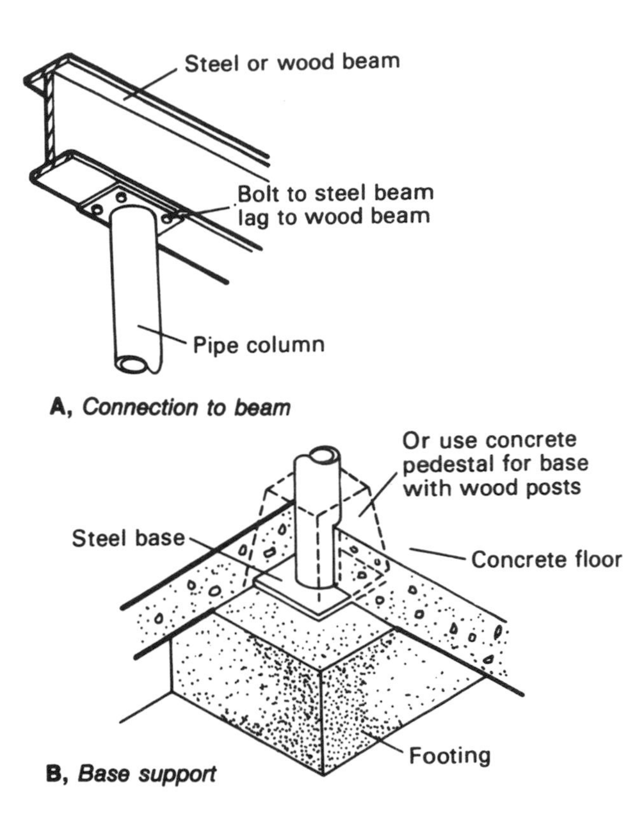

Figure 44: Steel post support for wood or steel beam: connection to beam.

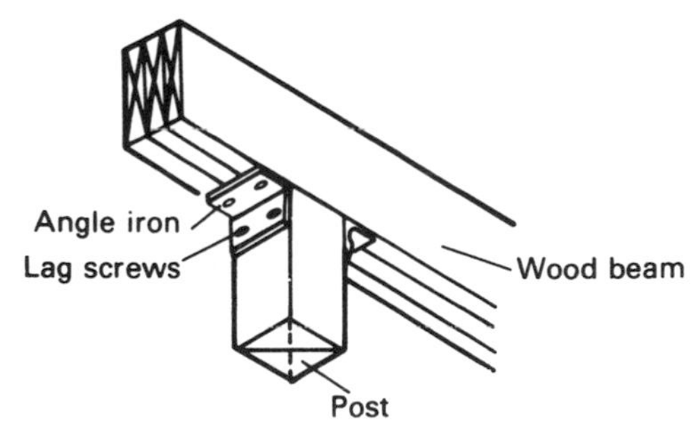

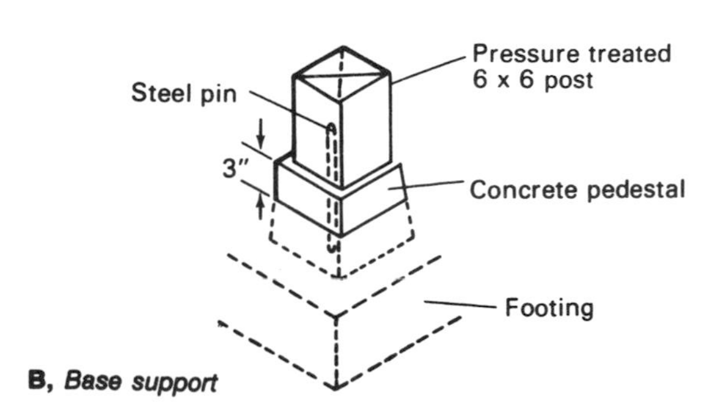

Figure 45: Wood post support for wood beam: A, connection to beam; B, base support.

walls of 2×4 depth. Buying special doors and windows or having custom setting done can cost a considerable sum and can negate many of the cost savings. However, many doors and windows today adjust to accommodate either wall size. Generally, 2×6 stud construction is slightly more expensive than conventional construction, but it does provide a significant energy savings. Make sure your carpenter is familiar with the inherent problems of this type of construction before deciding to use it.

Many of the cost savings of 2×6 construction can be preserved without any of the headaches by going to 2×4 construction spaced 24″ on center. This method is approved by all national building codes for single-story construction, and some local municipalities allow it on two-story construction, or at least on the second floor. Check your local building codes to be sure. Spacing studs 24″ on center saves time and materials and allows more space for insulation, since fewer studs are used. Ask your carpenter if he is comfortable with this method of construction. Look for window units that will match the 2′ framing width if you want to insert winoows between the studs.

ROOF FRAMING

The two basic types of roof framing consist of stick-built roofs and roof trusses. As with floor trusses, roof trusses have the advantage of being very sturdy for their size and weight. In most house designs, roof trusses require no load-bearing walls between the exterior walls. This gives you more flexibility in designing the interior of your home. Roof trusses are also generally cheaper than stick-built roofs because of significant labor savings in construction. Their only major disadvantage is the lack of attic space as a result of the cross member supports. This can be alleviated somewhat by ordering "space-saving" trusses that have the inner cross members angled to provide more attic space. This is called a *W* truss. Stick-built roofs are more appropriate for high pitches or odd-shaped or contemporary roof lines.

Your carpenter will install the roof sheathing, which usually consists of ½″ CDX plywood or pressed chipboard (also known as flake board and oriented strand board E—less expensive) especially made for this purpose. Request that plywood clips be used in

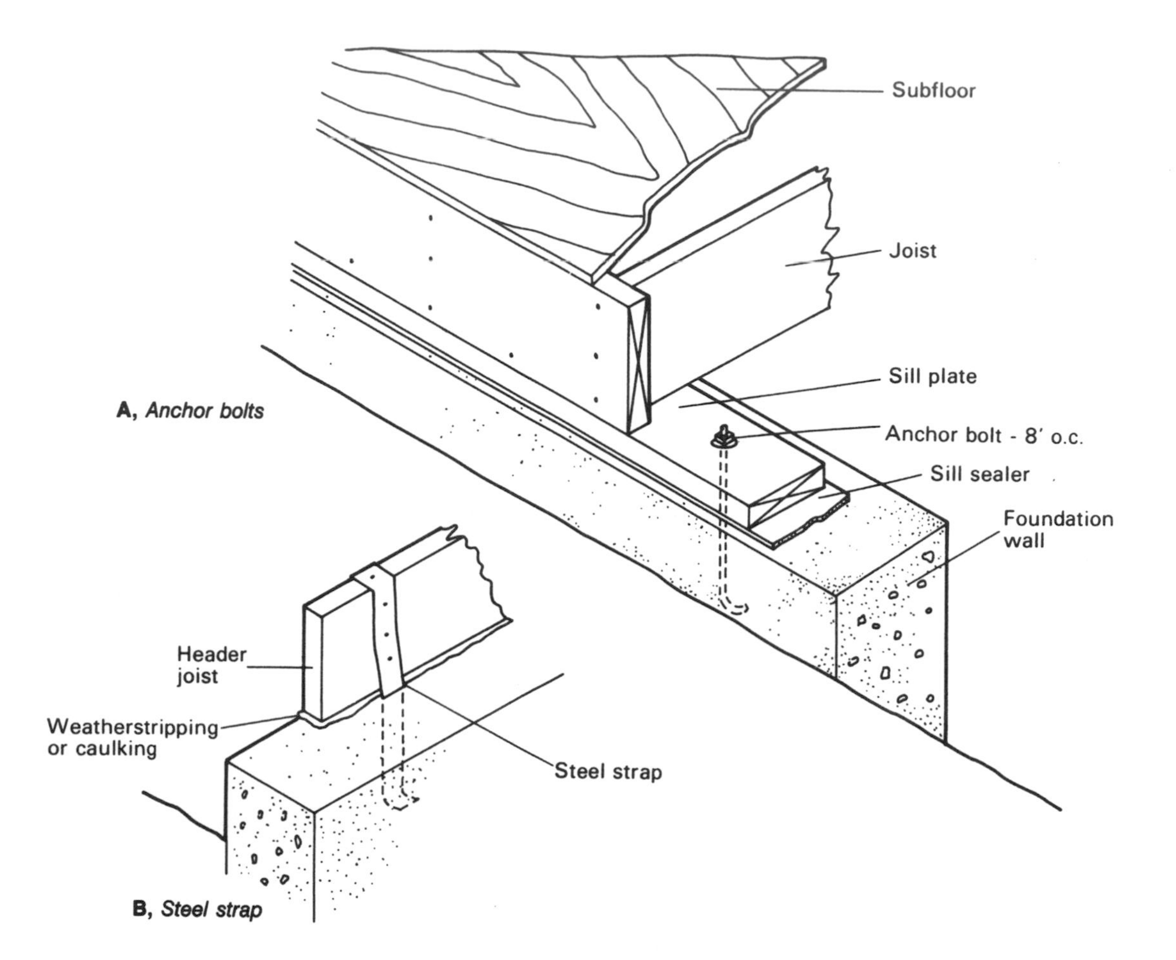

Figure 46: Anchoring floor system to foundation wall: A, anchor bolts; B, steel strap.

between each sheet of roofing material. These are small, *H*-shaped aluminum clips designed to hold the ends and sides of the plywood sheets together for added rigidity. This will help prevent warpage and "wavy" roof surfaces.

Due to the popularity of asphalt and fiberglass roofing shingles, this manual relates specifically to roof framing where these two types of roofing are used. Normal roof decking is not used with cedar shake, slate or tile roofs. If you are planning to have such a roof, consult with your framing and roofing subs. When installing the roof sheathing, the carpenter should be aware of the type of roof ventilation to be used. The most convenient and efficient type to use is continuous eave and ridge vents. The carpenter should leave a 1″ gap in the plywood at the roof ridge for installation of the ridge vent, if used.

SHEATHING

Sheathing comes in several different types. Asphalt-soaked fiber sheathing is the most common and the cheapest, but it has the least insulating value. The next in line would probably be "blue foam" insulation. It is a good compromise of value and insulation ability. Foil-backed urethane is by far the best insulator, but also the most expensive. If you plan to stucco, exterior gypsum board is most often used. Talk to your local supplier to see what types are available in your area, and their insulation ability. When installing the sheathing, your carpenter should take care not to rip or puncture the sheathing. If he does, ask him to replace the defective piece. Pay him only after all defective pieces have been replaced. He should use special nails with a plate on the shaft that holds the sheathing to the stud without puncturing the sheathing.

DOORS AND WINDOWS

After the roof and sheathing have been installed, the framing contractor will be ready to install the doors and windows and dry-in the structure. The proper

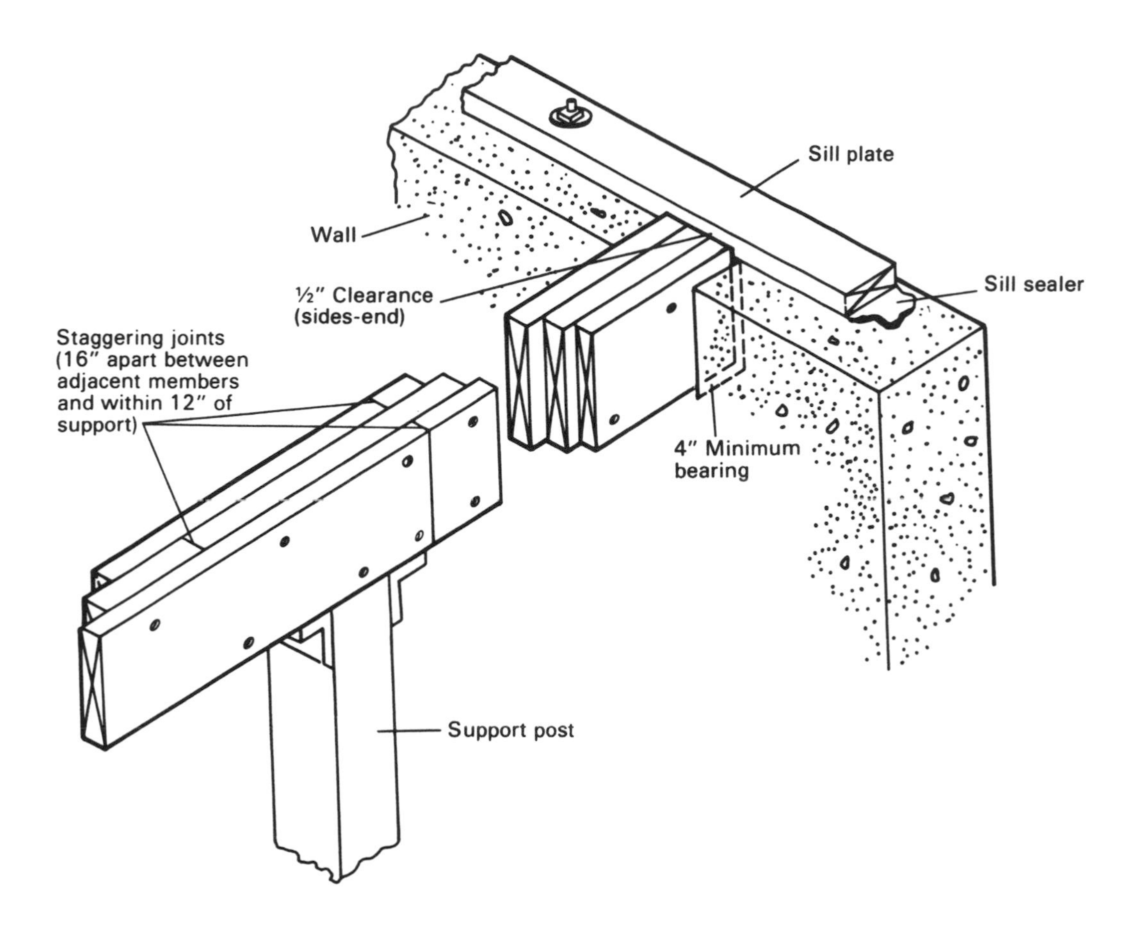

Figure 47: Built-up wood girder. Clearances for wood construction.

installation of doors and windows is probably the source of more irritation and hot tempers than any other job. A poorly hung door isn't worth much except as an argument starter at the end of a hard day's work. The advent of prehung doors has made the job quite a bit easier, but still not foolproof. With your level, check to make sure each door is completely square and that there is an even gap between door and frame from top to bottom. Make sure the door has plenty of room to swell in damp weather (if it is wood) and still open freely. Measure under the door to make sure there is plenty of room for subflooring, carpet pad and carpet without the door rubbing over the carpet when it is installed. The same amount of care should be taken with the windows to guarantee that they will slide freely within their frames. An ounce of prevention is worth a pound of pounding!

WASTE DISPOSAL

Few projects generate more scrap than frame construction. If your area allows burning, this may be your best option. Stack scrap lumber in convenient piles far away from the house and burn only on windless days. Consult with your local fire marshall to determine if the weather is conducive to burning. Monitor the fire constantly and keep a water hose available at all times. Don't burn plastic or other petroleum products. These items leave nondegradable by-products that will stay in your yard practically forever. Keep paint cans and aerosol cans out of the fire. These items can become explosive weapons when heated.

In winter, don't burn waste until the project is completed. You will usually find the subcontractors burning your scrap (and sometimes nonscraps!) in order to keep warm on the job site. Also, make sure the scraps you are burning are not usable. Many small pieces of lumber can be used as blocking and other items. Cornice subs use 2×4s 8″ and longer to run cornice framing. Electricians use 4″ to 6″ 2×4s for blocking around outlets and switch boxes. Plumbers use 2×4s 6″ and up to brace pipes, tubs and other fixtures.

Stack all usable scrap in a neat, accessible pile protected from the elements. This way, your subs can't help but notice the material. This will keep them from wasting lumber by cutting up new, larger pieces.

OTHER TRADES

Your framing sub must be familiar with the workings of the plumbing, HVAC and prefab fireplace subs so that he can anticipate for them where ductwork, pipes and heating equipment must be placed. Make sure his copy of the blueprints indicates the location of furnace and ducting.

FRAMING—Steps

FR1 CONDUCT standard material bidding process. Find the best package deal on good-quality studs, sheathing, pressboard, plywood, interior doors (prehung or otherwise), exterior doors and windows. Find out the return policy on damaged and surplus material and whether their price includes delivery.

FR2 CONDUCT framing labor bidding process. Ask to see their work and get recommendations. A good crew should have at least three people. Any fewer and the framing probably won't go as quickly as you would like. Consider a speed/quality incentive tied to payment.

FR3 DISCUSS all aspects of framing with the crew. This includes all special angles, openings, clearances and other particulars. Also make sure that the electrical service has been ordered. Discuss special-order materials,

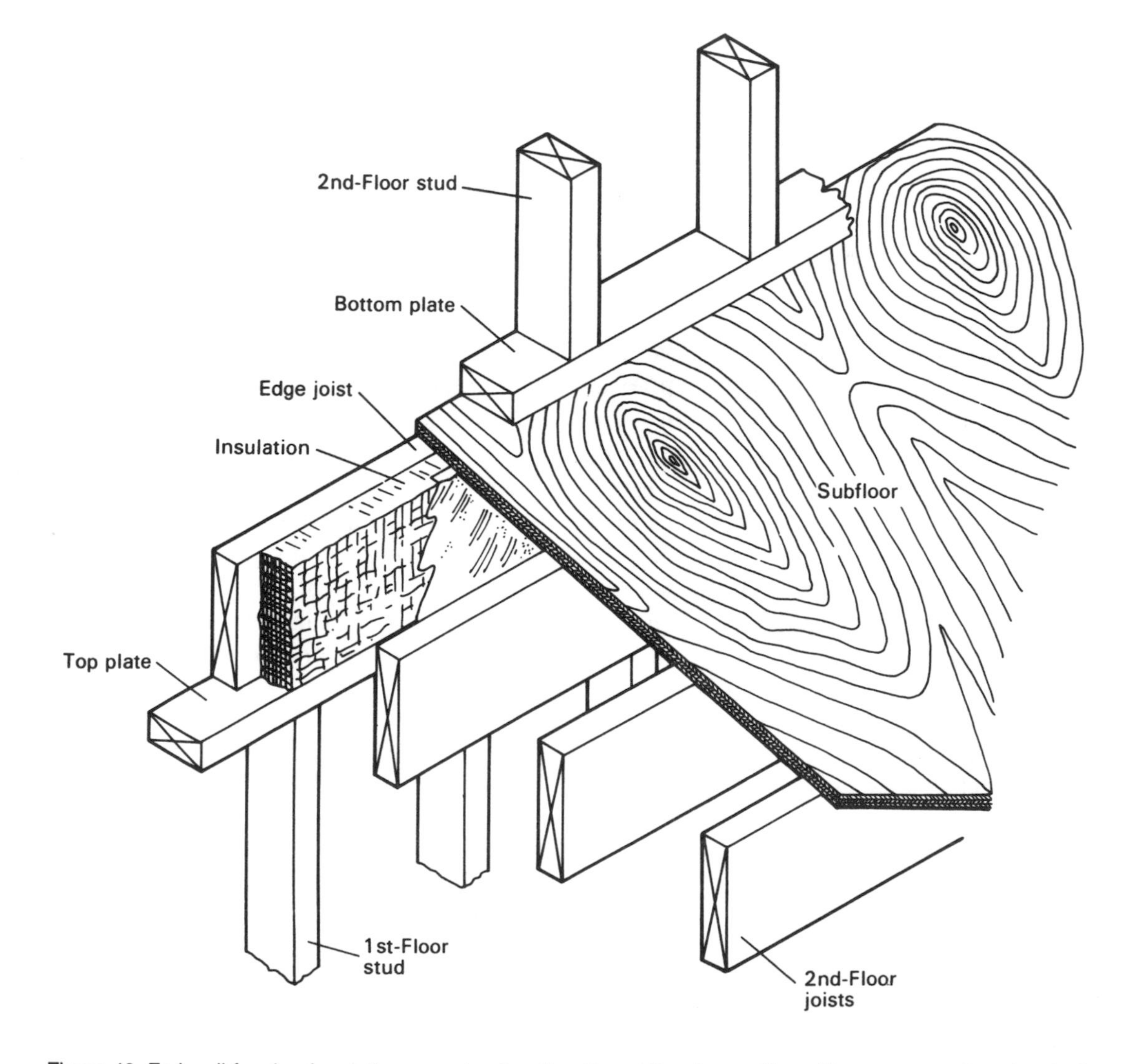

Figure 48: End-wall framing for platform construction (junction of first-floor ceiling with upper-story floor framing).

such as flitch plates or glue-lam beams, and order them as soon as possible.

FR4 ORDER special materials ahead of time to prevent delays later. This includes custom doors, windows, beams and skylights. This is also a good time to make sure your temporary electric pole works—you will need it for the power saws.

FR5 ORDER and receive first load of framing lumber. Have a flat platform ready to lay it on near the foundation. "Line out" framing walls with chalk lines on the foundation to

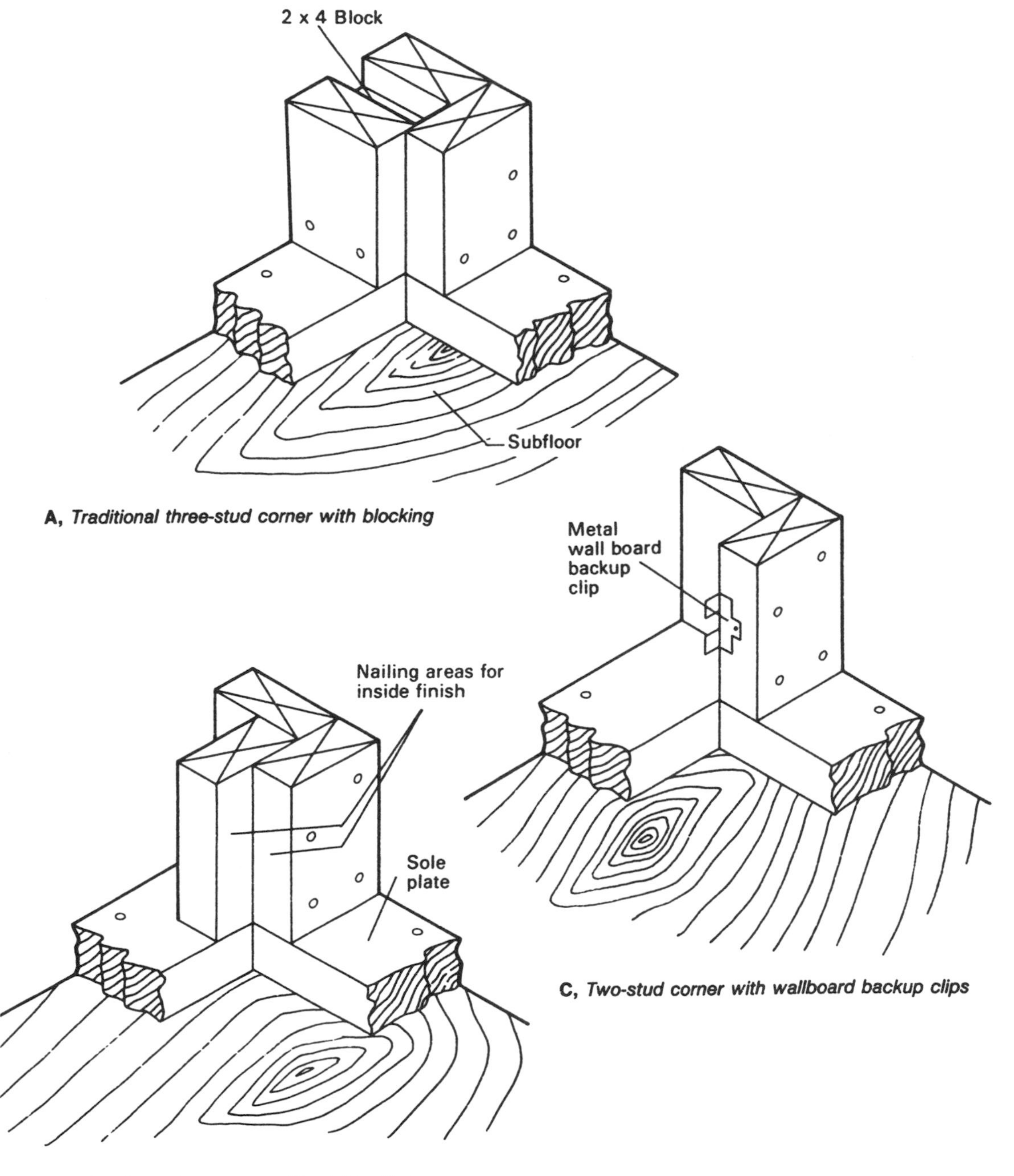

Figure 49: Corner stud assembly. Note use of wallboard backup clip to save additional stud normally needed for drywall backing.

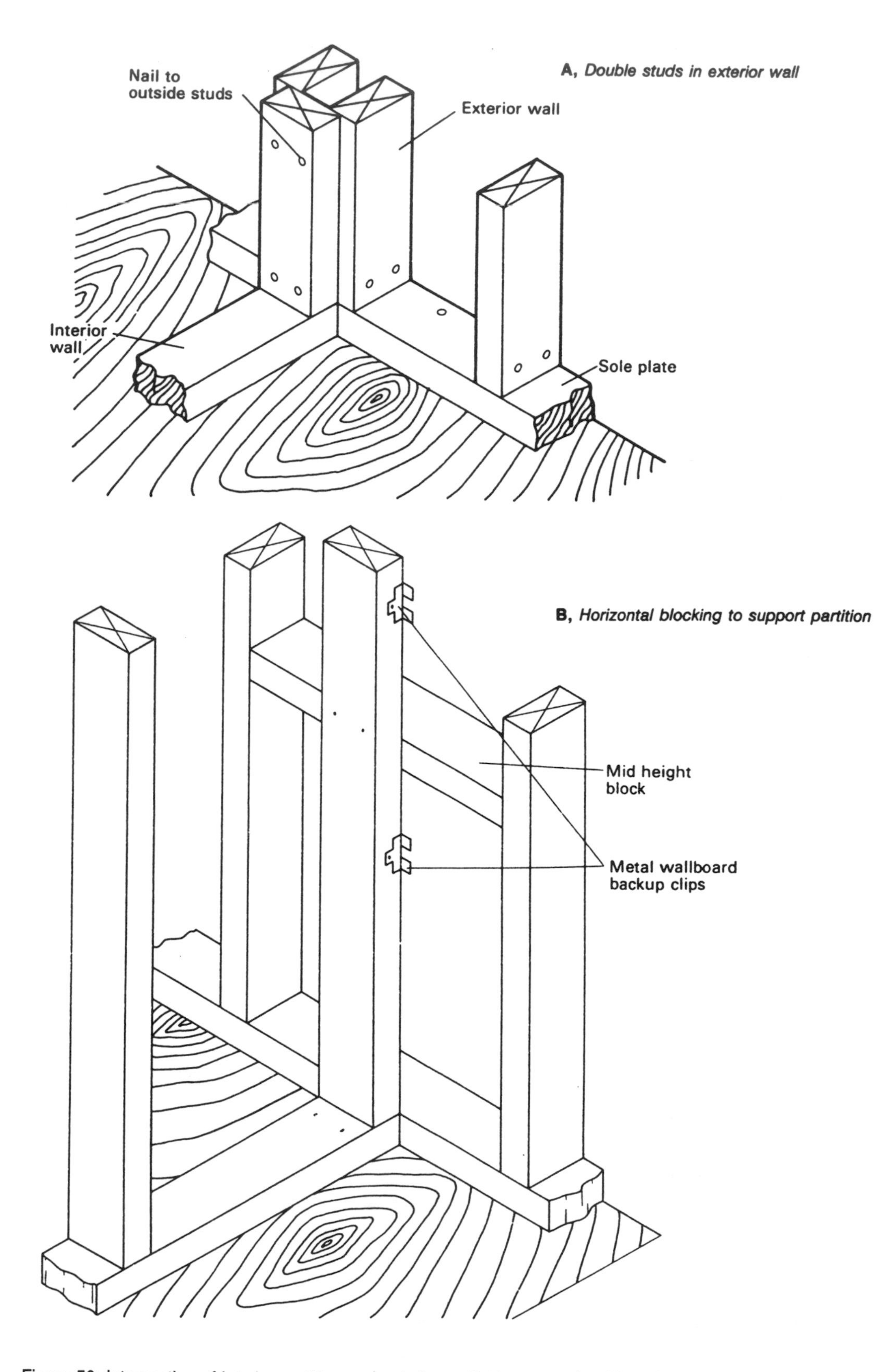

Figure 50: Intersection of interior partition and exterior wall. Note use of wallboard backup clips in place of backer stud.

act as a framing guide. Check chalk lines for accurate dimensions and squareness.

FR6 INSTALL sill felt, sill caulk, or both.

FR7 ATTACH pretreated sill plate to lag bolts embedded in the foundation.

FR8 INSTALL steel or wood support columns in basement.

FR9 SUPERVISE framing process. Check framing dimensions to be sure flooring and wall measurements are in the right places. This is an ongoing process. Check with framer in the morning each day to discuss progress and problems. Check for needed materials. Don't leave excess materials on-site—they have a tendency to disappear. Ask framers to cover lumber with plastic if rain is likely. Ask them to put nails and other valuable supplies in their trucks to prevent theft.

Check wall plates with a level. Also, run a

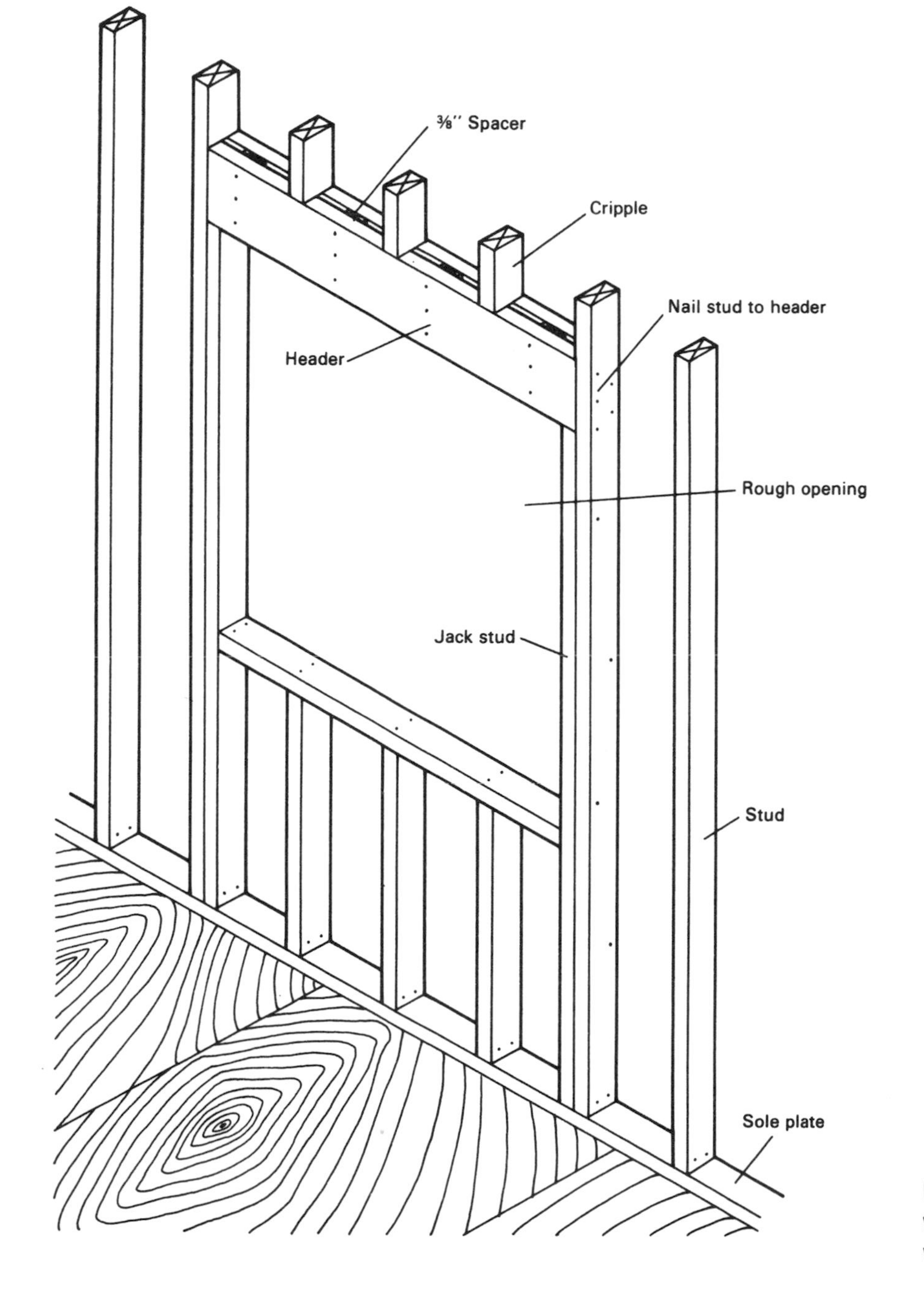

Figure 51: Traditional header assembly over window or door openings in load-bearing wall.

span board between two parallel walls and check with a level to insure that walls are all at the same exact height.

FR10 FRAME first-floor joists and install subfloor. Place these as close together as you can afford. If you can afford 2×10s instead of 2×8s, this will help give your floor a more solid feel. Tongue-and-groove plywood is a popular subfloor. Make sure to use exterior grade plywood, since it will be exposed to the weather during framing. At bay windows, joists will extend out beyond the walls.

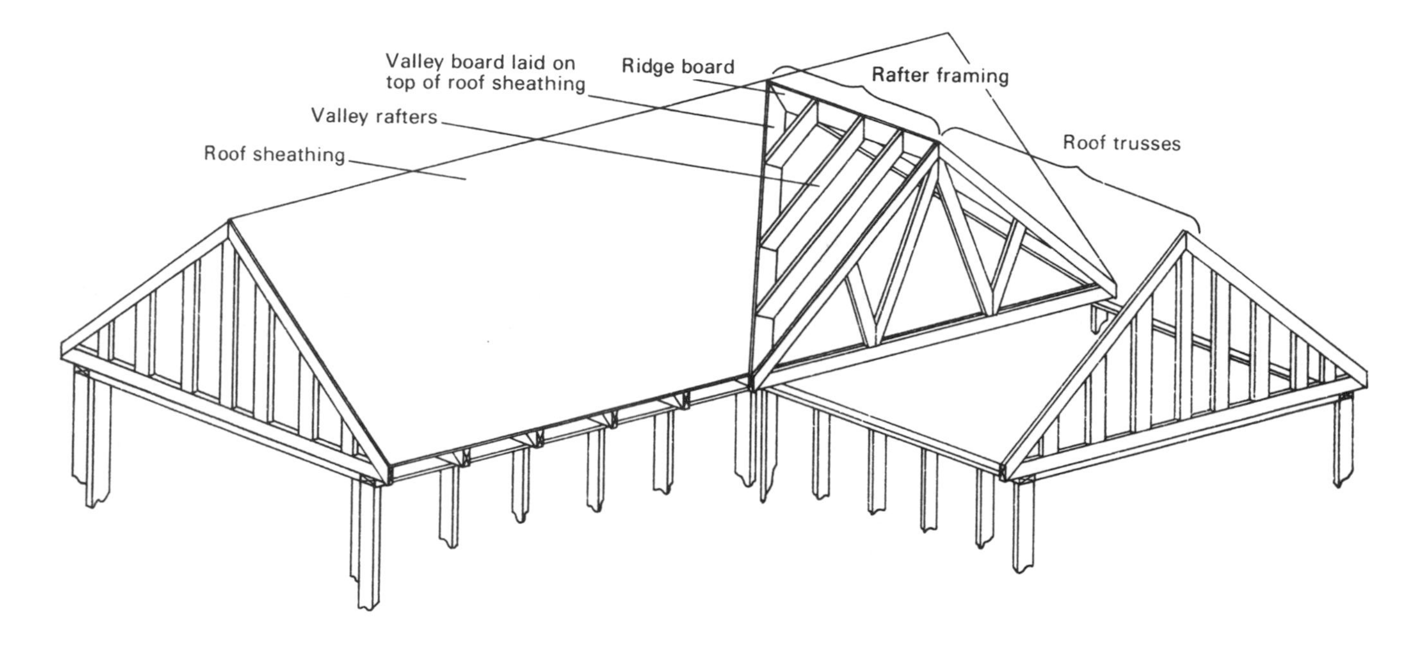

Figure 52: "Closed" rafter framing joining perpendicular truss roof segments.

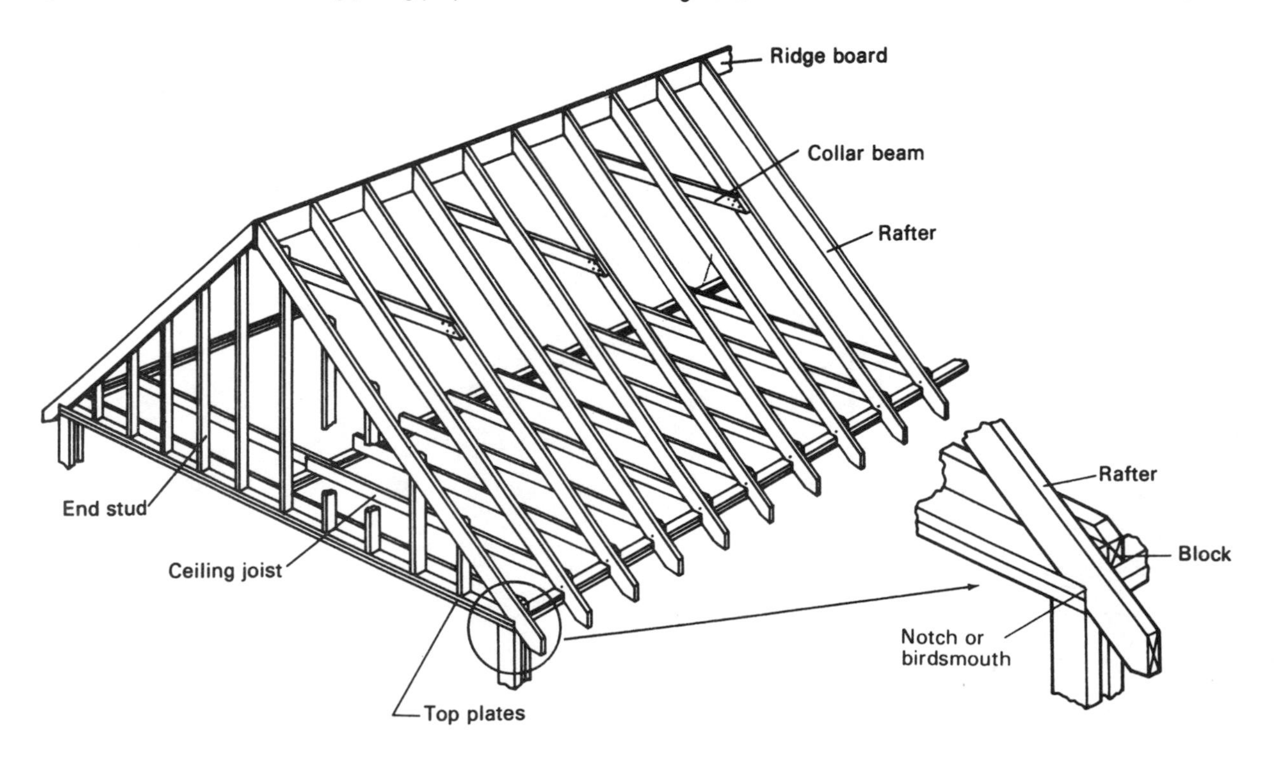

Figure 53: Typical rafter framing for pitched roof.

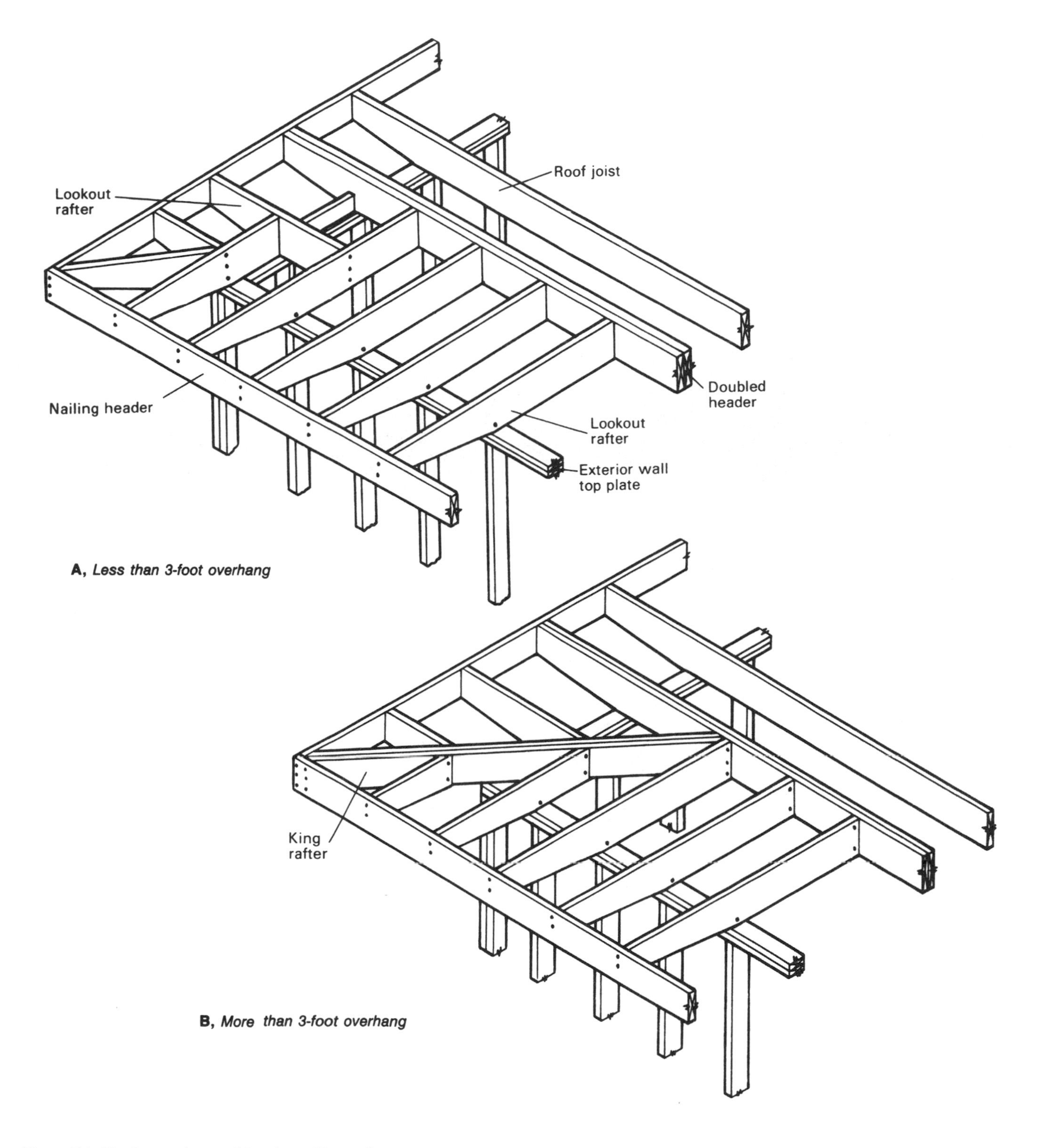

A, *Less than 3-foot overhang*

B, *More than 3-foot overhang*

Figure 54: Single-member roof framing with overhang.

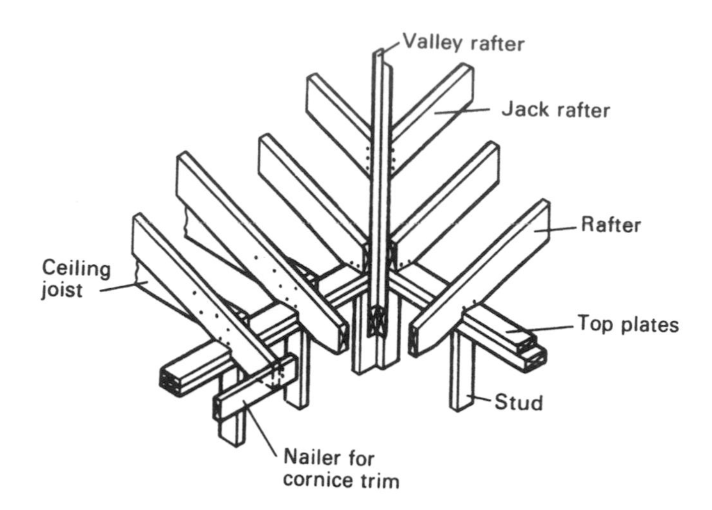

Figure 55: Framing at valley in rafter roof.

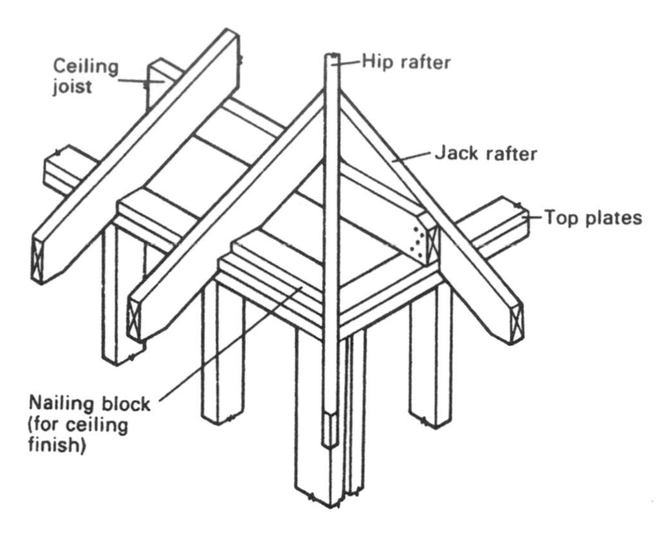

Figure 56: Framing at corner of hip-rafter roof.

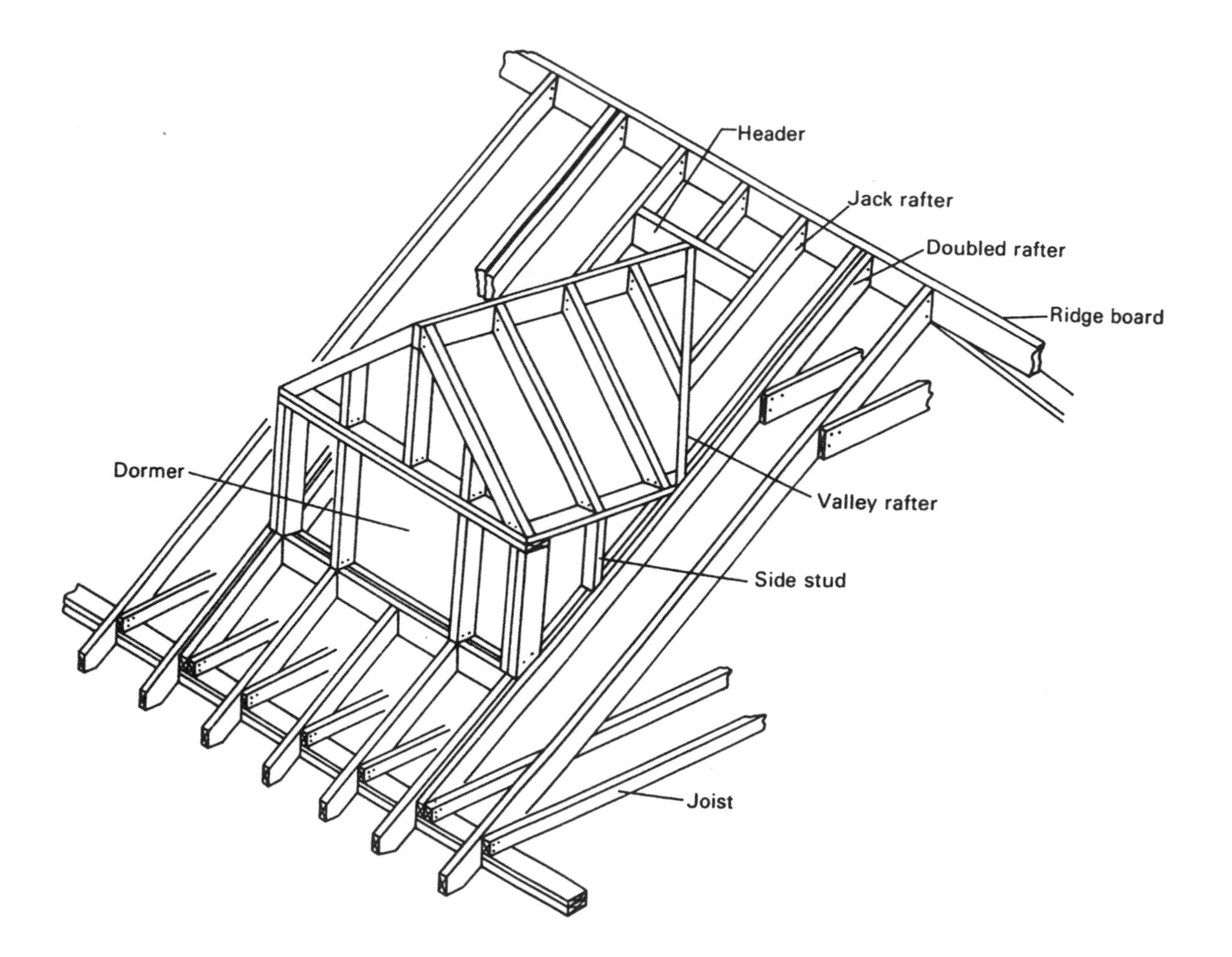

Figure 57: Gable dormer framing.

FR11 FRAME stairs to basement (if any). In many respects, a home is built around staircases. This is a critical step.

FR12 POSITION all large items that must be inserted before framing walls, prior to setting doorways and inner partitions. These include bathtubs, modular shower units, HVAC units and oversized appliances. Beware of installing oversized appliances that will not fit through the door when you move or when they have to be replaced.

FR13 FRAME exterior walls/partitions (first floor). Keep track of all window and door opening measurements. Constantly check squareness.

FR14 PLUMB and line first floor. This is critical to a quality job in drywall and trim. This involves setting the top plates for all first-floor walls with lapped joints while someone is checking to make sure that all the angles are plumb and perpendicular. This straightens the walls exactly—fine tuning. Walls are braced one by one. Start from one exterior corner and go in a circular direction around the exterior perimeter of the house, adjusting and finishing each wall.

FR15 FRAME second-floor joists and subfloor.

FR16 POSITION all large second-floor units prior to setting doorways and partitions. Again, these include bathtubs, modular shower units and other large items.

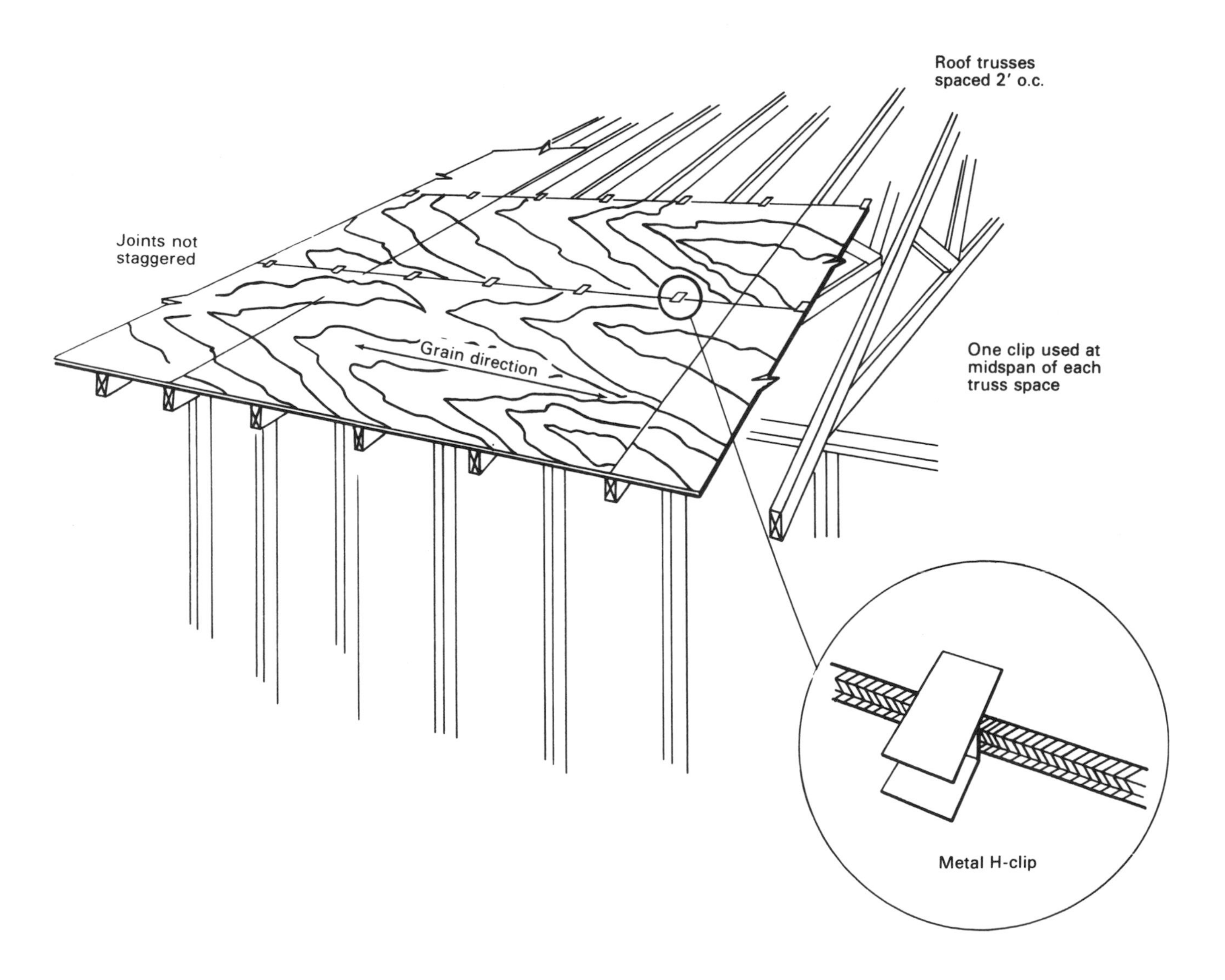

Figure 58: Plywood roof sheathing with H-clips for edge support.

FR17 FRAME second-floor exterior walls/ partitions.

FR18 PLUMB and line second floor. Refer to plumbing/lining first floor.

FR19 INSTALL second-floor ceiling joists if roof is stick built. Have your window and door supplier visit the site to check actual dimensions of all openings.

FR20 FRAME roof. If you want a space between your window trim and cornice, add a second plate on the top of the second-floor wall. Depending upon the size and complexity of your roof, you may either use prefab trusses or stick build your roof. Chances are, if your home is over 2,200 sq. ft. or has an 8/12 or steeper roof, your framers will probably end up stick building it.

FR21 INSTALL roof deck. Plywood sheathing will be applied in 4 × 8 sheets in a horizontal fashion. Each course of plywood should be staggered one-half width from the one below it. Your framers should use ply clips to hold the sheets together between rafters.

FR22 PAY first framing payment (about 45 percent of total cost). Check for missed items, such as framing for attic stairs, fir downs for kitchen cabinets, framing for house fan, hearth framing, stud supports around tubs, etc.

FR23 INSTALL lapped tar paper. This protects your roofing deck and entire structure from

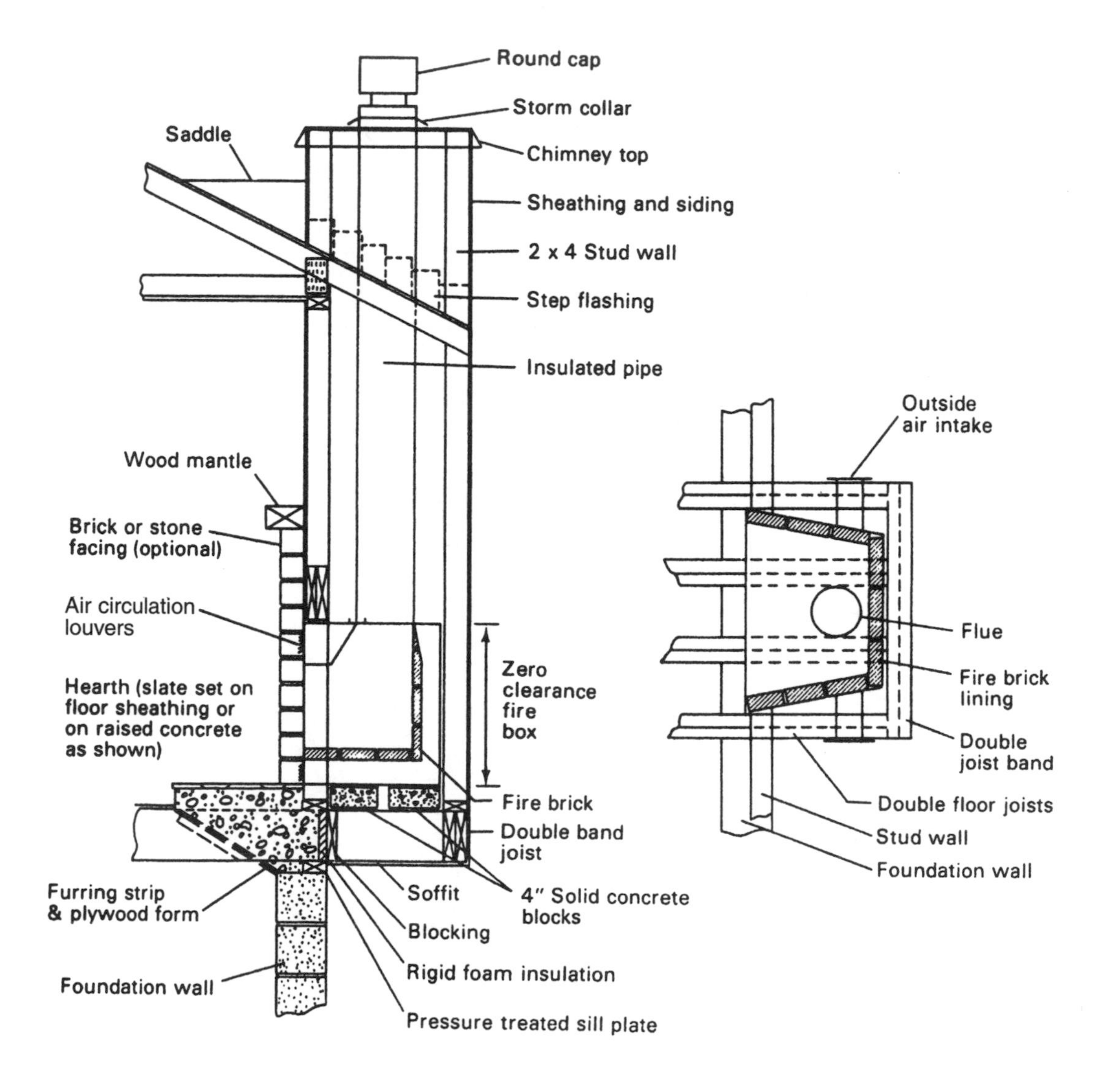

Figure 59: Zero-clearance fireplace framing details.

rain. At this point, the structure is dried-in, and a major race against the elements is over.

FR24 FRAME chimney chases, if you have them. Ask framer to cover top of chase with plywood or rain caps to protect against rain.

FR25 INSTALL prefab fireplaces. Install a fireplace with at least a 42″ opening and 24″ depth. Anything smaller will have problems taking full-size firewood. Prefab fireplaces are usually installed by the supplier or a separate sub. Since plans seldom indicate the specific type of fireplace, make sure framer is aware of the brand and requirements of your particular fireplace. Store your gas log lighter key in a safe place to avoid losing it.

FR26 FRAME dormers and skylights.

FR27 FRAME tray ceilings, skylight shafts and bay windows, if you have them. Skylight

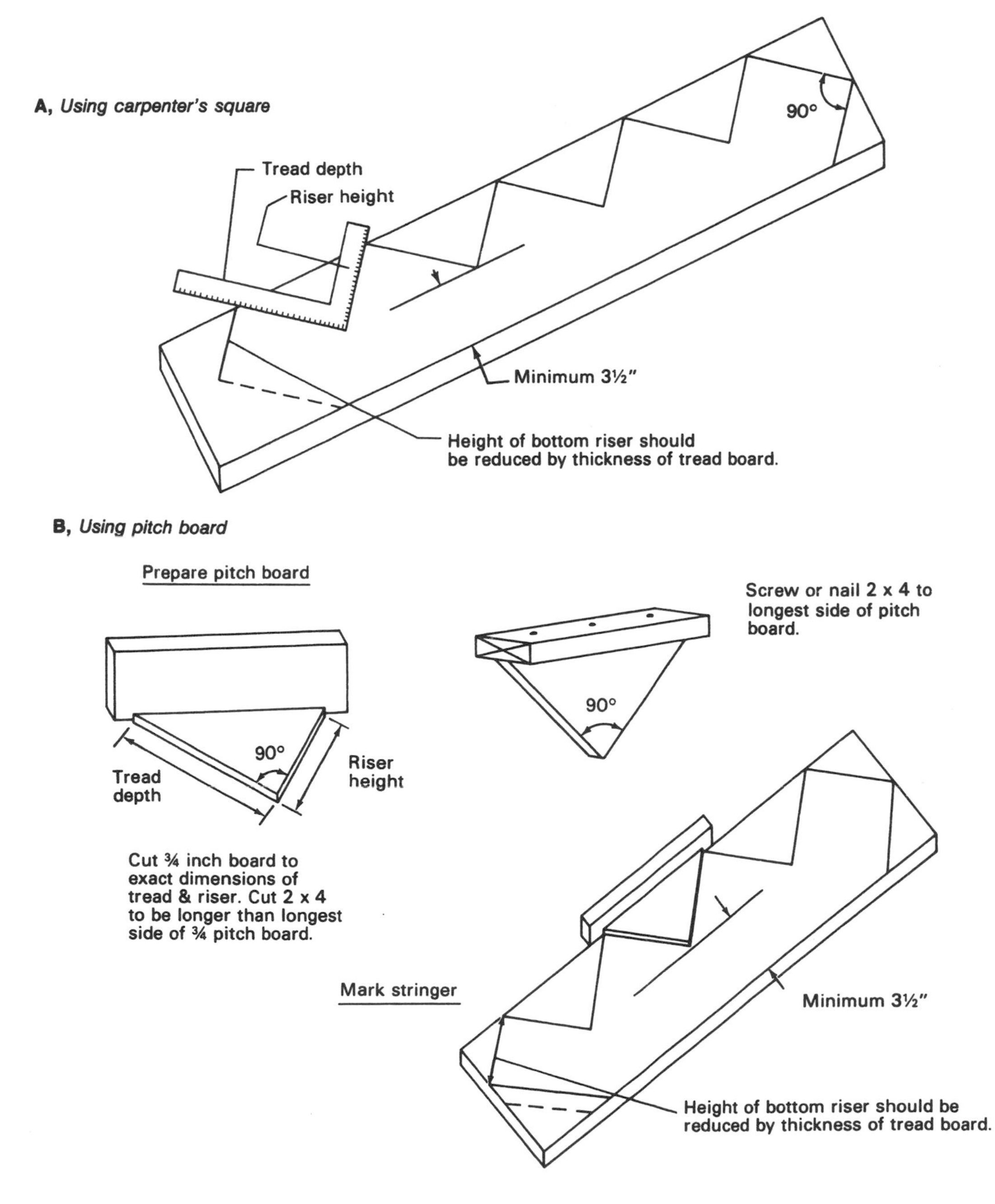

Figure 60: Laying out a stair stringer.

shafts should flare out as they come into the room. This allows more light to enter.

FR28 INSTALL sheathing on all exterior walls.

FR29 INSPECT sheathing. Watch for punctures and gaps in the insulation that can be fixed with duct tape.

FR30 REMOVE temporary bracing supports. Bracing supports installed prior to backfill and those used during the framing process are removed and used as scrap.

FR31 INSTALL windows and exterior doors. The siding and cornice sub may do this operation. If you are using prehung doors, this becomes an easier process. A prehung door comes with its own door frame, but without door knobs and locks. After installing door knobs and locks, the house can be locked up to prevent theft and vandalism. Just make sure to give your subs temporary keys. If your siding and cornice sub installs your windows, this is covered under another step.

FR32 APPLY dead wood. "Dead wood" consists of short pieces of lumber installed in areas that need backing to nail drywall on. These include tops and sides of windows where curtain hardware will be attached, small areas around stairs, locations of electrical boxes, and bracing for ceiling lights or fans.

FR33 INSTALL roof ventilators. This should only be done on the back of the roof.

FR34 FRAME decks using pressure-treated lumber to prevent rotting and termite infestation.

FR35 INSPECT framing. You should have been checking on the framing process all along, so there should be no big surprises at this point. This is the last chance to get things fixed before you make your last payment to the framing crew.

FR36 SCHEDULE and have your loan officer visit the site to approve rough-framing draw.

FR37 CORRECT any problems with framing job. This is an ongoing process that should occur throughout the framing phase. The earlier a problem is detected, the easier and cheaper it is to fix. Look for bowed wood that will make drywall attachment difficult. Correct this by shimming or by cutting a slice partway through the stud. At this point, you have one last bargaining tool before making the next payment to the framing crew.

FR38 PAY framing labor. Retain a percentage of pay until all work is completed and final inspection is approved. Remember to deduct any necessary worker's compensation.

FR39 PAY framing labor retainage. Get a signed affidavit when final payment is made.

FRAMING—Sample Specifications

- All materials are to be crowned (circular grain goes in the same direction). Where possible, all wall studs have crown pointing the same direction.
- All walls and joists are to be framed as specified in the blueprints.
- All joists are to be framed 16″ O.C.
- All vertical wall studs are to be framed 16″ O.C.
- All measurements are to be within ¼″ and plumb, perpendicular and level.
- Cross bridging is to be used at all joist midspans to increase strength and stability of floor.
- All lumber in contact with concrete must be pressure treated.
- Double joists are to be used under load-bearing partitions and bathtubs unless truss floors are used.
- Floor joists are not to interfere with tub, sink and shower drains.
- Four-inch overlaps on lap joists supported over girders are in place.
- All cuts necessary for HVAC and plumbing work are to be braced.
- Hip roof is to be 10/12 pitch, and gable roof is to be 12/12 pitch.
- No cuts are to be made in laminated beams.
- All chimney chases are framed with a cricket (also known as a saddle) installed to allow drainage of excess water.

- Worker's compensation insurance is to be provided by builder.

Framing Extras Include:

- Bay windows
- Stacked bay windows
- Staircases
- Prefab fireplace opening and chase
- Skylights
- Basement stud walls

SHEATHING

- Sheathing is to be installed vertically along studs.
- Sheathing is nailed every foot along studs.
- All gaps over ¼″ are to be taped.
- Skin on sheathing is not to be broken by nail heads, hammer or other means.
- Sheathing nails are to be used exclusively.

FRAMING—Inspection

Framing inspection will be an ongoing process. Errors must be detected as soon as they are made, for easy correction.

- ☐ All work is done according to master blueprints.
- ☐ No framing deviations exceed standard ¼″ leeway for error.
- ☐ Vertical walls are plumb.
- ☐ Opposite walls of rooms are parallel.
- ☐ Horizontal members, such as joists, headers and subfloors, are level.
- ☐ Window and door openings are proper dimensions and are square. Rough openings are 2″ wider than opening dimension.
- ☐ Adequate room exists at all door openings for trim to be attached. Opening is not flush against a corner.
- ☐ No loose or unreinforced boards or structures.
- ☐ Purlins are in all basement walls and load-bearing walls.
- ☐ Corner bracing is used in all wall corners.
- ☐ No room is framed before installing large fixtures or appliances that are larger than door openings.
- ☐ All framing is being done to within 1″ of on-center (O.C.) spacing requested.
- ☐ Wall corner angles are square. This is especially critical in kitchens (where cabinets and countertops are expressly designed for 90° angles).
- ☐ Check to see that vertical studs in stud walls are even and unwarped. This can be done by pulling a string along each stud wall. The string should just touch all studs at the same time.
- ☐ All joists, vertical studs and rafters are crowned. This means that the natural bow of the wood is running in the same direction for all adjacent members.
- ☐ No excessive scrap remains.
- ☐ Adequate lumber is available for framers. Have them notify you if shortages are about to occur.
- ☐ Site measurements match the blueprints. If something is to change, it should be recorded in red on the blueprints.

Subfloor

- ☐ Specified plywood or other subflooring is used. Proper spacing and nailing of sheets is done. Tongue-and-groove plywood is used where specified.
- ☐ No weak or crushed plywood in floor—bottom two sheets in a plywood load often are broken when sliding the load off the truck. Make sure they are not used in the floor.
- ☐ All sill studs are exterior grade and pressure treated.
- ☐ All subflooring is glued to studs with construction adhesive and nailed adequately. Check from underneath.
- ☐ Bracing between floor joists is as specified.

Stairs

- ☐ No squeaks or other noises come from treads or risers.
- ☐ Treads are deep enough for high-heeled shoes to step comfortably.
- ☐ Proper bracing in walls will support rails later.
- ☐ All risers are the same height.
- ☐ All treads are the same width or as specified.
- ☐ All framing is complete for trimwork to begin.
- ☐ Attic staircase works properly.

Fireplace

- ☐ Fireplace area headers are in proper position for prefab or masonry work.
- ☐ Framing for raised mantel is installed.
- ☐ Flue is framed with clearances according to building code.
- ☐ Raised hearth, if used, is proper height.

Roofing Deck

- ☐ Plywood sheets are staggered from row to row.
- ☐ Plywood sheets are nailed every 8″ along rafters or trusses.
- ☐ Ply clips are used on all plywood between supports.
- ☐ Proper exterior grade of plywood is used.
- ☐ Plywood sheets and roofing felt are cut properly for ridge and roof vents.

Deck or Porch

- ☐ Deck has galvanized joist hangers on joist ends.
- ☐ Pressure-treated lumber is used on outdoor decks.
- ☐ Galvanized decking nails or screws are used.

Miscellaneous Framing

- ☐ Tray ceiling areas are framed as specified.
- ☐ Tray ceiling angles are even.
- ☐ Skylight openings and skylights are according to specifications.
- ☐ Skylights are sealed and flashed.
- ☐ Access doors to crawl space and wall storage areas are installed according to specifications.
- ☐ Pull-down attic staircase is installed and operable.

Sheathing

- ☐ Proper sheathing is installed.
- ☐ Sheathing is applied vertically.
- ☐ Sheathing nails are used.
- ☐ No punctures or other damage have impaired sheathing.
- ☐ Perimeter of sheets is nailed adequately to prevent sheathing from blowing off in wind.
- ☐ Sheathing joints are tight and taped where necessary.

Roofing and Gutters

ROOFING

Roofing is measured, estimated and bid based on squares (units of 100 square feet roof coverage). Common roofing compositions include asphalt, fiberglass and cedar shakes, although special roofs are also composed of tile, slate and other more expensive materials. Fiberglass is a popular choice due to its favorable combination of appearance, price and durability. Roofing warranties are most often a minimum of 20 years. This pertains to material only.

Asphalt and fiberglass shingles, the most common and inexpensive forms, are of two basic types:

- Strip
- Individual

STRIP SHINGLES are normally 12″ tall and 36″ wide, with two or three tabs to make the units look like several individual shingles when installed. Weights vary from 160 lb. (per square) on the cheap side up to 280 lb. on the high end. The heavier they are, the more sound and long-lived the roof will be. Lighter shingles also have a tendency to get blown around in heavy wind. Strip shingles normally come in packages of 80 strips. About three packages cover one square of roof. When installed, about 5″ of the shingle's height remains exposed, hence a 7″ top lap. There are some fiberglass shingles out on the market that resemble slate roofing—from a distance. You just have to look around.

INDIVIDUAL SHINGLES are individual sheets normally measuring 16″×16″ or 12″×16″. Weights vary from 160 lbs. to 330 lbs. Be advised, your roofing sub will probably charge you more to install these than asphalt or fiberglass, and they are not nearly as popular as strip shingles. So make sure you really want them.

ROOFING SUBS should quote you a price with and without material. The steeper your roof pitch, the steeper his price per square installed. Prices start to get steep after about 10/12, which represents 10 feet of rise for every 12 feet of run (horizontal). Hence, 12/12 is a 45° angle. Appeal is growing for steeper and steeper pitches. Make sure the roofing sub has plenty of worker's compensation insurance, or deduct heavily from his fee if you have to add him to your policy. He's one fellow with a very high risk of injury. If you plan to supply the shingles, make sure they are at the site a day or so early.

NOTE: Since asphalt and fiberglass shingles are by far the most common used, the steps on the following pages relate specifically to them. If you use roofing materials other than these two, refer to a handbook and detailed construction methods. Due to the high asphalt content in asphalt and fiberglass shingles, it is highly recommended to apply them in 50° Fahrenheit or warmer weather, as they are highly susceptible to cracking in cold weather.

GUTTERS

Gutters are used to channel rainwater to downspouts and then to planned drainage paths. In some cases, gutters are used more for appearance than functionality. Typically, gutters are made of galvanized iron or aluminum. The number of downspouts will vary. Your gutter subcontractor will help you in determining the number and placement of downspouts and gutters. If you can, use underground drainpipes instead of simple splash blocks. Care must be taken that they are not completely buried during the backfill stage. A heavy-gauge PVC black tubing should be used. The primary gutter materials are:

ALUMINUM—These are good, but you must specify "paint grip" aluminum, or the paint will not stick.

VINYL—This maintenance-free alternative is fast becoming a favorite.

Gutter subs normally charge by lineal feet of gutters and downspouts installed. They may charge you a certain amount for any more than four corners. They will also charge you extra for installing gutter screens and drainpipe.

Get your gutters up as soon as possible to start good drainage and to keep mud from splashing on the side of the house. Gutters shouldn't go on until the cornice is primed and painted and the exterior walls finished (because of downspouts).

ROOFING—Steps

RF1 SELECT shingle style, material and color.

RF2 PERFORM standard bidding process for material and labor.

RF3 ORDER shingles and roofing felt (tar paper). This is done only if you have decided to use your roofer for labor only. Check with him to make sure he knows of the brand and how to install them.

RF4 INSTALL 3″ metal drip edge on eave, nailed 10″ on center. Roofing paper will go directly on top of this drip edge (optional).

RF5 INSTALL roofing felt. This should be done immediately after the roofing deck has been installed and inspected. Your framer will install the tar paper as the last thing before finishing the job. This step protects the entire structure from immediate water damage. Until this step is performed, the project should proceed as quickly as possible.

RF6 INSTALL 3″ metal drip edge on rake. This drip edge goes over the roofing paper. Also install aluminum flashing at walls and valleys (optional).

RF7 INSTALL roofing shingles. Roofs over 30′ wide should be shingled starting in the middle. Shorter roofs can be shingled starting at either rake. Starter strips, normally 9″ wide, are either continuous rolls of shingle material that start at the eave or doubled shingle strips. Shingles are laid from the eave, overlapping courses up to the highest ridge. Don't install shingles on very cold days if you have a choice.

RF8 INSPECT roofing. Refer to contract specifications and inspection checklist for guidelines. You will have to get up on the roof to check most of this.

RF9 PAY roofing subcontractor. Get him to sign an affidavit. Deduct for worker's compensation, if applicable.

RF10 PAY roofing sub retainage.

GUTTERS—Steps

GU1 SELECT gutter type and color.

GU2 PERFORM standard bidding process.

GU3 INSTALL underground drainpipe. Flexible plastic pipe is the type used most often. Make sure it is covered properly.

GU4 INSTALL gutters and downspouts.

GU5 INSTALL splashblocks and/or other water-channeling devices.

GU6 INSPECT gutters. Refer to inspection guidelines, related checklist, contract specifications and your building code.

GU7 INSTALL copper awnings and dormer tops.

GU8 PAY gutter subcontractor. Have him sign an affidavit.

GU9 PAY gutter sub retainage.

ROOFING—Sample Specifications

- Bid is to perform complete roofing job per attached drawings and modifications.
- Bid is to include all materials and labor and 10-year warranty against leakage.
- Apply 240 lb. 30-year fiberglass shingles (make/style/color) to roof, using galvanized roofing nails.
- Install two thermostatically controlled roof vents on rear of roof as indicated on attached drawings.
- Install continuous ridge vent per drawing specifications.
- Roofing felt is to be applied and stapled to deck as indicated below: no. 15 asphalt saturated felt over entire plywood deck.

Figure 61: Flashing at chimney.

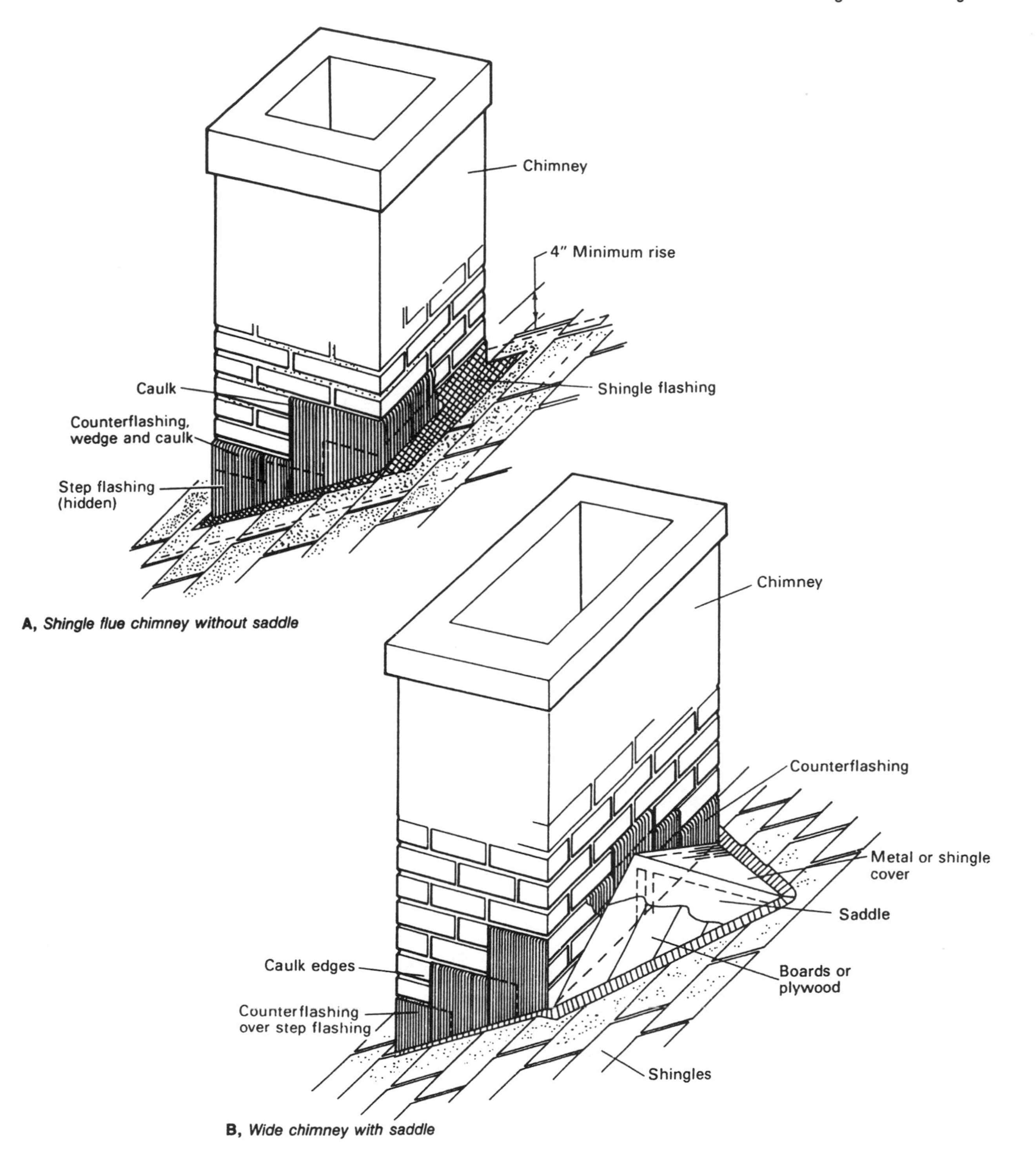

- All roofing felt is to have a vertical overlap of 6″.
- Install 3″ galvanized eave and rake drip edges nailed 10″ O.C.
- All work and materials are to conform to or exceed requirements of the local building code and must also be satisfactory to contractor.
- Roofer is to notify builder immediately of any condition which is or may be a violation of the building code.
- Nails are to be threaded and corrosion-resistant.

Figure 62: Flashing in roof valleys.

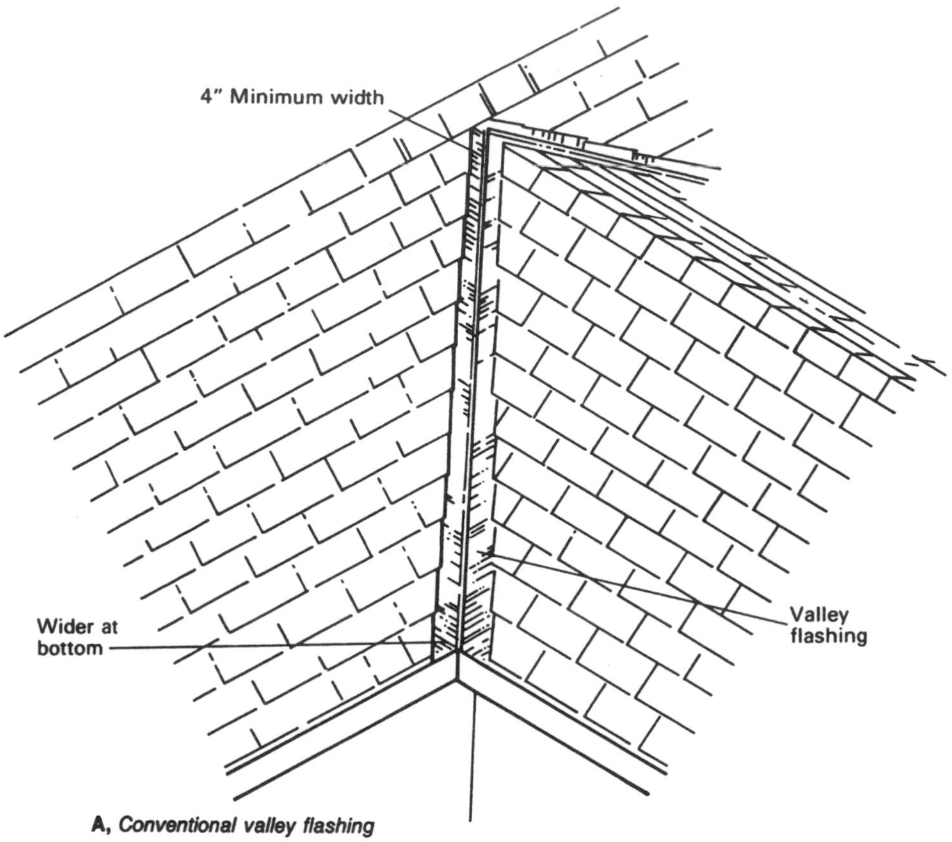

4" Minimum width
Wider at bottom
Valley flashing

A, *Conventional valley flashing*

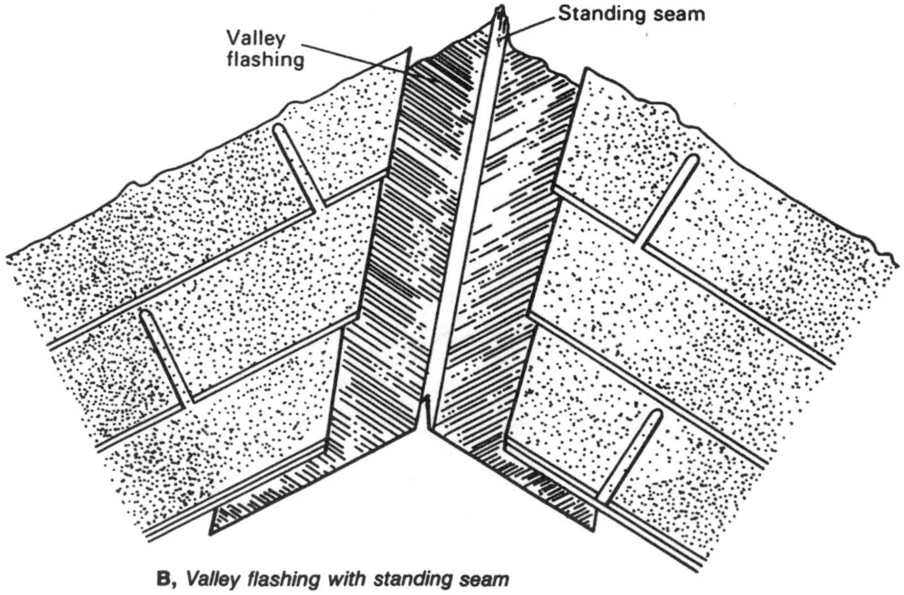

Valley flashing
Standing seam

B, *Valley flashing with standing seam*

Either aluminum or hot-dipped galvanized roofing nails are to be used, depending upon shingle manufacturer recommendations.

■ Nails are to be driven flush with shingles and installed per roofing specs.

■ Soil stacks are to cover all vents. All vents are to be sealed to shingle surface with plastic asphalt cement.

■ Flashing is to be installed at chimney, all roof valleys, soil and vent stacks and skylights.

Figure 63: Flashing at roof and wall intersection.

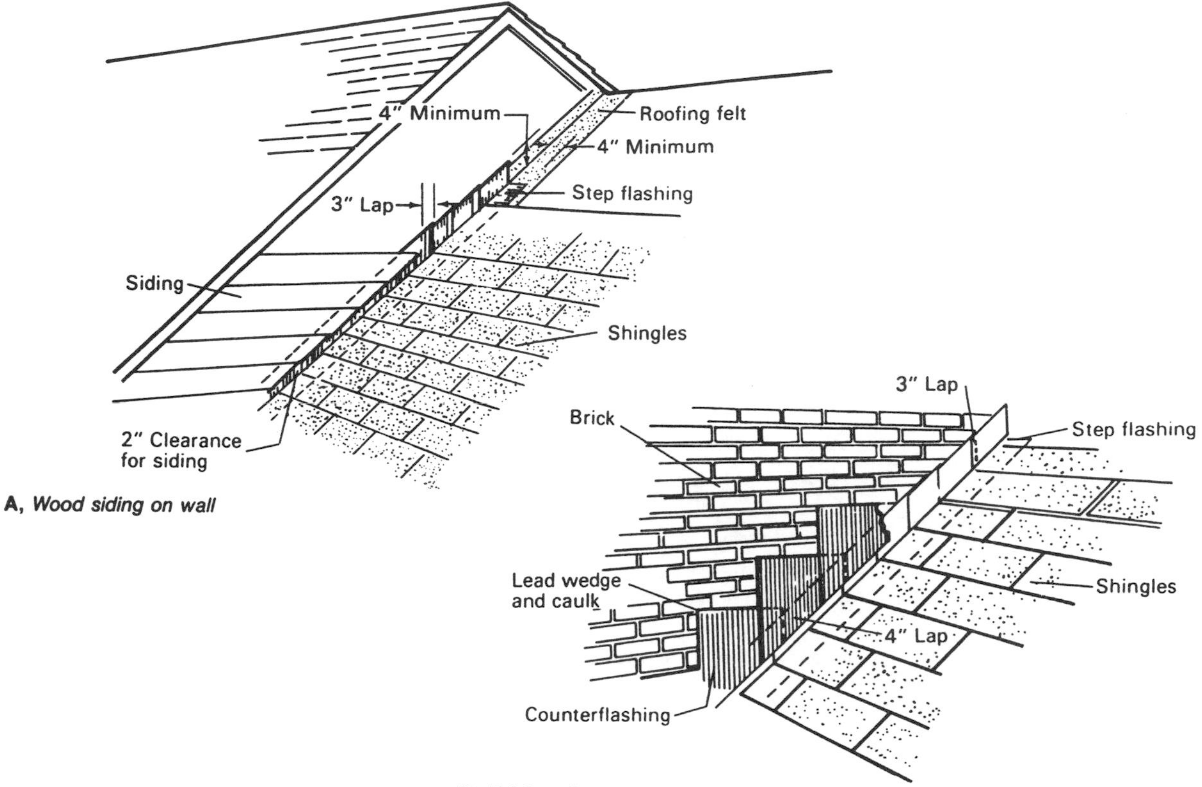

GUTTERS—Specifications

- Bid is to provide and install seamless, paint-grip aluminum gutters around entire lower perimeter of dwelling as specified in attached drawings.
- 5″ gutter troughs and 3″ corrugated downspouts to be used exclusively.
- Gutter troughs are to be supported by galvanized steel spikes and ferrules spaced not more than 5′ apart.
- Downspouts are to be fastened to wall every 6 vertical feet.
- Gutter troughs are to be sloped toward downspouts 1 (one) inch for every 12 lineal feet. Gutters should be installed for minimum visibility.

ROOFING—Inspection

☐ Shingle lines inspected for straightness with string drawn taut: Tips of shingles line up with string.

☐ Shingle pattern and color are even and uniform from both close up and afar (at least 80′ away).

☐ Shingles extend over edge of roofing deck by at least 3″.

☐ All roofing nails are galvanized and nailed flush with all shingles (random inspection).

☐ No shingle cracks are visible through random inspection.

☐ Hips and valleys are smooth and uniform.

☐ Roofing conforms to contract specifications and local building codes in all respects.

☐ Water cannot collect anywhere on the roof.

☐ Shingles fit tightly around all stack vents and skylights. Areas are well sealed with an asphalt roofing compound that blends with the shingles.

☐ Drip edges have been installed on eaves and rakes.

☐ No nail heads are visible while standing up on the roof.

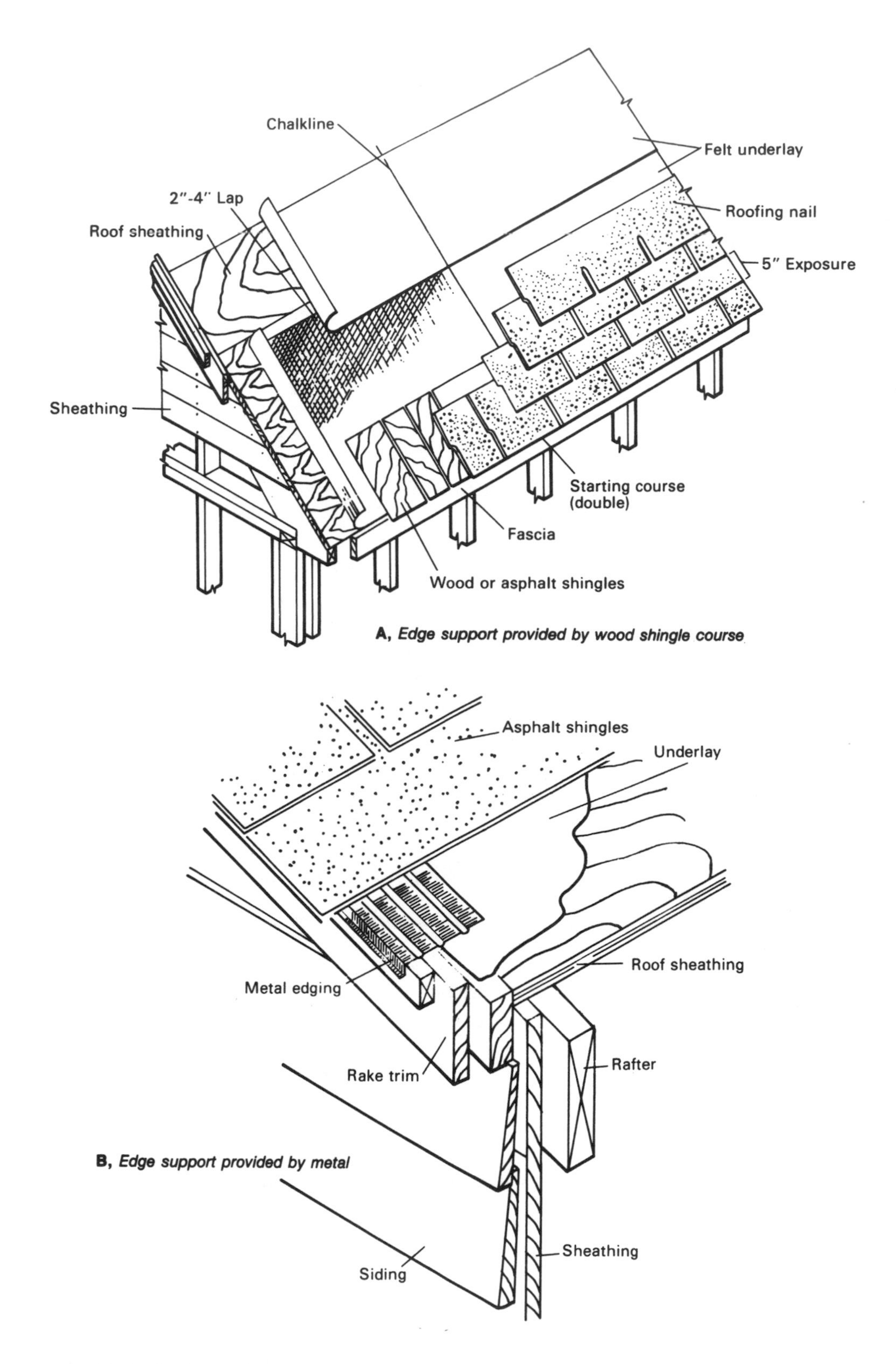

Figure 64: Application of fiberglass or asphalt shingles.

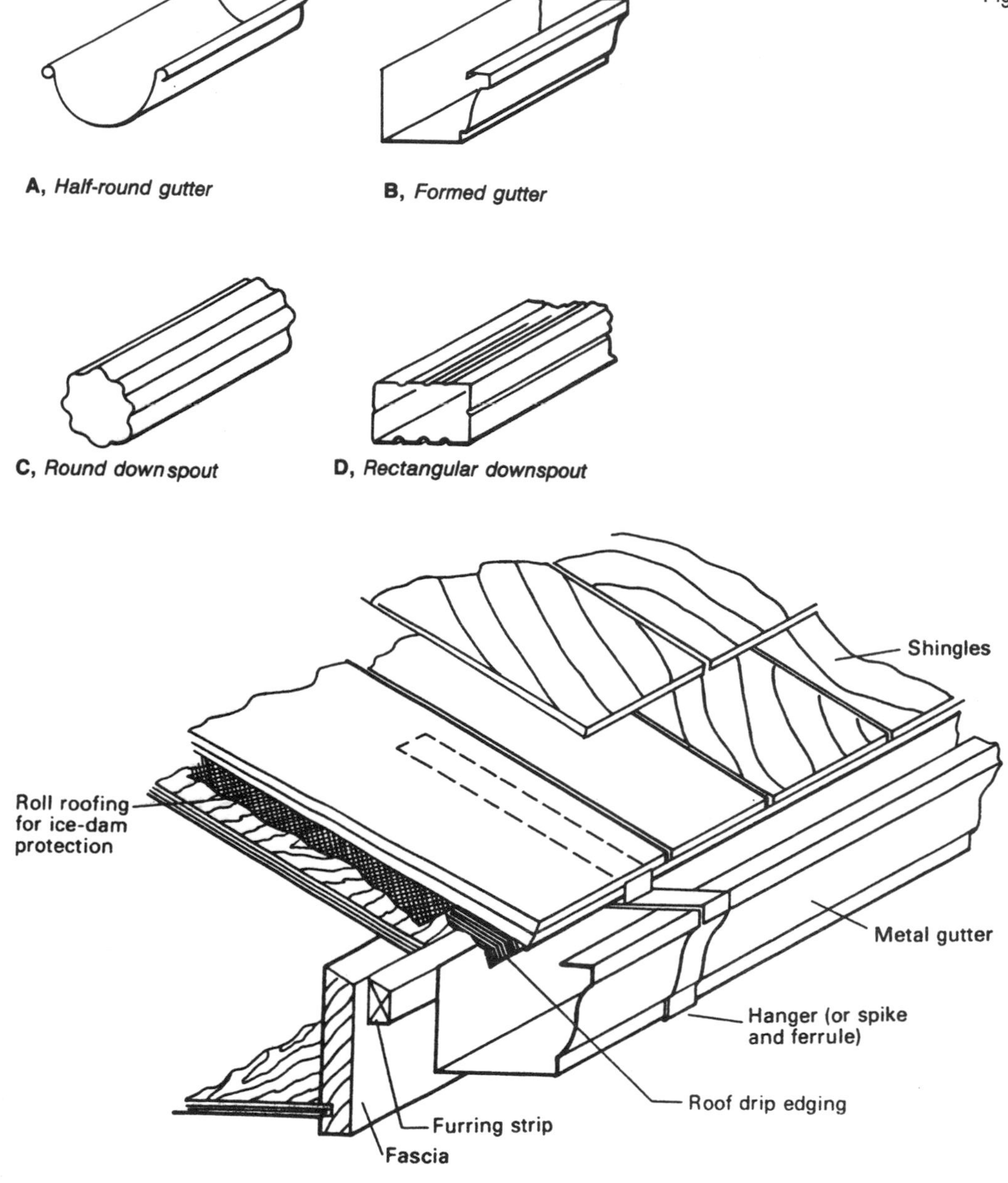

Figure 65: Gutters and downspouts.

Figure 66: Gutter installation.

- ☐ All shingles lie flat (no buckling).
- ☐ No visible lumps in roofing due to poor decking or truss work.
- ☐ Edges of roof are trimmed smooth and even.
- ☐ All vents and roof flashing are painted proper color with exterior grade paint.
- ☐ All garbage on roof removed.
- ☐ Shingle tabs have been glued down if this was specified. (Used only in very windy areas or where very lightweight shingles are used.)

GUTTERS—Inspection

- ☐ Proper materials are used as specified—aluminum, vinyl, etc.
- ☐ Proper trough and downspout sizes are used.
- ☐ Downspouts are secured to exterior walls.
- ☐ Adequate number of gutter nails support gutters securely.
- ☐ Water drains well to and through troughs when rain falls or water from a garden hose is applied.

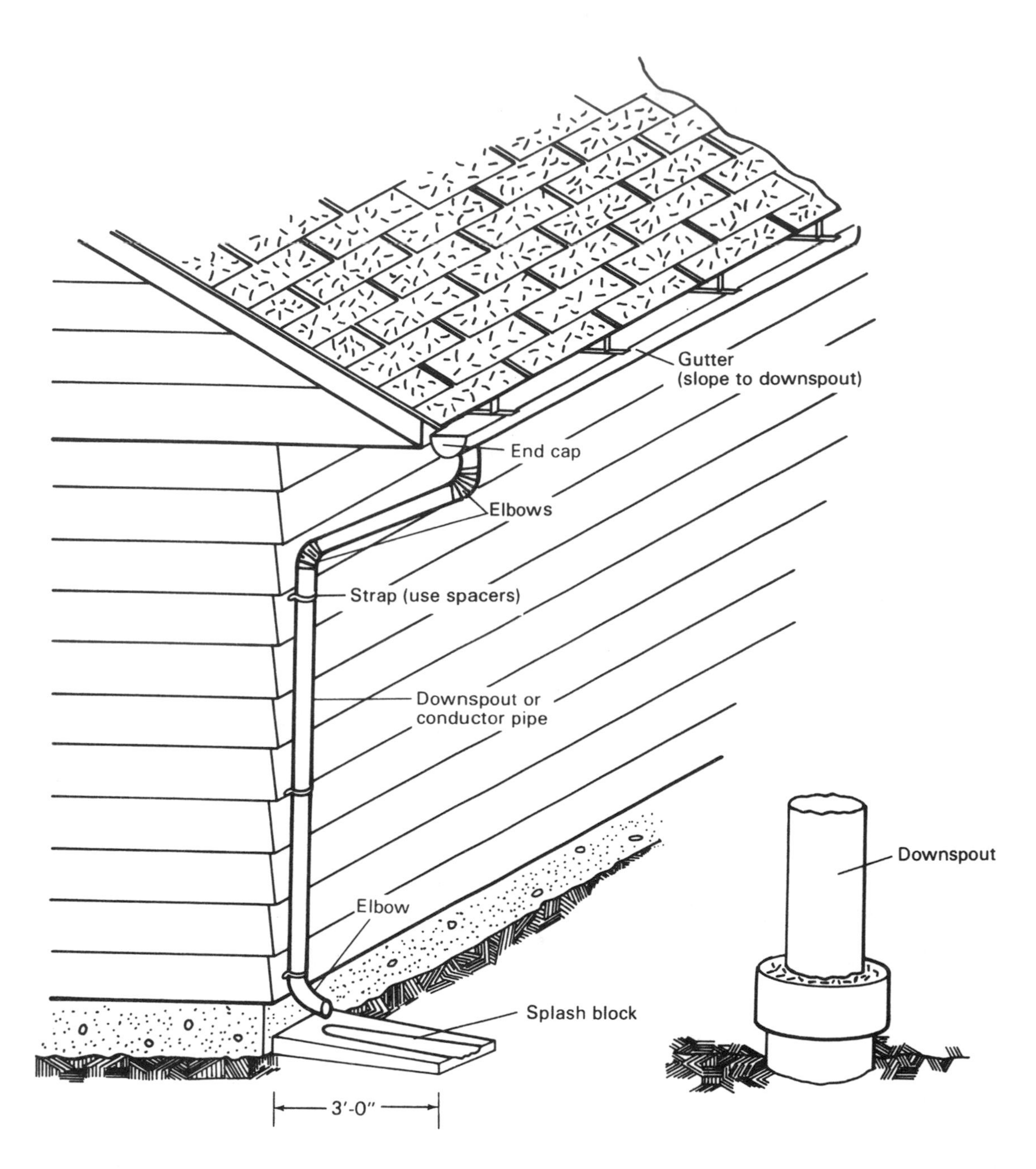

A, ***Downspout with splash block***

B, ***Downspout connected directly to storm sewer***

Figure 67: Downspout installation.

☐ No leaks appear in miters and elbows while water is running.

☐ Water does not collect anywhere in troughs and is completely drained within one minute.

Plumbing

Plumbing work can be started as soon as the house has been dried-in. Getting the plumbing right is essential—it is very difficult to change plumbing decisions later.

Plumbing is composed of three interrelated systems:

- The water supply
- The sewer system
- The (wet) vent system

THE WATER SUPPLY

Unless you have opted for an individual water supply (such as a well), you will tap into a public water supply. Unless you plan to live in your home many years (eight or more), the cost of drilling a well cannot be justified on cost alone. Remember, your water bill includes sewer service, too. You'll still pay for that (unless you opt for a septic tank or cesspool). If your local water supply is 80 psi or greater, you will need to install a pressure reduction valve at the main water service pipe. High water pressure can and will take its toll on most pipes and fixtures in a short period of time.

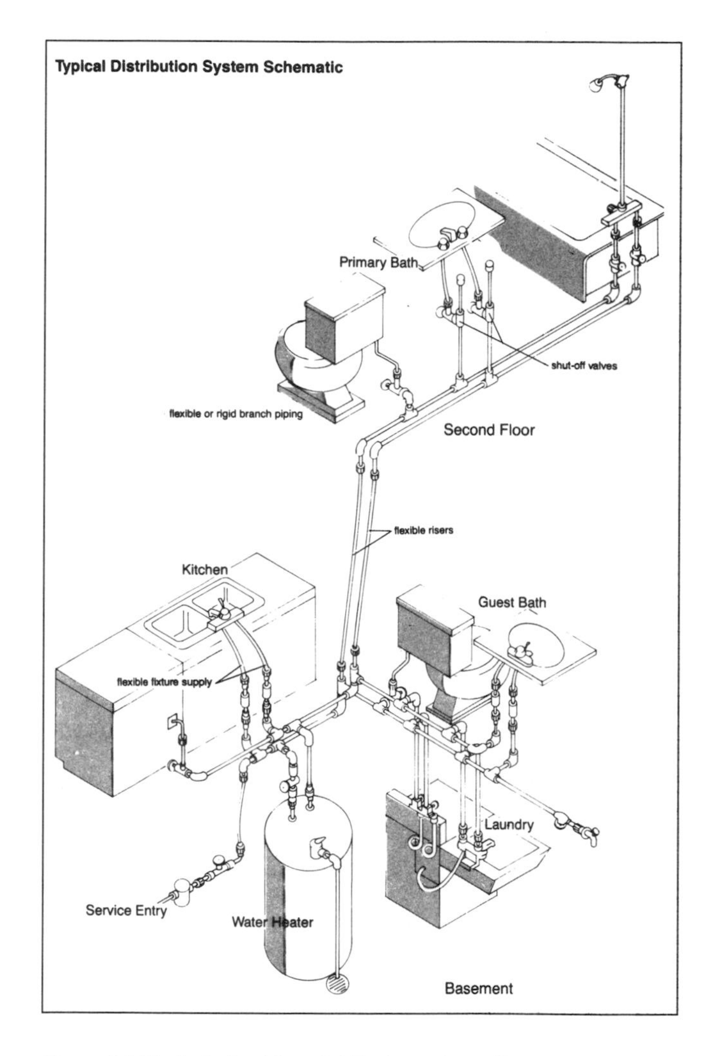

Figure 68: Typical plumbing distribution schematic.

Copper has been the material of choice of plumbers for water pipes for many years, but new materials available today have numerous advantages over copper. CPVC pipe is a plastic used for hot and cold supplies; its cousin, PVC, is only rated for cold water and sewer installations. CPVC offers several advantages over copper, including ease and speed of installation and some cost savings. Some people complain of a slight aftertaste in the water for the first year or so.

Polybutylene is a new semiflexible material that has taken the plumbing world by storm in recent years. It offers many advantages over copper and CPVC. It is totally inert and imparts no taste to the water. Its flexible nature allows it to be installed with a minimum of joints, which are the most expensive material in water systems. Polybutylene is also virtually freeze proof, a characteristic that can be appreciated by anyone who has had his pipes freeze in cold weather. It is very quiet in operation and reduces the effect of water hammer. Polybutylene joints are installed by one of two methods: crimping or grabber fittings. Crimp joints are the most cost-effective, but the grabber joints are ridiculously easy to install. You simply insert the pipe into the fitting and push—and that's it.

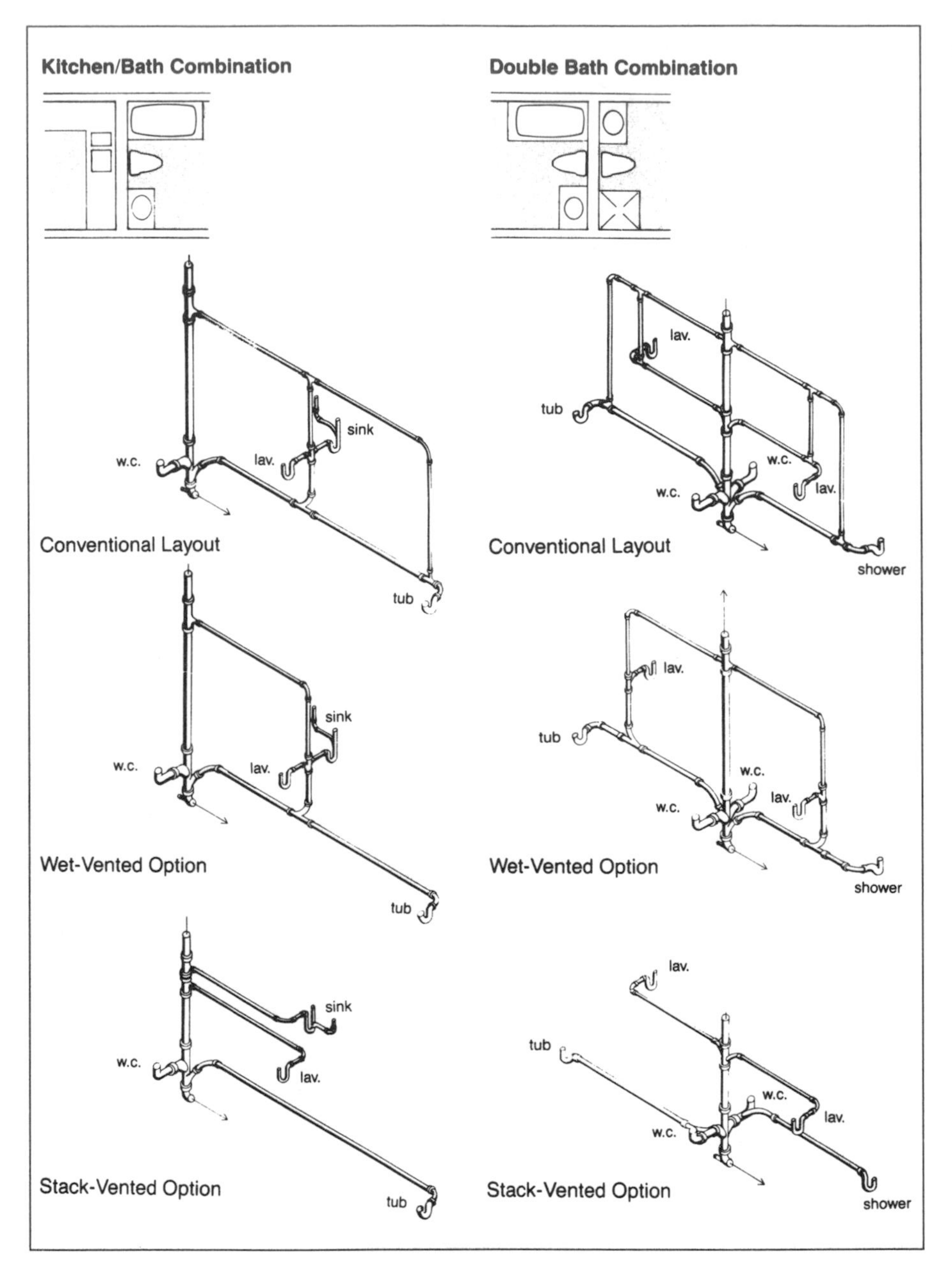

Figure 69: Different vent stack configurations based on back-to-back plumbing.

Polybutylene pipe is by far the least expensive method of installing plumbing, but you may have trouble finding plumbers willing to work with it. Plumbers are slow to accept innovations and may not pass the full savings of the installation to you. Many plumbers fail to realize the full cost savings available to them because they tend to install many more joints than are necessary. The flexibility of polybutylene allows it to be wrapped and molded around obstructions, eliminating the need for many joints. Check with your local plumbing inspector for plumbers familiar with this new material. It is well worth the research.

THE SEWER SYSTEM

The sewer system collects all used water and waste and disposes of it properly. Unless you opted for a private sewer (septic tank), you will connect to the public sewer system. In most cases, building codes will not allow you to build a private sewer if a public one is available. When designing the house, keep in mind that sewer systems are powered by gravity. Hence, your lowest sink, toilet or drain must be higher than the sewer line. The sewer line must have at least ¼″ drop per foot in order to operate properly. This may mean raising your foundation if necessary to accommodate proper drainage.

With sewer pipe materials, technology has created a new problem. PVC pipe is now the most common material used because of its ease of installation; however, plastic pipes transmit the noise of flushing and water draining much more than the old cast iron pipes. Install cast iron sewer pipe if possible to reduce this noise. If you use PVC pipe, take care to position the pipes in walls that are isolated from bedrooms or you will constantly hear water noises. It is critical to discuss with your plumber where the sewer pipe will run if you have a basement. Many finished basements have been ruined by post framing around poorly placed sewer pipes that breaks up the future living space.

The (Wet) Vent System

Also known as the vertical wet venting system, this permits your plumbing system to "breathe." This system protects against siphoning and backpressure. All drains are attached to one of several vents in a network. These vents are seen as pipes sticking out the back side of your roof.

PLUMBING SUBS

Plumbers will normally bid and charge you a fixed price per fixture to install all three systems described above. You are likely to pay extra for Jacuzzis, whirlpools and the like. The price of all fixtures (tubs, sinks, disposals) should be obtained from your sub. Fixtures above builder grade will usually cost more. If you use many custom fixtures, or if your house plan positions bathrooms and the kitchen close together to conserve plumbing material, consider asking for a bid based on actual material and labor costs. This can save you a considerable amount of money. If you purchase fancy fixtures, have the plumber deduct the standard fixture charge for the ones you have purchased.

After you have received the estimate, ask for the cash price. Ask your plumber how many hours it will take him to do the job. You will pay the plumber 40 percent after passing the rough-in inspection and 60 percent after passing the final inspection. If your plumbing code does not require individual shutoff valves on each fixture, do not use them on each fixture if you want to save a little money.

BUILDING CODES

Most likely, your local building code will impose a number of requirements on your plumbing system. This helps protect you from paying for an inadequate system. Your local building code will cover plumbing materials and allowable uses of each, water supply and sewer capacities, joints, drainage, valves, testing, etc. *Read this and know it well.* Although you should rely on your plumbing sub for advice, know your plumbing code!

ABOUT SEPTIC TANKS

If you are considering going this route:

- CONFIRM that septic tanks are permissible. Contact local inspectors and consult your local building code.
- CONDUCT a percolation test to determine how quickly the soil absorbs water. Certain minimum standards will be imposed on your system. It is necessary to hire a qualified engineer to perform this test.
- ESTIMATE your minimum septic tank capacity requirements.
- REFER to your local building code for additional guidelines and requirements for private sewer system.

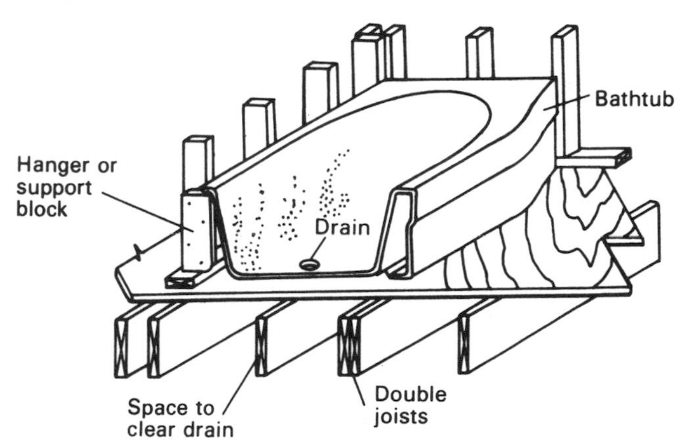

Figure 70: Framing for bathtub.

PLUMBING—Steps

Rough-In

PL1 DETERMINE type and quantity of plumbing

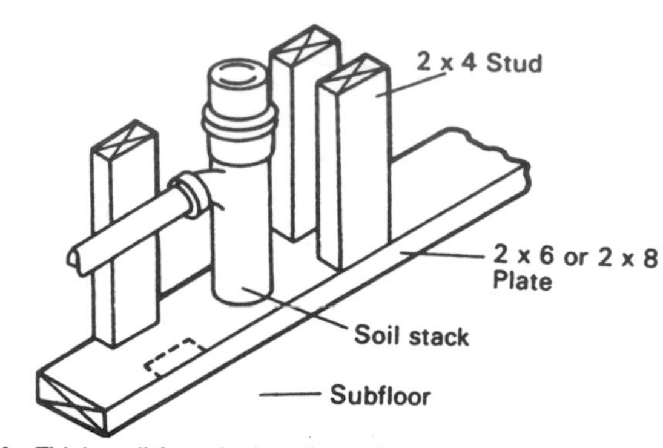

A, *Thick wall for 4-inch soil stack*

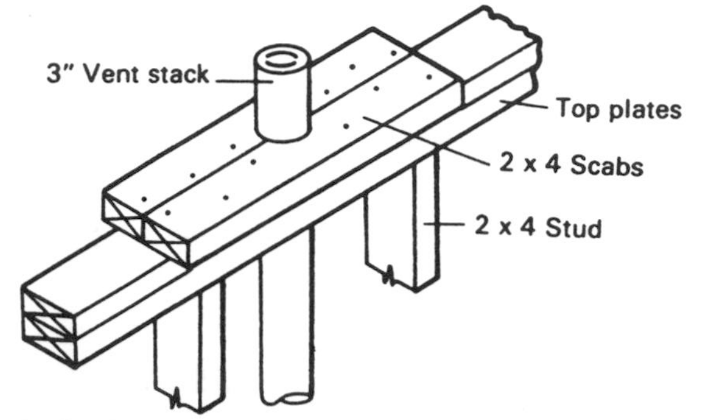

B, *Reinforcing scabs for 3-inch vent stack in 2 by 4 wall*

Figure 71: Plumbing stacks within partitions.

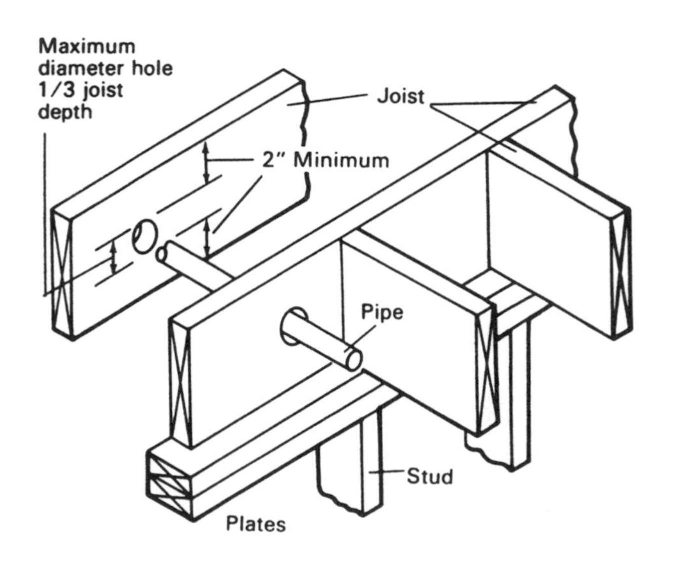

Figure 72: Drilled holes in joists.

fixtures (styles and colors). This includes:

- Sinks (kitchen, baths, utility, wet bar, etc.)
- Bathtubs (Consider one-piece fiberglass units with built-in walls requiring no tile. If you go this route, get a heavy-gauge unit.)
- Shower fixtures
- Toilets and toilet seats
- Water spigots (exterior)
- Water heater
- Garbage disposal
- Septic tank
- Sauna or steamroom
- Water softener
- Refrigerator ice maker
- Any other plumbing-related appliance

PL2 CONDUCT standard bidding process. Shop prices carefully; subs who bid on your job may vary greatly on their bid price, depending on what materials they use and their methods of calculating their fees. This is where you will decide what kind of pipe to use.

PL3 WALK THROUGH site with plumber to discuss placement of plumbing and any special fixtures needed. For instance, if you have a brick exterior, the plumber must extend faucets to allow for the thickness of brick veneer.

PL4 ORDER special plumbing fixtures. This needs to be done well in advance because supply houses seldom stock large quantities of special fixtures. These will take some time to get in.

PL5 APPLY for water connection and sewer tap. You will probably have to pay a sewer tap fee. Ask the plumber for an adjustment piece that will allow you to connect a garden hose to the water main. Record the number stamped on the water meter.

PL6 INSTALL stub plumbing. This applies to instances where plumbing is cast in concrete foundations. This must be done after batter boards and strings are set and before the concrete slab is poured. The plumber will want to know where the sewer line is located.

PL7 PLACE all large plumbing fixtures, such as large tubs, fiberglass shower stalls and hot tubs, before wall framing begins. These fixtures will not fit through normal stud or door openings. For security, you may want to chain expensive fixtures to the studs or pipes.

PL8 MARK location of all plumbing fixtures, including sinks, tubs, showers, toilets, outside spigots, wet bars, ice makers, utility tubs, washers and water heater. Make sure the plumber knows if you have a regular vanity or pedestal sink. Mark the ends of tubs where drains will be located. Mark areas in wall and ceiling where pipes must not be located, such as locations for recessed lights or medicine cabinets.

PL9 INSTALL rough-in plumbing. This involves the laying of hot (left) and cold (right) water lines, sewer and vent pipe. Pipe running along studs should run within holes drilled (not notched) in the studs. All pipe supports should be in place. Your plumber should use FHA straps to protect pipe from being pierced by drywall subs wherever cutouts are made. FHA straps are metal plates designed to protect the pipe from puncturing by nails. Exterior spigots should not be placed near the point where water mains or other pipes go through the exterior wall, as this will invite water into the home if the spigot freezes or leaks. Make sure plumber extends the spigot if brick veneer is used. The plumber will conduct a water pipe test using air pressure to ensure against leaks.

PL10 INSTALL water meter and spigot. This has to be done early, because masons will need a good source of water.

PL11 INSTALL sewer line. Mark locations of other pipes (water or gas) so diggers will avoid puncturing them.

PL12 SCHEDULE plumbing inspector.

PL13 INSTALL septic tank and line, if applicable.

PL14 CONDUCT rough-in plumbing inspection. *NOTE:* This is a very important step. No plumbing should be covered until your county inspector has issued an inspection certificate. Plan to go through the inspection with him so you will understand any problems and get a good interpretation of your sub's workmanship.

PL15 CORRECT any problems found during the inspection. Remember, you have the county's force behind you. You also have the specifications. Since your plumber has not been paid, you have plenty of leverage.

PL16 PAY plumbing sub for rough-in. Get him to sign a receipt or equivalent.

FINISH

PL17 INSTALL finish plumbing. This involves installation of all fixtures selected earlier. Sinks, faucets, toilets and shower heads are installed.

PL18 TAP into water supply. This is where your plumbing gets a real test. You will have to open all the faucets to allow air to bleed out of the system. The water will probably look dirty for a few minutes, but don't be alarmed. The system needs to be flushed of excess debris and solvents.

PL19 CONDUCT finish plumbing inspection. Call your inspector several days prior to the inspection. Make every effort to perform this step with your local inspector to see what he is looking for, so you will know exactly what to have your sub correct if anything is in error.

PL20 CORRECT any problems found during the final inspection.

PL21 PAY plumbing sub for finish. Have him sign an affidavit.

PL22 PAY plumber retainage.

PLUMBING—Specifications

- Bid is to perform complete plumbing job for dwelling as described on attached drawings.
- Bid is to include all material and labor, including all fixtures listed on the attached drawings.
- All materials and workmanship shall meet or exceed all requirements of the local plumbing code.
- No plumbing, draining and venting to be covered, concealed or put into use until tested, inspected and approved by local inspectors.
- All necessary licenses and permits for the project

are to be obtained by plumber.

- Plumbing inspection is to be scheduled by plumber.
- All plumbing lines shall be supported to ensure proper alignment and prevent sagging.
- Worker's compensation is to be provided by plumber.
- Floor drain pan to be installed under washer if located upstairs, or for water heater in attic.
- Plan for gas dryer.
- Install approved pressure reduction valve at water service pipe.
- Stub in basement, plumbing for one sink, tub and toilet.
- Drain pans are to be installed in all tile shower bases.
- Install construction water spigot on water meter with locked valve.
- Install refrigerator ice maker and drain.
- Plumb water supply so as to completely eliminate water hammer.
- Water pipes are to be of adequate dimension to supply all necessary fixtures simultaneously.
- Cutoff valves are to be installed at all sinks and toilets, and at water heater.

Sewer System

- Hook into public water system.
- Install approved, listed and adequately sized backwater valve.
- Horizontal drainage is to be uniformly sloped not less than 1″ in 4′ toward point of disposal.
- Drainage pipes to be adequate diameter to remove all water and waste in proper manner.

(Wet) Vent System

- Vent piping shall extend through the roof flashing and terminate vertically not less than 6″ above the roof surface and not less than 1′ from any vertical surface, 10′ from and 3′ above a window, door or air intake, or less than 10′ from a hot line.
- All vent piping is to be on rear side of roof without visibility from front.
- All vent piping is to conform to the local plumbing and building code.

PLUMBING—Inspection

Rough-In

- ☐ All supplies and drains specified are present and of proper material (PVC, CPVC, copper, cast iron, polybutylene).
- ☐ Sewer tap done.
- ☐ No pipes pierced by nails.
- ☐ Hot water fixture on left of spigot.
- ☐ No evidence of any leaks: This is particularly critical at all joints, elbows and FHA straps.
- ☐ All plumbing lines are inside of stud walls; must allow for a flush wall.
- ☐ FHA straps are used to protect pipes from nails, where necessary.
- ☐ All cutout framing done by plumber is repaired so as to meet local building code.
- ☐ Tub is centered in bath properly and secured in place.
- ☐ Tub levelers are all in contact with tub. Tub does not move or rock.
- ☐ Toilet drain is at least 12″ to center from all adjacent walls, to accommodate toilet fixture.
- ☐ Toilet drain is at least 15″ to center from tub, to accommodate toilet fixture.
- ☐ Exterior spigot stub-outs are high enough so that they will not be covered over by backfill. Minimum of 6″.
- ☐ Ice maker line for refrigerator is in place (¼″ copper pipe).
- ☐ Roof stacks have been properly flashed with galvanized sheet metal.
- ☐ All wall-hung sinks have metal bridge support located with center at proper level.
- ☐ Water heater is firmly set and connected.

- ☐ Attic water heater has a floor drain pan.
- ☐ Water main shutoff valve works properly.
- ☐ Water lines to washer are installed.
- ☐ Rough-in inspection is approved and signed by local building inspector.

Finish

- ☐ Tub faucets (hot and cold) operate properly, with no drip. Drain operates well.
- ☐ Sink faucets (hot and cold) operate properly, with no drip. Drain operates well.
- ☐ Toilets flush properly. Filled to proper line, action stops completely with no seepage.
- ☐ Kitchen sink faucets (hot and cold) operate properly, with no drips. Drain operates well.
- ☐ Garbage disposal operates properly. Test as needed.
- ☐ Dishwasher operates properly (hot and cold), with no leaks. Drain operates well. Run through one entire cycle.
- ☐ No scratches, chips, dents or other signs of damage are on any appliances.
- ☐ No evidence of water hammer exists in entire system. Turn each faucet on and off very quickly and listen for a knock.
- ☐ All water supplies: hot on the left and cold on the right.
- ☐ Turn on all sinks and flush all toilets at the same time and check for significant reduction in water flow. Some is to be expected.
- ☐ Cutoff valves on all sinks, toilets and water heaters.
- ☐ All exterior water spigots are freeze-proof and operating properly.
- ☐ All roof and exterior wall penetrations test waterproof.
- ☐ Pipe holes in poured walls are sealed with hydraulic cement.
- ☐ *NOTE:* It is highly recommended that you obtain the assistance of your local inspector when performing your plumbing inspections. Your plumbing inspector is well trained for such tests.

Heating, Ventilation and Air Conditioning (HVAC)

Heating, Ventilation and Air Conditioning is one of the big three trades in terms of cost and impact. Heating and air-conditioning systems share the same ventilation and ductwork. Cold air for air conditioning is heavier than hot air, hence it normally takes a higher-speed fan. For this reason, see if you can get a two-speed fan—one speed for cold air in the summer and one for hot air in the winter. The fan unit, normally a "squirrel cage" type, looks just like its name implies. If you have a two-story home, a large split-level or a long ranch, consider a zoned (split) system. Zoned systems work on the premise that two smaller, dedicated systems operate more efficiently than one larger one. For large homes (2,500 sq. ft. or more), this becomes a *strong* consideration. Heating is measured in BTUs (British Thermal Units), while cooling is normally measured in tons (no relationship to 2,000 lbs.). Normally, it takes 30,000 BTUs of heat and one ton of A/C for each 800 sq. feet of area.

The efficiency of heating and cooling systems varies widely. Select your systems carefully with this in mind. Contact your local power and gas companies for information regarding energy efficiency ratings used to compare systems on the market. When shopping for systems, check the efficiency ratings, and be prepared to pay more for more efficient systems. Your power and gas company may also conduct, free of charge, an energy audit of the home you plan to build, and help suggest heating and cooling requirements. For efficiency reasons, your HVAC system should be located as close to the center of your home as possible, to minimize the length of ductwork required. Also, ductwork should have as few turns in it as possible. Air flow slows down every time a turn is made, decreasing the system's overall efficiency and effective output. Seven-inch-diameter round ductwork is fairly standard.

Air-conditioning units operate by transferring heat from the air inside your home to the outside via a fluid called Freon. Several grades of Freon exist. Freon-16 is used in most A/C units. Freon is measured by the pound (weight)—not volume. Make sure your system is fully charged before operating, to avoid damage to the compressor.

Other Climate Control

At your option, your HVAC sub will also install other climate control devices. Electric air filters attach to your HVAC system air intake, eliminating up to 90 percent of airborne dust particles when cleaned regularly. Humidifiers, popular in dry areas, add moisture to the air. Special water supplies can be attached for automatic water dispensing. Dehumidifiers, popular in the Southeast and other humid areas, draw water out of the air. These units may be attached to a water drain for easy disposal of accumulated water.

HVAC SUBS

HVAC subs charge based on the size of your system (BTUs and tonnage). There is a tendency for subs to overestimate your heating and cooling requirements for three reasons:

- More cost equals more profit.
- Nobody ever complains about having too much heating and cooling.
- They adhere to the adage, "Better safe than sorry."

Have your HVAC sub submit a separate bid for the ductwork. Your HVAC sub will also install your natural gas line (if you use natural gas) and hook it up to the gas main. The gas company will install your gas meter. Shop carefully for your HVAC sub: A wrong decision here can cost you increased utility bills.

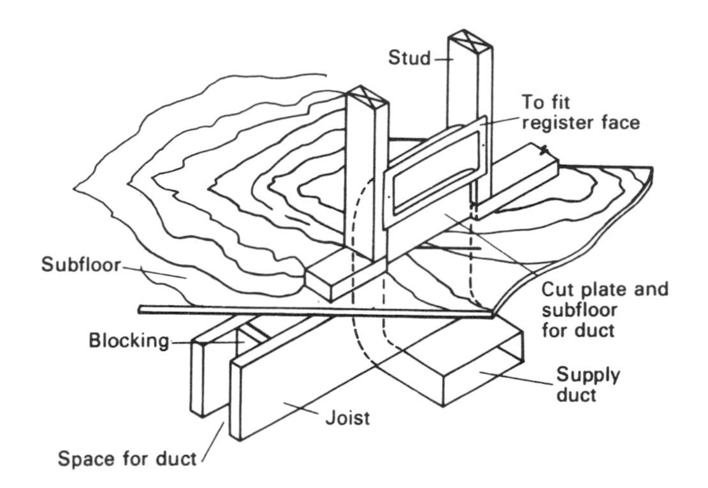

Figure 73. Spacing joists to allow installion of ductwork in load-bearing partitions.

Ductwork

The ductwork circulates conditioned air to specific points in the home; it is split between air supplies that deliver conditioned air and air returns that recirculate the air back to the HVAC system. Ducts will usually be constructed of sheetmetal or insulated fiberglass cut and shaped to fit at the site. Fiberglass ducts are rapidly gaining in popularity due to their ease of fabrication at the site. The duct itself is made of insulated fiberglass and can be cut with a knife. This extra insulation makes for a more efficient and quieter air system. Sheetmetal ducts are notorious for developing mysterious noises caused by expansion and air movement—noises that may be difficult or impossible to eliminate.

Register Placement

Proper placement of supplies and returns is critical to making your total HVAC system efficient and economical to run. Most registers are placed near doors or under windows. This type of placement puts the conditioned air near the sources of greatest heat loss and promotes more even distribution of air.

SUPPLIES should be located primarily along the exterior walls of the home. This is known as a radial duct system. Where possible, hot air supplies should be positioned directly below windows and alongside exterior doors near the floor, since heat rises. Cold air supplies should be located up high in rooms since cool air falls.

RETURNS should be located low for warm air returns since the cooled air drops. Cool air returns should be located high to remove the air as it warms up. It is a good idea to have two supplies (high and low) and two returns (high and low) in critical rooms in order to handle either season. Your HVAC sub will be able to help you with the proper quantity and placement of your supplies and returns. As a general rule, you will need at least one supply and one return for each room.

Thermostats

Many elaborate thermostats are available on the market today. Many automatically shut heating systems down or set them at a low level during "off" hours. In any event, do not put the thermostat within 6′ of any air supply register or facing one on an opposite wall. Never place a thermostat in a room that has a fireplace. Consider placing a thermostat in the hallway.

Get the Most Out of Your Subs

If you are building in the winter, subs will normally be used to working in a cold, unheated house. This means subs may not show up; or when they do, they might spend a lot of time just getting warm by a fire somewhere. If you get your heating system approved early, normally by passing a quick test, you can turn it on before the house is finished. This will also save you some money when other builders may have gone out and rented a space heater. Drywall and paint will dry faster, and subs will really appreciate the additional comfort.

HVAC—Steps

HV1 CONDUCT energy audit of home to determine HVAC requirements. Your local gas and electric companies can be very helpful with this, sometimes providing computer printouts at little or no cost. You may only have to drop off a set of blueprints. Decide whether you will have a gas or electric dryer and determine where it will be situated.

HV2 SHOP for the best combination of cost, size and efficiency in heating and cooling systems. The higher the efficiency, the higher the price.

HV3 CONDUCT bidding process on complete HVAC job. Have subs give a separate bid for the ductwork.

HV4 FINALIZE HVAC design. Have a representative from the local gas company or an inspector look over your plan to make sure you aren't doing anything that could be a problem. Specify location of dryer exhaust vent.

HV5 ROUGH-IN heating and air-conditioning system. Do not install any external fixtures, such as compressors, at this time. Compressors are a favorite theft item. If you frame floor joists 12″ O.C., you may need to use floor registers.

HV6 INSPECT heating and air-conditioning rough-in. This should be done with your county inspector. Call him in advance.

HV7 CORRECT all deficiencies noted in the rough-in inspection.

HV8 PAY HVAC sub for rough-in work. Have him sign a receipt.

Finish Work

HV9 INSTALL heating and air conditioning (finish work). This is where the thermostat and all other HVAC electrical work are hooked up. All registers are installed. The A/C compressor is installed and charged with Freon. Discuss the placement of the A/C compressor location with the HVAC sub. He may supply the concrete pad if you do not have one.

HV10 INSPECT heating and air conditioning final. Again, this should be done by your county inspector. Call him several days in advance so you can be present for the inspection. Some builders like to have their HVAC sub present, too. Make a note of any deficiencies. Be sure electrician has run service to the unit and that it is functional.

HV11 CORRECT all deficiencies noted in the final inspection.

HV12 CALL gas company to hook up gas lines.

HV13 DRAW gas line on plat diagram.

HV14 PAY HVAC sub for finish work. Get him to sign an affidavit.

HV15 PAY HVAC sub retainage after system is fully tested.

HVAC—Sample Specifications

- Bid is to install heating, ventilation and air-conditioning system (HVAC) as indicated on attached drawings.
- Install zoned forced-air gas-fired (FAG) heating units with 5-year warranty—one in basement and one in attic.
- Install two A/C compressors.
- Install gas range hookup.
- Install 50-gallon gas water heater.
- Install downdraft line for island cooktop.
- Connect 1½″ copper gas line to gas main, a minimum of 12″ underground.
- Install sloped PVC tubing for A/C drainage.
- Install gas grill line and grill unit on patio.
- Install one fireplace log lighter and gas line.
- Install gas line to porch and mailbox gaslights.
- All work and materials are to meet or exceed requirements of the local building code.
- Provide and install all necessary air filters.
- Leave opening to furnace large enough to permit its removal and replacement.
- Charge all Freon lines with Freon.

HVAC—Inspection

Rough-In

- ☐ All equipment is UL-approved with warranties on file.
- ☐ Heating and air units are installed in place and well anchored.
- ☐ Heating and air units are proper make, model and size.
- ☐ Zoned systems have proper units in proper locations.
- ☐ Air compressors are firmly anchored to footings.
- ☐ All ductwork is installed according to specifications.

- ☐ Proper number of returns and supplies have been installed.
- ☐ All ductwork meets local building codes.
- ☐ All ductwork joints are sealed tightly and smoothly with duct tape.
- ☐ No return ducts are in baths or kitchen.
- ☐ Bathrooms vent to outside if specified.
- ☐ All ductwork in walls is flush with walls, so as not to interfere with drywall.
- ☐ All ductwork is in ceiling tied off so as not to interfere with drywall or paneling.
- ☐ Attic furnace has a floor drain pan.
- ☐ All heat exhaust vents are isolated from wood or roofing by at least 1″.
- ☐ No exhaust vents are visible from the front of the home.
- ☐ Air-conditioning condensate drain is installed.
- ☐ All ductwork outlets are framed properly to allow installation of vent covers.
- ☐ Vent hood or downdraft is installed in kitchen according to plan.
- ☐ HVAC rough-in inspection is approved and signed by local building inspectors.
- ☐ Gas meter is in place.
- ☐ Gas line is connected to gas main.

Finish

- ☐ Line to gas range is hooked up. Range is installed.
- ☐ Line to gas dryer is hooked up. Dryer vent is installed properly.
- ☐ Line to gas grill is hooked up. Gas grill is installed.
- ☐ HVAC electrical hookup is completed.
- ☐ Thermostats operate properly and are installed near the center of the house. Thermostats are located away from heat sources (fireplaces and registers), doors and windows to provide accurate operation.
- ☐ The Manual/Auto switch on the thermostat works properly.
- ☐ A/C condensate pipe drains properly. This will be difficult or impossible to test in cold weather. The pipe should at least have a slight downward slope to it. You cannot test A/C if the temperature is below 68°.
- ☐ Fan noise is not excessive.
- ☐ Furnace, A/C and electronic air filters are installed.
- ☐ Water line to humidifier is installed and operating.
- ☐ Water line from dehumidifier is installed and operating.
- ☐ All vent covers are on all duct openings and installed in proper direction.
- ☐ All holes and openings to exterior are sealed with exterior grade caulk.
- ☐ Downdraft vent for the cooktop is installed and operating.
- ☐ All HVAC equipment papers are filed away (warranty, maintenance, etc.).
- ☐ Air flow is strong from all air supplies.
- ☐ Air returns function properly.

Electrical

Electricians will perform all necessary wiring for all interior and exterior fixtures and appliances. Electricians will often charge a fixed price for providing and wiring each switch, outlet and fixture. Double switches will count as two switches, and so on. You can save money by purchasing your own fixtures; the markup is incredible. Code will normally require at least one outlet per wall of each room and/or one for each specified number of running wall feet. The electrical code most often used and referred to is the National Electrical Code (NEC), but the one in your county should prevail. Your electrician needs to be a licensed master electrician in your county.

Your electrical sub will help you determine where switches and outlets should be placed. There are a few standards, such as minimum spacing of outlets and height of switches and outlets, which should be indicated in your local building code. Consider multiple switches for a single light (such as at the top and bottom of stairs), floodlight switches in the master bedroom and photoelectric cells for driveway and porch lights. Make sure to install ground-fault interrupters (GFIs) in all bathroom outlets. GFIs are ultrasensitive circuit breakers that provide an extra margin of safety in areas where the risk of shock is high. Hobby shop and garage areas are also good sites for these devices. Key areas for additional outlets:

- Plugs in breakfast areas
- Along kitchen counters
- Switched outlets in attic
- Foyer area

Wire gauges used for circuits will vary. Electric stoves, ovens, refrigerators, washers, dryers and air compressors are each normally on separate circuits and use heavier gauges of wire. Your local building code will describe load limits on each circuit and the appropriate wire types and gauges to use.

After your home is dried-in, the electrical sub will rough-in all the electrical wiring, fuse box and electrical HVAC connections. Make sure to have the HVAC and plumbing subcontractors complete the rough-in before calling in the electrician. This will prevent any wiring from having to be cut or rerouted in order to make room for plumbing or ducting. It is much easier for the electrician to work around obstructions. If all ducting is in place, the electrician will also know precisely where the furnace will be located and will be able to wire for it properly.

Many house plans include a wiring diagram, but make sure to examine it carefully. These diagrams are usually done as an afterthought and are produced by the draftsman, not an electrician. It is best to walk through the house with the electrician and mark the locations of outlets and switches. A good electrician will make suggestions on the location of fixtures.

Additional Wiring

While the electrical wiring is being installed, consider installing other necessary wiring for appliances such as phones, alarm systems, TV antennas, doorbells, intercoms and computers. Ask your electrician for bids on installing this other wiring, although his price may be quite high. This type of wiring makes an excellent do-it-yourself project if you are eager to get involved. Installing this wiring before drywall is installed is easy and saves considerable installation expense later.

STEPS—Electrical

Rough-In

EL1 DETERMINE electrical requirements. Unless your set of blueprints includes a wiring diagram, you will want to decide where to place lighting fixtures, outlets and switches. Make sure no switches are blocked by an open door. You may want to consider furniture placement while you are doing this. Even if your blueprint has an electrical diagram, don't assume it is correct. Most of the time, these can be improved. You may wish to investigate the use of low-voltage and fluorescent lighting. Remember that you cannot dim fluorescent lights.

EL2 SELECT electrical fixtures and appliances. Visit lighting distributor showrooms. Keep an eye out for attractive fluorescent lighting, which will be a long-term energy saver (but cannot use a dimmer switch). Special orders should be done now due to delivery times. You should listen to the doorbell sound before purchasing.

EL3 DETERMINE whether phone company charges to wire home for modular phone system. Find out what it charges for, and how much. Even if you plan wireless phones, install phone jacks in several locations.

EL4 CONDUCT standard bidding process. Have subs give a price to install each outlet, switch and fixture. They will normally charge extra for wiring the service panel and special work. Use only a licensed electrician.

EL5 SCHEDULE to have phone company install modular phone wiring and jacks. If your phone company charges too much for this service, just have your electrician do the job or do it yourself. Also try to arrange to have your home phone number transferred to your new house later so that you can keep the same phone number.

EL6 APPLY for and obtain permission to hook up your temporary pole to public power system. You may need to place a deposit. Perform this early so that hookup will take place by the time the framers are on-site.

EL7 INSTALL temporary electric pole. Make sure the pole is within 100′ of the center of the house foundation (preferably closer).

EL8 PERFORM electrical rough-in. This involves installing wiring through holes in wall studs and above ceiling joists. All the wiring for light switches and outlets will be run to the location of the service panel. To ensure that your outlets and switches are placed where you want them, mark their locations with chalk or a marker. Otherwise, the electrician will place them wherever he desires. It is important to get lights, switches and outlets exactly where you plan to place furniture and fittings. Provide scrap 2×4s for blocking next to outlets.

EL9 INSTALL modular phone wiring and jacks and any other wiring desired. You may also wish to run speaker, cable TV, security and computer wiring throughout the house. It's much easier to do it now than later. Staple wires to studs where wire enters the room so the drywall crew will notice them. Many security system contractors will prewire your home for their security system at their cost if you decide to purchase their system.

EL10 SCHEDULE electrical inspection. The electrician will usually do this.

EL11 INSPECT electrical rough-in. The county inspector will sign off on the wiring. Remember that this *must* be done before final drywall is in place.

EL12 CORRECT any problems noted during rough-in inspection.

EL13 PAY electrical sub for rough-in work. Get a receipt or use canceled check as a receipt.

EL14 INSTALL garage door(s). Typically this is not done by the electrician, but by a specialist. The garage door must be installed so that the opener can be hooked up by the electrician.

EL15 INSTALL electric garage door opener. Store remote control units in a safe place. Wire electrical connection.

Finish

EL16 PERFORM finish electrical work. This includes terminating all wiring appropriately (switches, outlets, etc.). Major electrical appliances such as refrigerators, washers, dryers, ovens, vent hoods, exhaust fans, garage door openers, doorbells and other appliances are installed at this time. The air compressor will also be wired.

EL17 CALL phone company to connect service.

EL18 INSPECT finish electrical. Again, your county inspector must be scheduled. Have your electrician present so he will know just what to fix if there is a problem. Store all appliance manuals and warranties in a safe place.

EL19 CORRECT electrical problems if any exist.

EL20 CALL electrical utility to connect service.

EL21 PAY electrical sub, final. Have him sign an affidavit.

EL22 PAY electrical sub retainage after power is turned on and all switches and outlets are tested.

ELECTRICAL—Sample Specifications

- Bid is to perform complete electrical wiring per attached drawings.
- Bid is to include all supplies, except lighting fixtures and appliances and includes installation of light fixtures.
- Bid is to include furnishing temporary electric pole and temporary hookup to power line.
- Bid is to include brass switch plates in living room, dining room and foyer.
- Bid may include wiring and installation of the following items:
  - 200 amp service panel with circuit breakers
  - Light switches (one-way)
  - Light switches (two-way)
  - Furnace
  - Dishwasher
  - Garbage disposal
  - Microwave (built-in)
  - Door chime set
  - Door chime button(s) with lighted buttons
  - Bath vent fans
  - Bath heat lamp with timer switch
  - Jacuzzi pump motor with timer switch
  - Light fixtures (all rooms and exerior areas)
  - Three-prong (grounded) interior outlets
  - Three-prong (grounded) waterproof exterior outlets
  - Electric range
  - Double-lamp exterior flood lights
  - Electric garage door openers with auto light switch
  - Washer
  - Dryer
  - Central air-conditioning compressor
  - Climate control thermostats
  - Electric hot water heater
  - Hood fan
  - Central vacuum system
  - Intercom, radio units, speaker wires
  - Coaxial cable TV lines and computer cables
  - Sump pump
  - Electronic security system with sensors
  - Ceiling fans
  - Whole-house attic fan with timer switch
  - Automatic closet switches
  - Humidifier/dehumidifier
  - Time-controlled heat lamps
  - Thermostatic roof vents
  - Phone lines in all rooms as shown
  - Time-controlled sprinkler system
- All work and materials are to meet or exceed all requirements of the National Electrical Code unless otherwise specified.
- Natural earth ground is to be used.
- Dimmer switches are to be installed on fixtures as shown on plan.
- Rheostat is to be installed on ceiling fans.

ELECTRICAL—Inspection

Rough-In

☐ Outlets and switch boxes are offset to allow for drywall and base molding.

☐ All outlets and switches are placed at 42″ height as indicated on plans. Outlets and switches are unobstructed by doors.

☐ All special outlets, switch boxes and fixtures are in place where intended. Refer to your specifications and blueprints.

☐ Electrical boxes for chandeliers and ceiling fans are adequately braced to hold the weight of the fixtures.

☐ All lines are grounded.

- ☐ All visible splices have approved splice cap securely fastened.
- ☐ Bath ventilator fans are installed.
- ☐ Attic power ventilators are installed.

Finish

- ☐ All outlets and fixture wires measure 117 V with circuit tester or voltmeter. Test between the ground and each socket.
- ☐ Dimmer switches are installed where specified (dining room, den, foyer and master bath).
- ☐ Air conditioner cutoff switch is installed at A/C unit.
- ☐ No scratches, dents or other damage has marred electrical appliances and fixtures.
- ☐ Furnace is completely connected and operational.
- ☐ All bath vents are connected and operational. They should be on a separate circuit from the light switch.
- ☐ Electric range is connected and operational.
- ☐ Electric pilot light starter on gas range is connected and operational.
- ☐ Range hood, downdraft and light are connected and operational.
- ☐ Built-in microwave is connected and operational.
- ☐ Garbage disposal is connected and operational.
- ☐ Trash compactor is connected and operational.
- ☐ Dishwasher is connected and operational.
- ☐ Refrigerator outlet is operational.
- ☐ Door chimes (all sets) are connected and operational.
- ☐ Power/thermostatic ventilators on roof are connected and operational.
- ☐ Washer and dryer outlets are operational.
- ☐ Garage opener is installed and operational.
- ☐ Electronic security system is connected and operational. All sensors need to be tested one at a time.
- ☐ Phone outlets work at all locations.
- ☐ Service panel is installed properly with sufficient load-carrying capacity. Breakers are labeled properly.
- ☐ Exterior lighting fixtures are installed and operational.
- ☐ All recessed lighting, ceiling lighting, heat lamps and other lighting fixtures are connected as specified and operational.
- ☐ All intercom/radio units are connected and operational.
- ☐ Jacuzzi pump motor is connected and operational. For safety reasons, the switch should not be reachable from the tub.
- ☐ Central vacuum system is connected and operational.
- ☐ All other specified appliances and fixtures are connected and operational.
- ☐ All switch plate and outlet covers are installed as specified (in rooms to be wallpapered, these will be temporarily removed).
- ☐ Finish electrical inspection is approved and signed by local building inspector.

Masonry and Stucco

MASONRY

Masonry is the trade involved with the application of concrete blocks, clay or tile bricks and stone. Masonry is a true art: Watch a bricklayer sometime and you will see why. A skilled bricklayer possesses a strong combination of good workmanship and efficiency—key ingredients to producing a strong, well-built brick structure on schedule. Your brick mason and stonemason may be two different subs.

Concrete Blocks

Concrete blocks are normally used below ground as an alternative to a poured foundation. Although a poured foundation is three times stronger, concrete blocks are less expensive, while being of adequate strength. The standard size of a concrete block is 7⅝″ × 7⅝″ × 15⅝″. Blocks are normally laid with ⅜″ flush mortar joints. Concrete blocks are applied on concrete footings.

Clay Bricks

Often called brick veneer, brick facing is normally composed of brick 3⅝″ × 7⅝″ × 2¼″ set in a ⅜″ mortar joint. For estimating purposes, brick veneer, laid in running bond (a typical pattern), requires seven bricks per square foot of wall. This does not apply for such items as chimneys or fancy brickwork. Your mason charges for laying units of 1,000 bricks (one skid). Additional expenses may be charged for multistory or fancy brickwork, where scaffolding and/or additional skill and labor will be required. Bricks are anchored to walls with galvanized metal wall ties. Once bricks are laid, brick walls are self supporting.

In some styles of architecture, brick or stone veneer is used for all or part of the exterior finish. It is good practice, when possible, to delay applying the masonry finish over platform framing until the joists and other members reach moisture equilibrium. Waterproof paper backing and sufficient wall ties should be used. It is normal practice to install the masonry veneer with a 3″ space between the veneer and the wall sheathing. This space provides room for the bricklayer's fingers when setting the brick.

STUCCO

Once popular and widely used throughout Western Europe, stucco has come back in style to bring back that old flair. Stucco is really just mortar applied over a metal lath or screen used for support and adhesion. Stucco finishes are applied over a coated, expanded metal lath and, usually, over some type of sheathing. In some areas, where local building regulations permit, such a finish can be applied to metal lath fastened directly to the braced framework.

DURABILITY. Some builders complain of maintenance problems (cracking) of traditional stucco after ten years or so of service. If stucco is applied properly with proper expansion joints, it should provide decades of sound performance. Expansion joints should be placed approximately every 300 square feet. Some stucco cracking should be expected.

INSULATION. Stucco is not the best insulator, so you should go with heavier batt insulation in your exterior walls. Higher grade sheathing is also in order.

NEW IMPROVED. Several brands of artificial stucco are on the market. Although they boast of higher insulating qualities and lower maintenance, they are also more expensive. Some say it pays off in use of cheaper insulation and lower utility bills. It is also easier to do fancier trim work, such as lintels, quoins and other work, because it uses Styrofoam cutouts as a base.

COSTS. Stucco is charged by the installed square yard. Normally, area is not deducted for doors and windows except for the garage door. The rationale is that it is at least as much trouble to work around

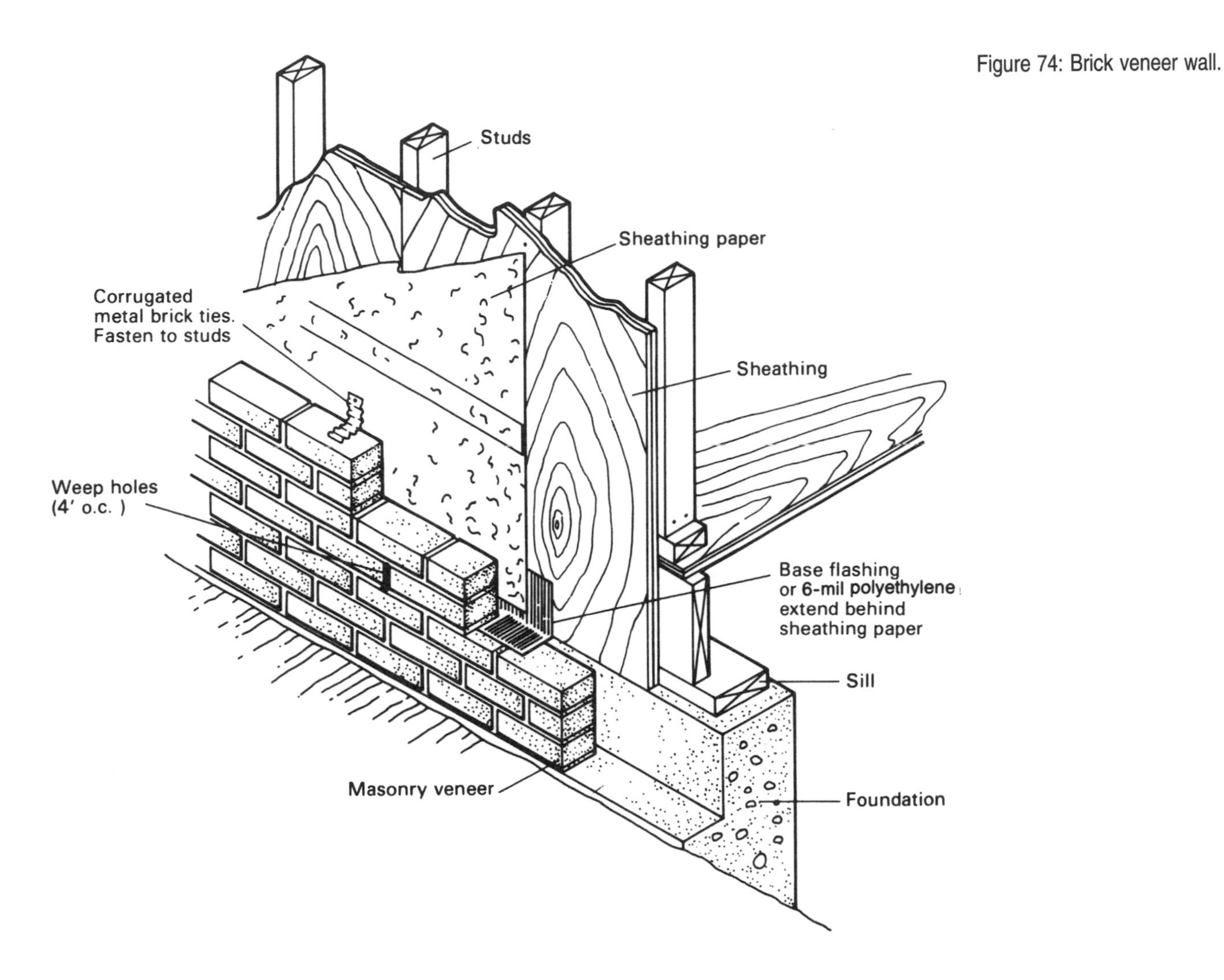

Figure 74: Brick veneer wall.

an opening as it is to cover the area. Extras will probably include:

- Smooth finish
- Decorative corner work (quoins)
- Decorative window and door work (keystones and bands)
- Corner bead—all edges

Stucco can be applied rough, smooth or anywhere in between. Subs will charge you more for a smooth finish because it takes longer and imperfections show up easier. If you use a dyed stucco, it may not require painting.

For the Best Stucco Job

- Don't attempt to stucco foil- or polyethylene-backed sheathing due to poor adherence. Use either Styrofoam (⅝″ to 2″) or exterior gypsum (½″). Check to see what sheathing your stucco sub recommends.
- Avoid mixing pigments with stucco mortar. This can lead to uneven drying and coloration. Paint the stucco after it is dry.
- To avoid cracking, be liberal with expansion joints. They are hardly noticeable and do not detract from overall appearance.
- Don't attempt to apply stucco if there is *any* chance of rain in the forecast.

About Fireplaces

If you go with a custom masonry fireplace, the critical thing to get is a proper draft. If your fireplace doesn't "breathe" properly, you will always have smoke in your home during its use. If it "breathes" too well, you may not get too much effect from the fireplace. This is not a big problem with prefabricated fireplace units, since they are predesigned to

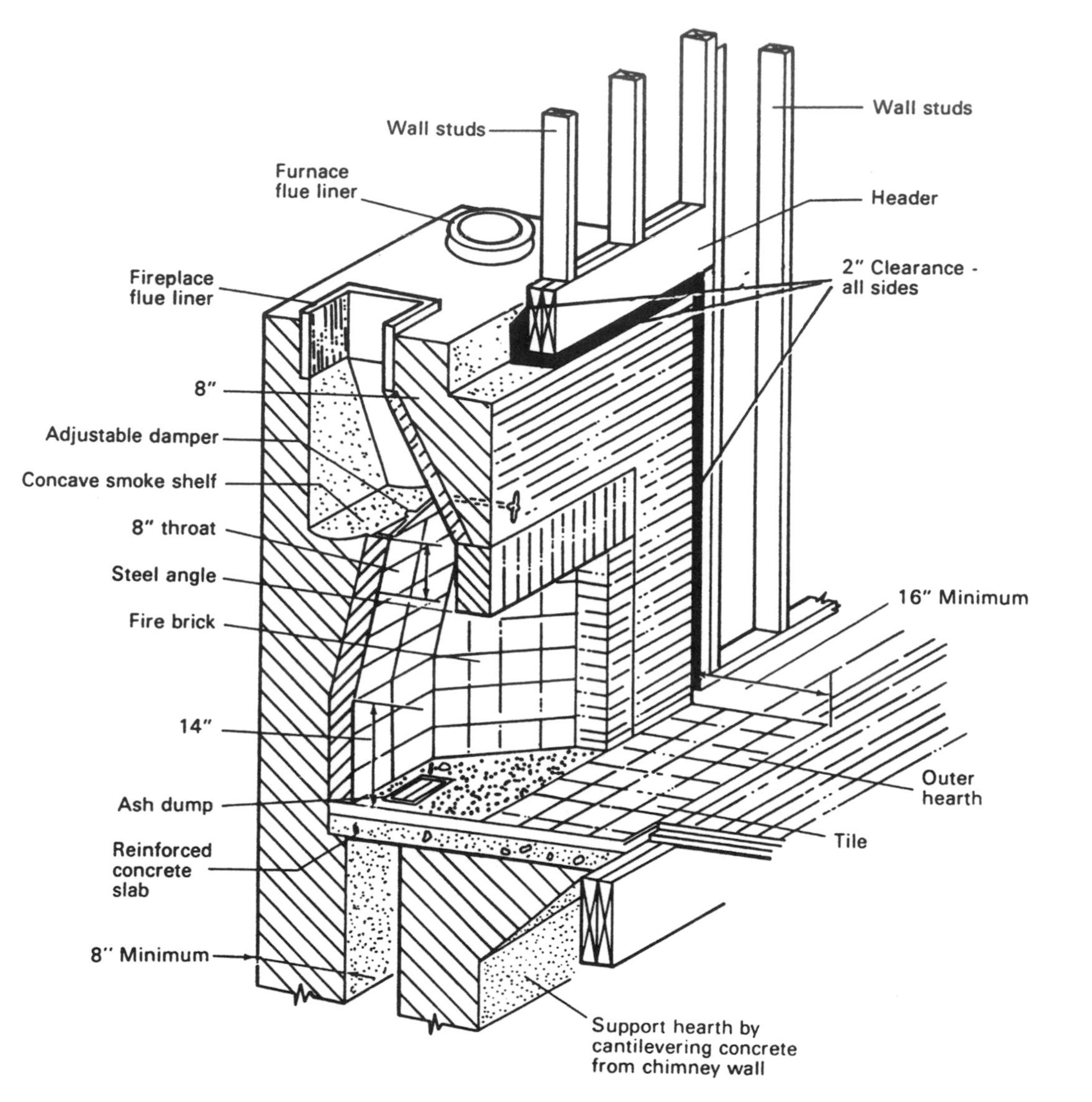

Figure 75: Masonry fireplace components.

operate properly. Below are a few options you may want to consider regardless of what type fireplace you have:

FRESH AIR VENT—Allows the fire to get its combustion air from outside your home so that it doesn't rob you of all your warm air.

ASH DUMP—Makes fireplace cleaning a lot easier.

LOG LIGHTER—Easier log starting.

RAISED HEARTH—Easier access to working on the fire.

BLOWER—May provide more room heating.

HEAT RECIRCULATOR—May also be able to provide more heating.

One important safety issue is the flue. Avoid burning pine whenever possible, especially if you are using a prefab fireplace. The resin buildup in the flue can cause problems later (fire hazard). If you plan to have a stucco home, consider a prefab fireplace. A masonry fireplace can look out of place unless covered over with stucco, which isn't very cost effective!

MASONRY/STONE—Steps

MA1 DETERMINE brick/stone pattern, color and coverage and joint tooling.

MA2 DETERMINE stucco pattern, color and coverage.

MA3 PERFORM standard bidding process. Contact material suppliers, local brick companies, masonry unions and other subs for the names of good masons. Do not select a mason until you have seen his work. Ask the suppliers if there is a penalty for returning unused bricks.

MA4 PERFORM standard bidding process for stucco subs. You really need to see their work, too.

MA5 APPLY flashing above all windows and doors if not already installed by roofer.

MA6 ORDER brick, stone and angle irons.

MA7 LAY brick/stone veneer exterior, chimney, fireplace and hearth. This is the time to have the mason throw in a brick mailbox if you want one. Remember to provide a "postal approved" metal mailbox insert.

MA8 FINISH bricks and mortar.

MA9 INSPECT brickwork. Refer to specifications, local code and checklist on this and the following page.

MA10 CORRECT any brick/stonework requiring attention.

MA11 PAY mason. If you have extra bricks that either can't be returned or are not worth returning, turn them into a brick walkway or patio later.

MA12 CLEAN up excess bricks and dried mortar. Spread any extra sand in the driveway and sidewalk area. Masons charge too much by the hour to have them do this work.

MA13 INSTALL base for decorative stucco. This includes keys, bands and other trimwork. Use exterior grade lumber or Styrofoam if using synthetic stucco.

MA14 PREPARE stucco test area. This should be at least a 3′×6′ sheet of plywood covered with test stucco to make sure you agree with the actual color and texture. Make sub finish mailbox if desired. If you wish to protect the stoops from stucco, cover them with sand.

MA15 APPLY stucco lath and decorative formwork to sheathing.

MA16 APPLY first coat of stucco. The color of the first coat does not matter.

MA17 APPLY second coat of stucco. You need to wait a few days for the first coat to dry completely. Place a thick mat of straw at base of house to prevent mud from staining stucco. While the scaffolding is set up, have the painter work on cornice and second-story windows.

MA18 INSPECT stucco work.

MA19 CORRECT any stucco work requiring attention.

MA20 PAY stucco subs.

MA21 PAY mason's retainage.

MA22 PAY stucco sub's retainage.

MASONRY—Specifications

- Brick veneer coverage is to be as indicated on attached drawings.
- All brick, sand and mortar mix are to be furnished by builder.
- All necessary scaffolding is to be supplied and erected by mason.
- Brick entrance steps are to be laid as indicated in attached drawings.
- Brick patio is to be laid as indicated in attached drawings.
- All excess brick is to remain stacked on-site.
- Wrought iron railings are to be installed on entrance steps as indicated in attached drawings. Wrought iron is to be anchored into surface with sufficient bolts.
- Bricks are to be moistened before laying to provide superior bonding.
- Window, door, fascia and other special brick patterns are to be done as indicated in attached detail drawings.
- Moistened brick is to be laid in running bond with ⅜″ mortar joints.

- ■ Mortar is to be medium gray in color; Type C mortar is to be used throughout job.
- ■ Mortar joints are to be tooled concave.
- ■ Steel lintels are to be installed above all door and window openings.
- ■ Masonry walls are to be reasonably free of mortar stains as determined by builder. Mortar stains are to be removed with a solution of one part commercial muriatic acid and nine parts water to sections of 15 square feet of moistened brick, and then washed with water immediately. Door and window frames are to be protected from the cleaning solution.
- ■ Flashing is to be installed at the head and sill of all window openings.
- ■ Flashing is to be installed above foundation sill and below all masonry work.
- ■ Bid is to include all necessary touch-up work.

STUCCO—Specifications

- ■ Bid is to provide all material, labor and tools to install lath and stucco per attached drawings.
- ■ Stucco is to be finished smooth, sand texture. (Specify color.)
- ■ All trim areas are to be 2″ wide unless otherwise specified.
- ■ Quoins are to have beveled edges; alternating 12″ and 8″ wide.
- ■ Two coats of stucco are to be applied; pigment is to be in second coat.
- ■ Stucco work is to come with a 20-year warranty for material and labor against defects, chipping and cracks.
- ■ Expansion joints are to be as indicated on blueprint or as agreed upon.

MASONRY & STUCCO—Inspection

- ☐ Brick and stone style and color are as specified.
- ☐ Mortar color and tooling are as specified.
- ☐ All wall sections are plumb, checking with plumb bob and string. No course should be out of line more than ⅛″. Overhang distance between the top course and bottom course should not exceed ⅛″.
- ☐ Flashing is installed above all windows, exterior doors and foundation sill.
- ☐ No significant mortar stains are visible from up close.
- ☐ Stoops and patio steps drain properly.
- ☐ No bricks or stones are loose.
- ☐ All quoins are smooth, even and as specified.
- ☐ Touch-up brickwork is finished.
- ☐ All wrought iron railing is complete per drawings and is sturdy.

Stucco

- ☐ Stucco coverage, colors and materials are as specified.
- ☐ Stucco is finished as specified.
- ☐ Stucco finish is even, with no adverse patterns in surface. Surface is level, with no dips or bumps. Sight down surface from corners.
- ☐ Stucco is applied to proper depth.
- ☐ All decorative stucco work such as quoins, keys and special trim are in place, complete as specified and well-finished, with sharp edges and corners.
- ☐ Stucco has adequate number of expansion joints (one approximately every 300 square feet). Expansion is symmetrical between window openings.
- ☐ Openings for exterior outlets and fixtures are not covered up.
- ☐ Stucco has no large cracks.

Siding and Cornice

The siding and cornice sub will handle all siding application and cornice work (soffit, fascia, frieze and eave vents). This sub can also install windows and exterior doors, if needed. Usually, the siding sub will first install thc sheathing, then the windows and doors. Finally, the siding will be installed flush to the window and door frames.

SIDING

Siding is the most economical exterior covering. The most popular materials include cedar, vinyl, aluminum and Masonite. There are advantages and disadvantages to each:

Type	Advantages	Disadvantages
Cedar	Natural appearance, moderate cost	High maintenance, may raise fire insurance premium, may warp or crack
Texture 111	Inexpensive	Poor appearance
Masonite	Inexpensive	High maintenance, requires painting
Vinyl	Washes easily, low maintenance	Expensive
Aluminum	Washes easily, low maintenance	Expensive

WOOD SIDING provides some of the greatest variety in styles and textures, as well as cost-effectiveness. It can be installed horizontally, vertically, diagonally or a combination of these to provide a crcativc and varied exterior.

Some of the wood types available include pine, cedar, cypress and redwood. The full range of types and styles exceeds the scope of this book. Check with your local supplier for the styles available in your area.

Regardless of the type of wood siding you choose, the important properties required for quality are freedom from warpage, knots and imperfections. Wood siding is prone to shrinkage after installation due to drying. If painted right after installation, lap siding will continue to shrink and leave an unpainted strip of wood where the pieces overlap. Try to delay painting until the siding has an opportunity to stabilize.

Drying will also cause boards to warp and crack as they shrink. Always retain a portion of your siding sub's payment so that you can get him to return later to replace these defective pieces.

The introduction of several plywood siding types, such as Texture 111, has simplified siding installation. These plywood sheets simulate several styles of wood siding, such as board and batten, but are much easier to install and do not shrink. Labor costs to install are also much lower than traditional wood siding.

Wood siding is prone to leakage around windows and doors, especially on contemporary designs that have many angled walls. Therefore, it is important that your siding sub or painter caulk all butt joints. Specify a tinted caulk that matches the final color of the siding.

Ask the siding sub to hand-pick the best siding pieces for the front and back of the house. The areas bordering decks and porches are the most visible.

MASONITE SIDING has many of the aesthetic qualities of wood, with the added advantage of more uniform and stable material. Masonite is usually free of imperfections and will not shrink. It also has the added advantage of being preprimed, which may eliminate one paint coat. Most traditional styles that use siding have gone to Masonite; however, the introduction of new textures that simulate wood has made Masonite more popular for rustic and contemporary designs as well.

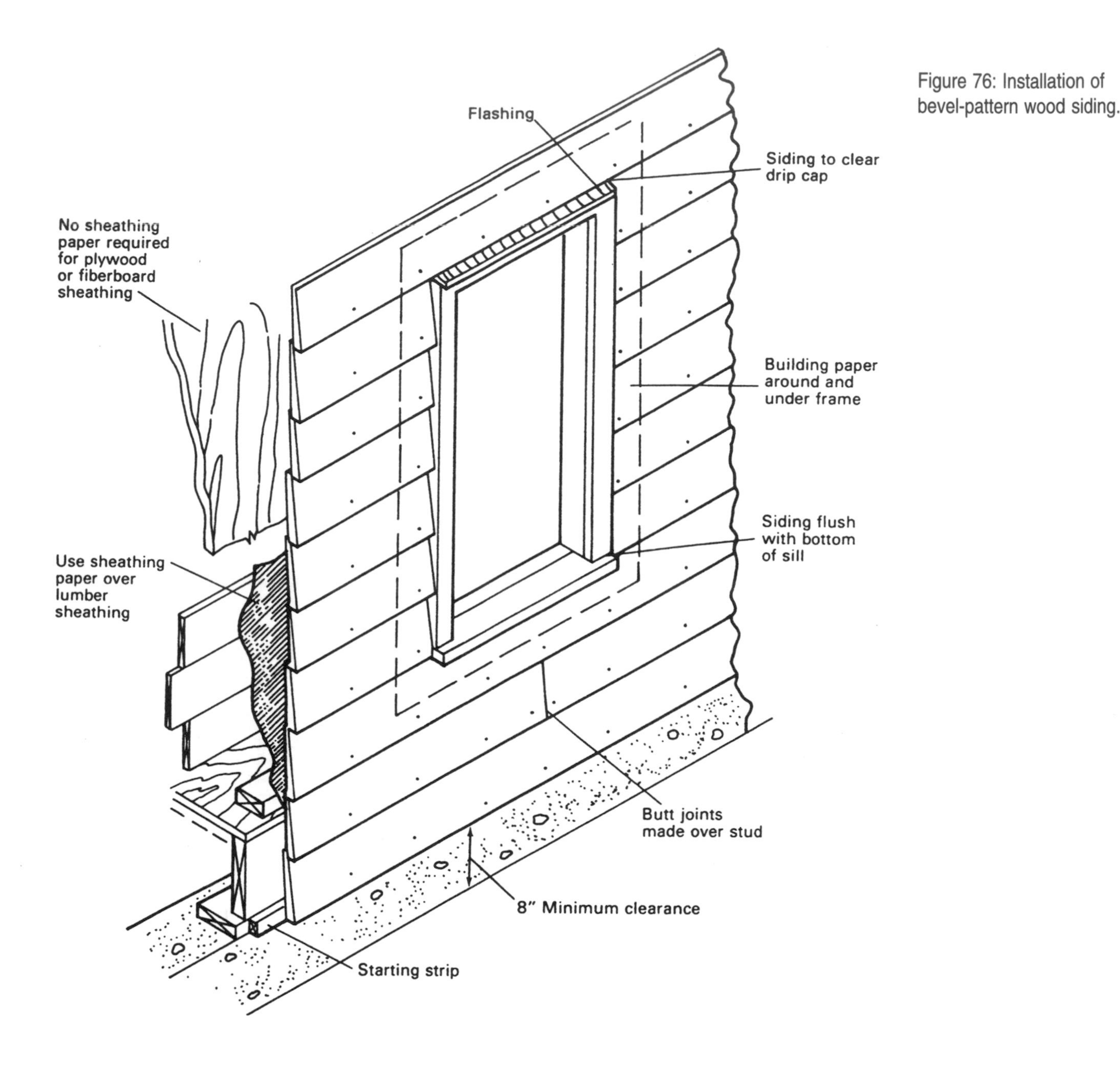

Figure 76: Installation of bevel-pattern wood siding.

The nail holes in Masonite are much more visible than in wood, so watch your sub carefully during installation. Specify that the nails to be used are galvanized and countersunk.

VINYL AND ALUMINUM SIDING is usually installed by a subcontractor who supplies labor and materials for the job. Although expensive, these siding types will last for many years with little or no maintenance. Many come with a 20-year unconditional guarantee.

CORNICE

The cornice is the finishing applied to the overhang of the roof, called the eave. The overhang of the roof serves a decorative function, as well as protecting the siding from water stains and leakage. Generally, the greater the overhang, the more expensive the look of the house.

The cornice will be installed by your siding or trim sub. If you plan to use fancy trim around your cornice, such as dentil molding, you may want to trust

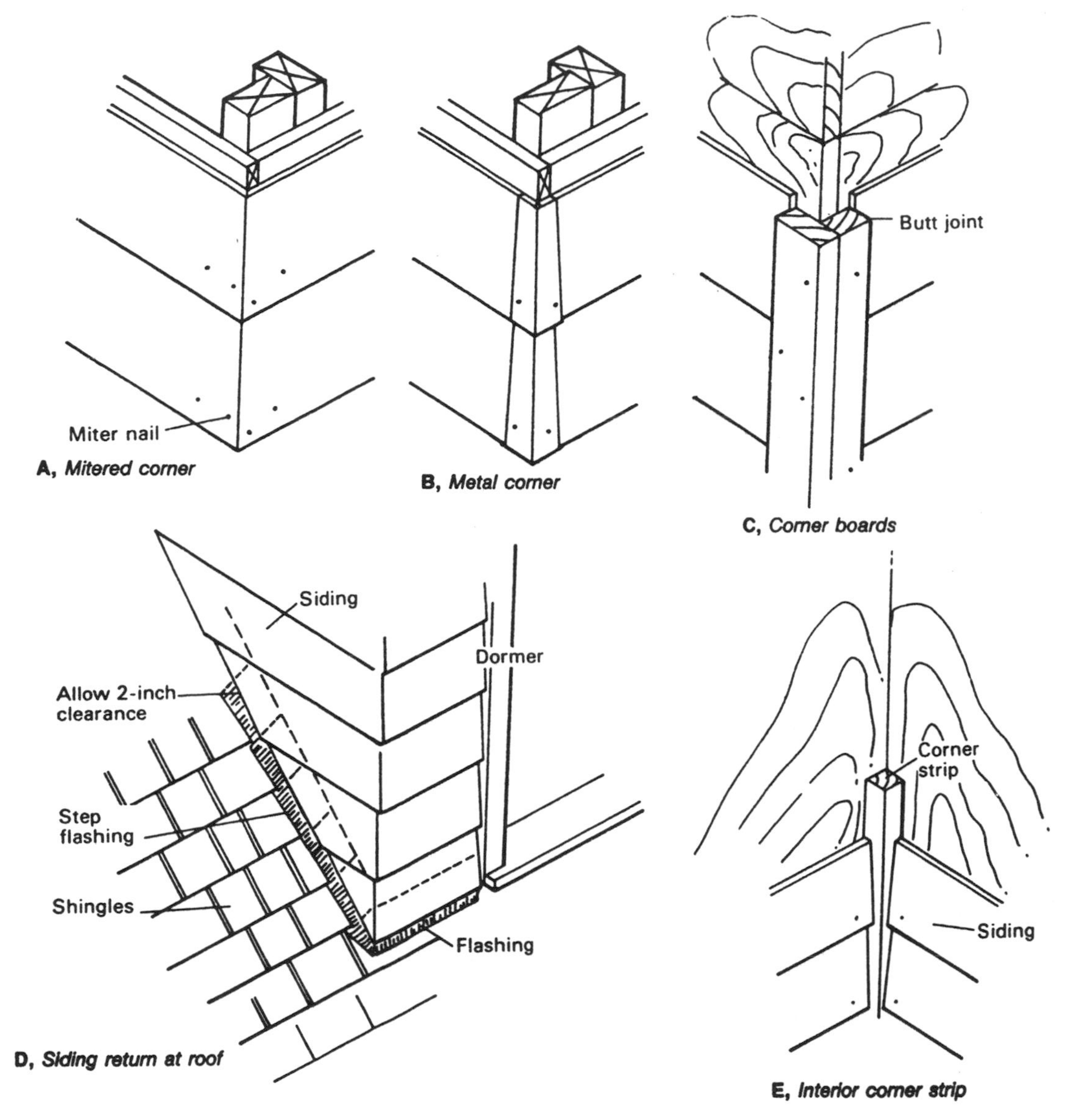

Figure 77: Siding installation details.

the job to your trim sub. He is more used to working with decorative trim.

The two types of cornices most commonly used are the open cornice and the box cornice. The open cornice is the simplest type and is more appropriate for contemporary styles. The open cornice looks just as it sounds. The rafter overhang is left unfinished and open and requires care in the choice of roof sheathing, since it will be visible from below. The box cornice has the overhang boxed in with plywood and is finished off with a variety of trim types and styles. This style is more commonly used on traditional styling and has the added advantage of providing a place to install eave ventilation, which will improve the energy efficiency of the house.

The style and quality of the cornice can have a big effect on the appearance of your house, so take care in the design. If you use a draftsman or an architect, ask him to provide a cornice detail drawing so that its design will be clear to the cornice sub. This will also help you in estimating materials for the job.

Cornice work is installed before brick or stucco; if you have that type of finish, be sure that the cornice sub builds the cornice out (usually with 1×6 or 2×6) so that the cornice sits out beyond the brick or stucco. Once installed, the cornice should be primed with exterior paint as soon as possible. For convenience, you may wish to have your painters prime the cornice material prior to installation.

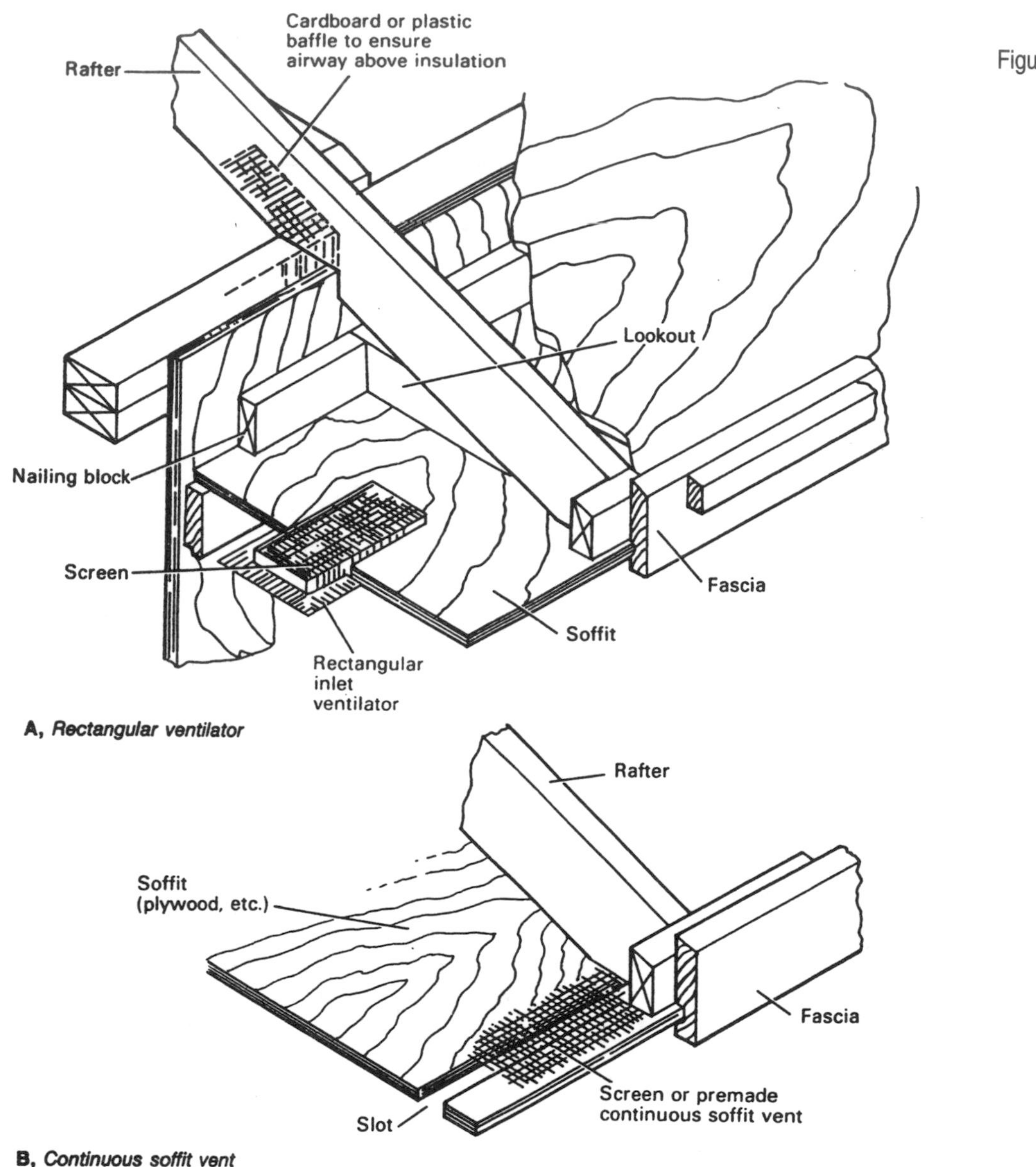

Figure 78: Soffit inlet ventilators.

SIDING AND CORNICE—Steps

SC1 SELECT siding material, color, style and coverage.

SC2 CONDUCT standard bidding process. Most often, the siding and cornice will be bid as a package deal.

SC3 ORDER windows and doors. Before ordering, have the window supplier walk through the site to check openings. All window and door brands differ in their real dimensions. The supplier can help to provide more exact dimension information. Mark the proper dimensions on the openings if adjustments need to be made by the framing or siding sub.

SC4 INSTALL windows and doors. Inspect windows and doors for bad millwork *before* installing. Have wood shims available to adjust openings properly. If installing brick veneer, nail a 2×4 under exterior door thresholds for additional support until brick is installed. Inspect window and door positioning. They should be plumb and secure. Galvanized finish nails should be used, with the heads of the nails set below the surface of the trim. Install locks and hardware on doors and windows.

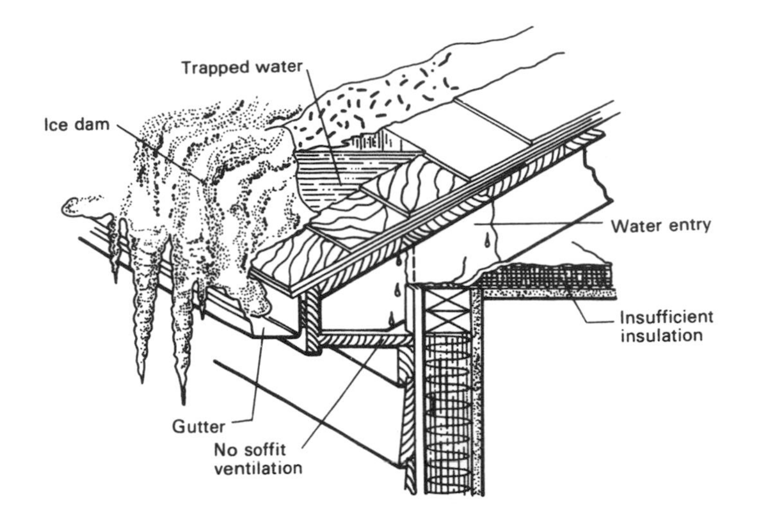

A, Ice dam forming at roof overhang causing melting snow water to back up under shingles and fascia board which damages ceilings inside and paint outside

Figure 79: Eave protection for snow and ice dams. Lay smooth surface 45-lb. roll roofing on roof sheathing over the eaves, extending upward well above the inside line of the wall.

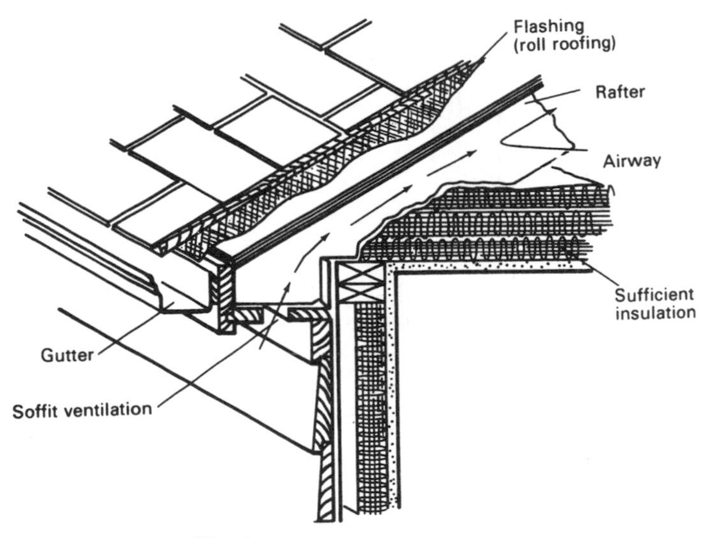

B, Eave protection for snow and ice dams

SC5 INSTALL flashing above windows, exterior door openings and around bottom perimeter of house.

SC6 INSTALL siding.

SC7 INSTALL siding trim around corners, windows and doors.

SC8 CAULK all areas where siding butts against trim or another piece of siding.

SC9 INSPECT siding. Refer to related checklist and specifications.

SC10 PAY sub for siding work. Make sure to retain a portion of payment for callbacks.

SC11 INSTALL cornice. This involves installing the cornice, fascia, frieze, soffit and eave vents. If you have a traditional front, you may want to use a dentil molding. Stucco fronts normally use an exterior grade of crown molding ranging from 3″ to 6″ thick. Frieze bed molding runs in size from 4″ to 12″ and should be made of clear wood. Soffit should be made of an AB grade of ½″ plywood.

SC12 INSPECT cornice work.

SC13 CORRECT any problems noted with cornice work.

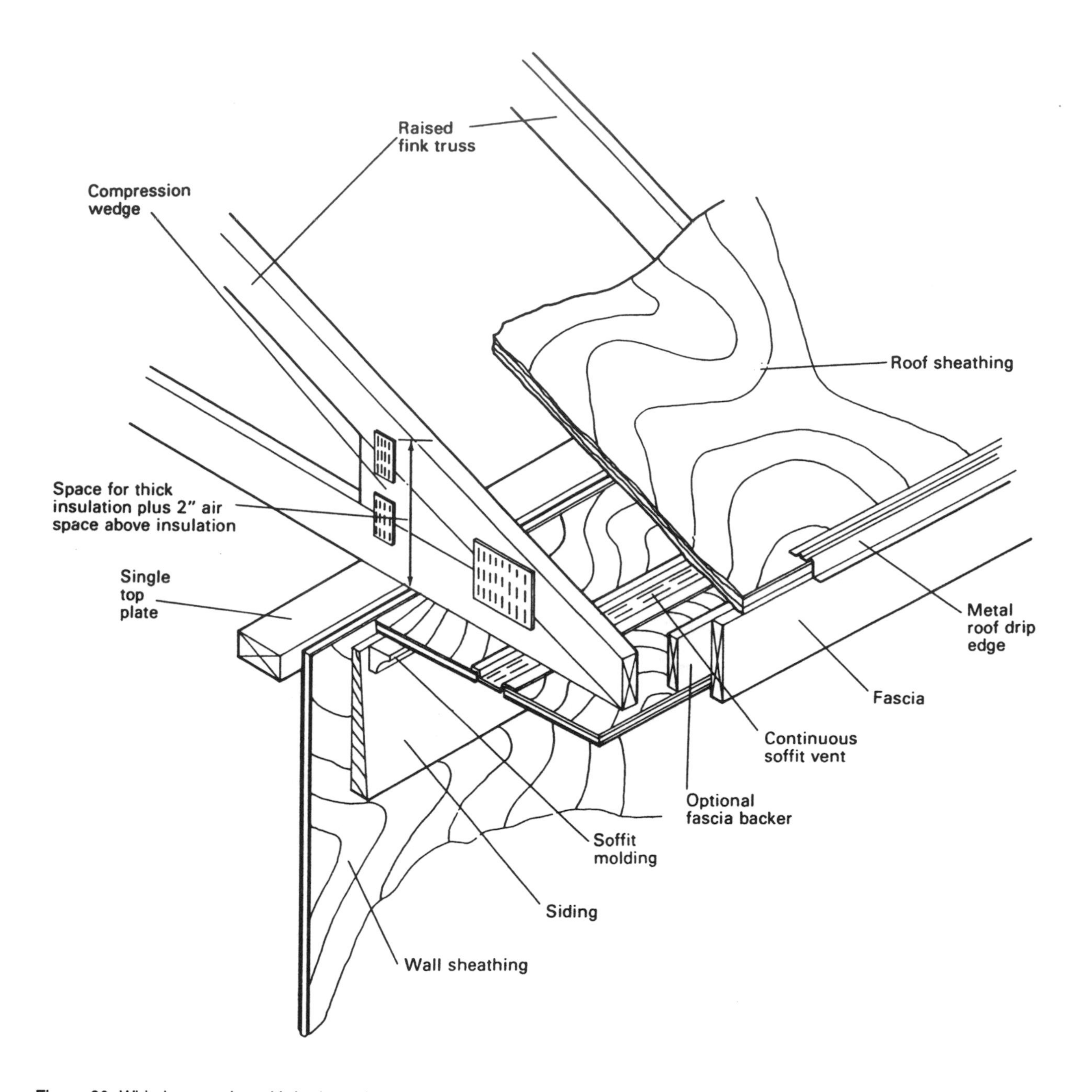

Figure 80: Wide box cornice with horizontal return and raised fink trusses.

SC14 PAY for cornice work.

SC15 CORRECT any problems noted with siding after it has had time to shrink.

SC16 ARRANGE for painter to paint and caulk trim.

SC17 PAY siding retainage.

SC18 PAY cornice retainage.

SIDING AND CORNICE—Specifications

Siding

- Exterior galvanized finish nails are to be used exclusively.
- All nails are to be flush or countersunk.
- All laps of siding are to be parallel.

- Joints are to be staggered between courses.
- All work and materials are to conform to the local building code.
- All siding edges are to be terminated in a finished manner and caulked.
- All openings and trim are caulked and flashed.

Cornice

- All soffit and fascia joints are to be trimmed smooth and fit tight.
- All fascia is to be straight and true.
- Soffit vents are to be installed.
- All soffit, fascia, frieze and trim materials are to be paint grade fir or equivalent.

Insulation and Soundproofing

INSULATION

With the cost of most energy forms on the rise, insulation is gaining importance. The insulating property of a material is measured as an R-value. The thickness of a material has little to do with its R rating. Don't buy thick insulation, buy high R-rated insulation. Insulation normally comes as fiberglass or cellulose. Either can be blown into most void areas, while fiberglass also comes in batts.

Insulation requirements vary across the country. Contact your local power company; representatives will assist you with recommended R ratings for ceilings, floors, walls, attics, etc. Observe that their recommendations may exceed your minimum local code requirements. Don't worry. You can pay for insulation now or pay for energy later. R-value requirements are not necessarily the same for all walls. Exposures that receive heavy sunlight or wind should be more heavily insulated. This is another job that you can do yourself. If you choose to install insulation yourself, ask a supplier for a copy of the manufacturer's recommended installation instructions. Use gloves and place an air filter over your nose.

Sheathing

Sheathing is a lightweight, rigid panel nailed to the outer side of the exterior walls. Several different types of sheathing are now available, each with different properties and costs. Foil-covered sheathing has several advantages, including higher insulation values, a finish more resistant to abrasion during installation and greater airtightness. It's also one of the most expensive types. Sheathing can add considerably to the insulation value of your house, so research your insulation needs carefully before deciding which type to use.

Fiberglass Batts

These are rolled sections of fiberglass bonded to a sheet of heavy duty kraft paper. Fiberglass, as well as most other forms of insulation, loses its effectiveness when wet, moist, compressed or otherwise not installed in accordance with the manufacturer's instructions. Batts should fit snugly between studs and joists, with no gaps. Many subs staple them in place.

Blown-In Materials

Blown-in materials are very popular as attic insulation because of ease of installation and the ability to fill in around odd-shaped areas and framing. It is usually not suitable as wall insulation, however, because of its tendency to settle. Fiberglass, rock wool and cellulose can be blown in (as a loose material). This type of insulation is normally applied with depths ranging from 6″ to 12″. Cellulose is the best bargain, but is susceptible to rodents and moisture. Beware of cellulose that does not conform to local fire treatment standards. Rock wool is a good compromise between quality and cost. Blown insulation makes a good do-it-yourself project. You can usually rent the blowers from the insulation supplier.

Insulation Subs

Insulation subs should give you a price for material and labor. Insulation will normally be charged by the square foot for batts and by the cubic foot for blown-in insulation. Investigate whether you qualify for any energy credit.

SOUNDPROOFING

This is a refined touch. Consider soundproofing bathrooms and the HVAC area to quiet down the system. Rooms where you plan to have stereo systems are also good candidates. Soundproofing material comes in 4×8 sheets and can be applied between the drywall and studs.

INSULATION—Steps

IN1 DETERMINE insulation requirements. Your

local energy company can help with guidelines and suggestions.

IN2 PERFORM standard bidding process.

IN3 INSTALL wall insulation. Hand-pack insulation in small nooks and crannies first, then around chimneys, where framing offsets occur, and where pipes come through walls. Make sure the sub uses plenty of staples when attaching vapor barrier, to avoid gaps.

IN4 INSTALL soundproofing. This is a nice touch to deaden noise. Use in bathroom walls to deaden sound.

IN5 INSTALL floor insulation on crawl space and basement foundations. Metal wires cut to the length of joist spacing are used to hold the insulation in place.

IN6 INSTALL attic insulation. This normally consists of batts or blown-in insulation. If blown-in, attic insulation will be installed after drywall has been nailed into place on the ceiling. If you use two layers of fiberglass batts, place the second layer perpendicular to the first, to cut off any major air leaks. The second layer should have no vapor barrier. Plug up all HVAC vent openings with leftover insulation.

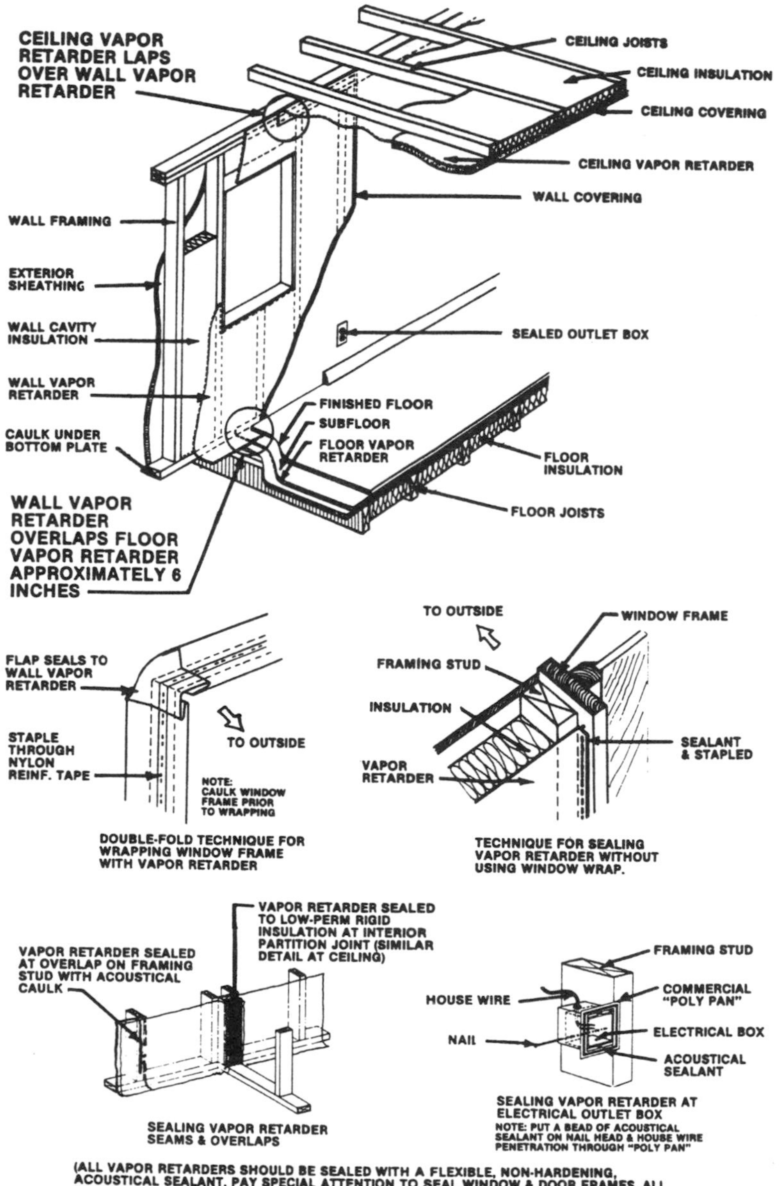

Figure 81: Precautions to be taken in insulating around openings.

This keeps drywall dust out of the HVAC vents during the dusty drywall installation and sanding process.

IN7 INSPECT insulation. A kraft paper or poly vapor barrier should be installed on the warm side of the insulation. All areas around plumbing, electrical fixtures, doors and windows should be stuffed with insulation to prevent air infiltration.

IN8 CORRECT insulation work as needed.

IN9 Pay insulation sub and have him sign an affidavit.

IN10 PAY insulation sub retainage.

INSULATION—Sample Specifications

- Bid is to provide and install insulation as indicated below:

Location	R-Value	Type
Walls	R13	Fiberglass batts
Ceiling	R19	Blown rock wool

- Bid is to include labor and materials for wall and ceiling insulation.
- All insulation is to include a vapor barrier, free of rips and reasonably sealed against air infiltration, on the warm side of the insulation.
- All joints and gaps around door frames, window frames and electrical outlets are to be packed with fiberglass insulation.
- All insulation used will be of thickness sufficient to meet R values specified after insulation has settled.

INSULATION—Inspection

- ☐ All specified insulation has been installed per specifications. Check labels on insulation.
- ☐ All insulation is installed tightly, with no air gaps.
- ☐ No insulation is packed down. That reduces the effective R value.
- ☐ Vapor barrier faces the interior of the home (the side heated in the winter).
- ☐ Any rips or tears have been repaired.
- ☐ No recessed lighting fixtures are covered with insulation, to allow heat to flow.
- ☐ Pull-down attic staircase is weatherstripped and well insulated.
- ☐ No punctures have been made in vapor barrier.
- ☐ Insulation adheres firmly to all adjoining surfaces.
- ☐ No eave vents have been covered by insulation.
- ☐ Ductwork in basement is insulated.
- ☐ Plumbing in basement is insulated.
- ☐ All gaps around doors and windows are stuffed with insulation.
- ☐ Gap between wall and floor is caulked completely.
- ☐ All openings to the outside for plumbing, wiring and gas lines are sealed with spray foam insulation.
- ☐ All gaps in siding around windows and corners are caulked thoroughly.
- ☐ All fireplaces are properly insulated.

Drywall

Drywall is composed of gypsum sandwiched between two layers of heavy-gauge paper. This material, normally used for interior walls, comes in 4′×8′, 4′×10′ and 4′×12′ sizes. When planning your home, it may be helpful to plan around these dimensions to some extent, to minimize waste.

Drywall comes in two common thicknesses: ⅜″ and ⅝″. Five-eighths inch is the thickness most often used because it does not warp as easily, especially where studs are 24″ on center (O.C.). Drywall sheets have the long edges tapered and the short edges full thickness. This is to allow room for the drywall mud and tape. Sheets should be installed so that the full edges butt against studs. After drywall is applied to studs, the joints between sheets of drywall are smoothed using special drywall tape. Moisture affects drywall adversely, so store it in a dry place.

Drywall installation normally involves the following steps:

- Apply drywall to studs with either nails or screws, and glue.
- Apply fill coat of joint compound and drywall tape.
- Sand all joints smooth (if necessary).
- Apply joint compound (second coat).
- Sand all joints smooth (if necessary).
- Apply finish coat of joint compound (third coat).
- Sand all joints smooth.

Application should not be done in extreme heat or cold. Ceilings are treated just as walls, unless you decide to stipple them. Stipple is a coarse-textured compound that eliminates the need for repeated sanding and joint compound application. This is a time- and money-saver.

Drywall Subs

Your drywall sub is a key player in the final, visible quality of most surfaces in the interior. Drywall subcontractors normally bid and charge for drywall work by the square foot, although they install it by the sheet. Drywall subs charge extra for special projects such as:

- tray ceilings
- vaulted ceilings
- open foyers
- curved walls and openings
- high ceilings
- unstippled ceilings
- water-resistant drywall

Of all your subcontractors, quality work here is important. Drywall is one of the most visible parts of your home. Get the best drywall sub you can afford. Below are signs of a good drywall sub:

- Finishing coats are very thin and feathered.
- Sanding the first two coats is unnecessary because they are so smooth.
- Pieces are cut in place to ensure proper fit.

Your drywall sub can quickly tell you the quality of the framing job. Knots, warped lumber and out-of-square walls will make the drywall sub's job more difficult. If the problems are major, have the framer fix them before continuing. Your drywall sub can't fix these underlying problems effectively.

WARNING: Your drywall sub may use stilts to reach ceilings and high walls. Stilts are dangerous, but many times virtually unavoidable. Some states will not award worker's compensation to subs injured while using stilts. If this is true in your state, have your sub sign a waiver of liability to protect you from such accidents. Scaffolding should be used for stairway work. If stilts are used, the floor should be cleared of scrap wood and trash to avoid tripping.

DRYWALL—Steps

DR1 PERFORM drywall bidding process (refer to standard bidding process). Refer to drywall contract specifications for ideas. You may wish to ask painters for names of good drywall subs; they know good finishing work—they paint over it for a living!

DR2 ORDER and receive drywall materials, if not supplied by drywall sub.
NOTE: Before drywall is installed, mark the location of all studs on the floor with a builder's pencil. This will make it easier for the trim sub to locate studs when nailing trim to the wall.

DR3 HANG drywall on all walls. Outside corners must be protected with metal edging. Inside

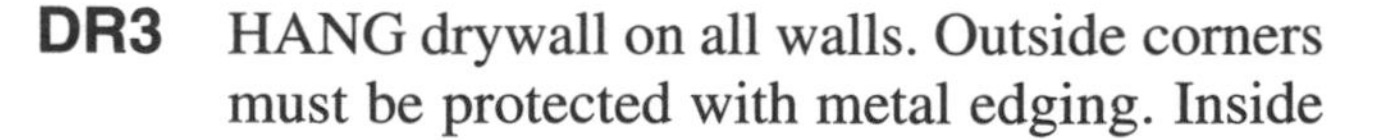

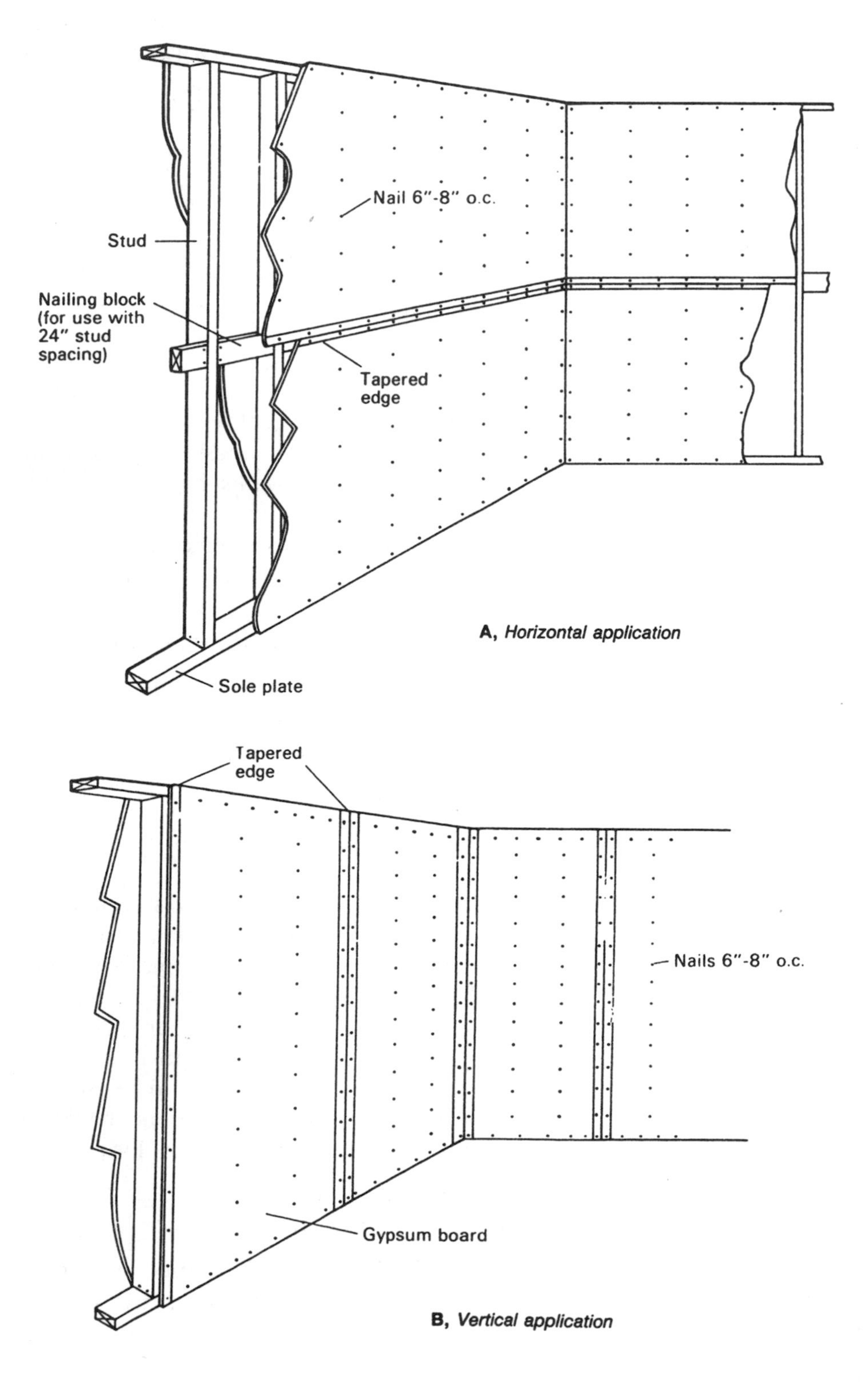

Figure 82: Application of gypsum board finish.

corners must also be taped.

DR4 FINISH drywall. Drywall must be finished in a series of steps, as outlined below:

- Spackle all nail dimples.
- Sand nail dimples smooth.
- Apply tape to smoothed nailed joints.
- Spackle tape joints (fill coat).
- Sand tape joints (as needed).
- Spackle tape joints (second coat).
- Sand tape joints (as needed).
- Spackle tape joints (finish coat).
- Sand tape joints (third time).

DR5 INSPECT drywall. Refer to specifications and inspection guidelines. To do this properly, turn out all lights and look at the wall while shining a light on it from an angle. Slight shadows will appear if there are imperfections on the surface. Mark them with a pencil so the drywall sub will know where to fix walls. Don't use a pen, as it will leave a mark that will show through paint.

DR6 TOUCH up and repair imperfect drywall areas. There are always a few. Place large drywall scraps in basement if you want to finish the basement later.

DR7 PAY drywall sub. Get him to sign an affidavit.

DR8 PAY drywall sub retainage. You may want to wait until the first coat of paint has been applied so that you can get a good look at the finished product.

DRYWALL—Sample Specifications

- Bid is to provide all material, labor and equipment to perform complete job per specifications of attached plans. This includes:
    - Drywall, tape and drywall compound
    - Metal corner bead
    - All nails
    - Sandpaper
    - Ladders and scaffolding
- Apply ⅝″ gypsum board (drywall), double-nailed at top; four nails per stud.
- All joints are to be taped with three separate coats of joint compound, each sanded smooth.
- All outside corners are to be reinforced with metal corner bead.
- All inside corners are to be reinforced with joint tape.
- All necessary electrical outlet, switch and fixture cutouts are to be made.
- All necessary HVAC ductwork cutouts are to be made.
- All ceilings are to be stippled as specified.
- All ceiling sheets are to be glued and screwed in place.
- Wall adhesive is to be applied to all studs prior to applying drywall.
- Moisture-resistant gypsum is to be used along all wall areas around shower stalls and bathtubs.
- Use drywall stilts at your own risk.

DRYWALL—Inspection

Before Taping

- ☐ No more than a ⅜″ gap exists between sheets.
- ☐ Nails are driven in pairs (2″ apart). Nail heads are dimpled below the surface of the drywall.
- ☐ Nails are not hit hard enough to break the surface paper. All joints shall be double-nailed or glued.
- ☐ No nail heads are exposed to interfere with drywall.
- ☐ No sheet warpage, bowing or damage exists. Sheets are easier to replace before taping.
- ☐ Rough cuts around door and window openings are cut close, so that trim will fit properly.
- ☐ Waterproof drywall, or Wonderboard, is installed in shower stalls and around bathtubs. No taping is necessary here.
- ☐ Metal bead is installed flush on all outside corners.

During Finishing

- ☐ Three separate coats of mud are applied to all joints. Stippling will hide any imperfections. Each successive coat should leave a wider track and a smoother finish.

After Finishing

- ☐ Look down the length of installed drywall; there should be no warpage or bumps. If any are found, circle the area with a pencil (ink may show through when painted). Have the drywall subcontractor fix all marked areas before final payment.
- ☐ All joints are feathered smooth, with no noticeable bumps, either by sight or touch.
- ☐ All electrical wiring remains exposed, including bath and kitchen vent fans, garage door opener switches and doorbell in garage.
- ☐ Proper ceilings are smooth and stippled, as specified.
- ☐ Cuts are clean around register openings, switches and outlets, so that covers will cover exposed areas.
- ☐ Nap of paper is not raised or roughened by excessive or improper sanding.
- ☐ All touch-up work is completed and satisfactory.
- ☐ No corner bead is exposed.

NOTE: Some drywall imperfections will not appear until the first coat of paint has been applied. This is the first time you will see the wall as a single, uniform color. If you can, wait until this point to pay retainage.

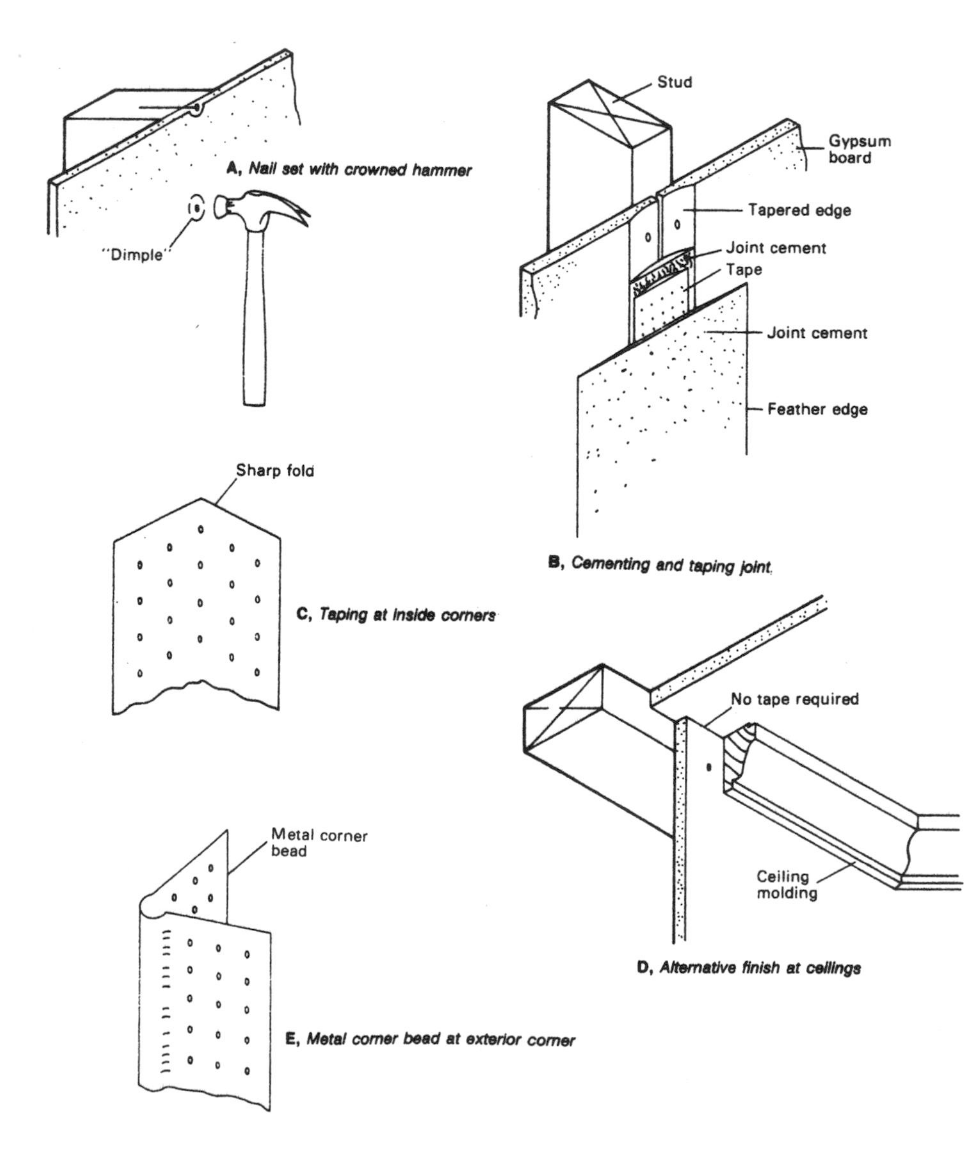

Figure 83: Finishing gypsum drywall.

Trim

Good trim requires some of the most precise work to be found in a home. If you are not very experienced in this type of work, you may not want to tackle this job. Interior trim subcontractors usually will do the following:

- Set interior/exterior doors and door sills
- Set windows and window sills
- Install base, crown, chair rail, picture and other molding
- Install paneling, raised paneling and wainscoting
- Install stairway trim
- Install fireplace and main entry door mantels
- Install closet shelves and hanger rods
- Install any other special trim
- Install door and window hardware, door stops, etc.

Trim subs charge for work in many ways:

- By the opening (for doors, windows and entryways)
- By the cut—the more cuts involved, the greater the cost
- By the hour or day
- By the job

It is advisable to get the bid for the job. Have subs give you a bottom-line figure—let them do the figuring. Go with the bottom-line price and quality you are most comfortable with.

Exterior trim (fascia and cornice work) is covered under a separate section, since it is usually done by the siding and cornice sub.

Decorative trim is one of the most visible items in your house, and can make the difference between an ordinary interior and one that stands out. The workmanship of your trim sub will be very visible and hard to repair if not done properly. Make sure your trim sub is very quality-conscious, and watch his work closely. If you want to keep a close eye on his work, volunteer to assist him in doing the trimwork.

The critical part of door installation is making sure that the door is set level and square, so that doorknob hardware will work properly and the door will operate without binding. Special care should also be taken with bifold and sliding-track doors. Improper installation is hard to fix and results in binding.

Weather stripping is essential in all exterior doors. This has become a specialty trade in itself. As such, your trim sub may not necessarily be the one who does this small but critical job.

Trimwork has become somewhat easier with the advent of preset doors and window trim kits. To save money, consider using prehung hollow-core interior doors. These doors install easily and look as good as or better than custom-hung doors.

In choosing your trim material, you must first decide whether you plan to paint or stain the trim. You may be able to use lesser grades of millwork or Masonite doors if they will be painted. Finger-joint trim, consisting of scraps glued together, is fine if you plan to paint. Masonite panel doors are becoming very popular and are inexpensive. If you plan to stain, you must use solid wood trim and birch-paneled wood doors, but the fine appearance and low maintenance of stained trim may be worth the expense.

Protect trim from abuse prior to installation, to avoid dents, cracks, scratches and excessive waste. Lay trim on the floor in neat stacks. Don't purchase trim from different suppliers—the different pieces may not match.

When installing base molding, consider whether trim will go around wall registers or merely be interrupted by the register. The first example is more expensive but yields a nicer appearance. If you use a tall base molding, you can build your registers and receptacles into the base molding.

Modern houses are making extensive use of decorative trim such as paneling and wainscoting. Ask your trim sub for suggestions on decorative approaches. It is amazing what can be done with a minimum of materials.

TRIM—Steps

TR1 DETERMINE trim requirements. Millwork samples can help. Mark all walls that get different trim treatments.

TR2 SELECT molding, window frames and door frames.

TR3 INSTALL windows. Windows and exterior doors should be installed as soon after framing as possible. Don't have windows and doors delivered until just before installation. Remind the trim sub to use the best trim pieces around the openings in the most visible locations.

TR4 CONDUCT standard bidding process.

TR5 INSTALL interior doors. If prehung doors, install and adjust door stops and doors.

TR6 INSTALL window casings and aprons.

TR7 INSTALL trim around cased openings.

TR8 INSTALL staircase molding—treads, risers, railings, newels, baluster, goosenecks, etc.

TR9 INSTALL crown molding. This includes special-made inside and outside corners, if specified. This is installed first, so that the ladder legs will not scratch the base molding.

TR10 INSTALL base and base cap molding. Where the final floor level will be higher than the subfloor (when hardwood is installed), the base should be installed a little higher.

TR11 INSTALL chair rail molding.

TR12 INSTALL picture molding.

TR13 INSTALL, sand and stain paneling.

TR14 CLEAN all sliding-door tracks.

TR15 INSTALL thresholds and weather stripping on exterior doors and windows.

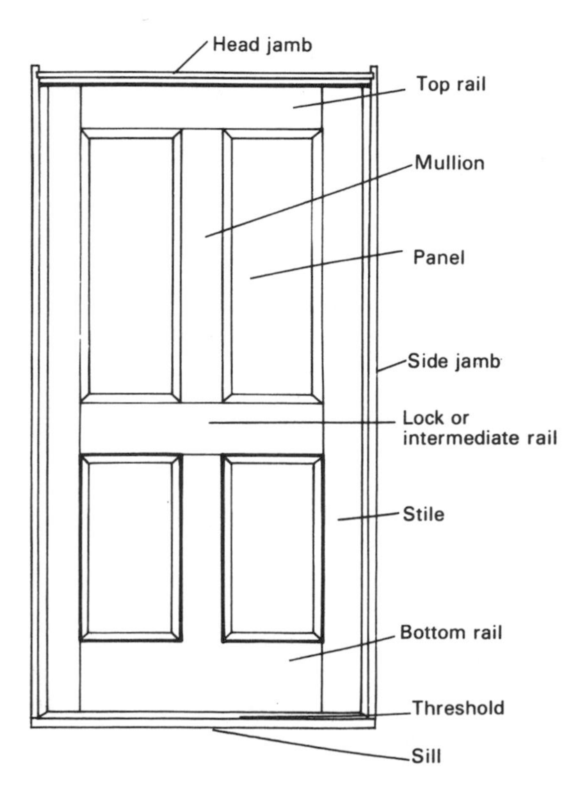

Figure 84: Panel door components.

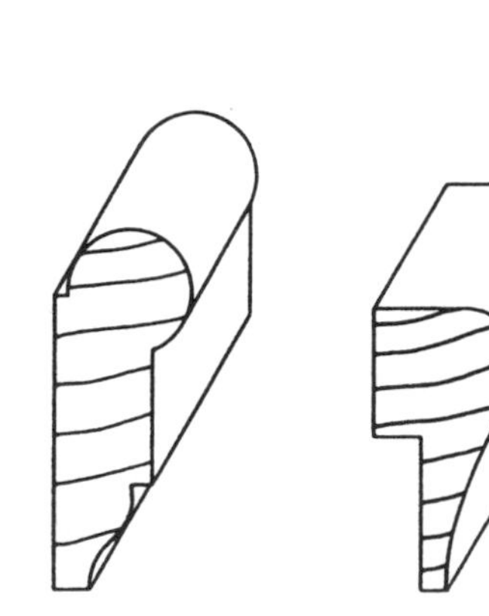

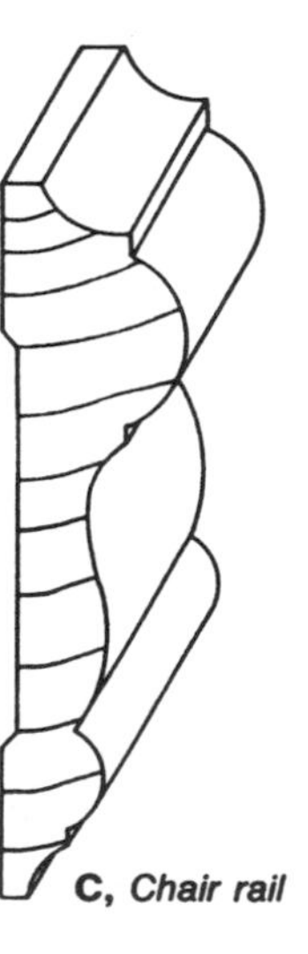

Figure 85: Wall moldings.

TR16 INSTALL shoe molding after flooring is installed. Painting the shoe molding before installation will reduce time and touch-up work.

TR17 INSTALL doorknobs, dead bolts, door stops and window hardware. Rekey locks, if necessary. Consider keying all the locks to just one key. Install a dead bolt-type lock on sliding glass doors for improved security.

TR18 INSPECT trimwork. Refer to your specifications and the checklist that follows.

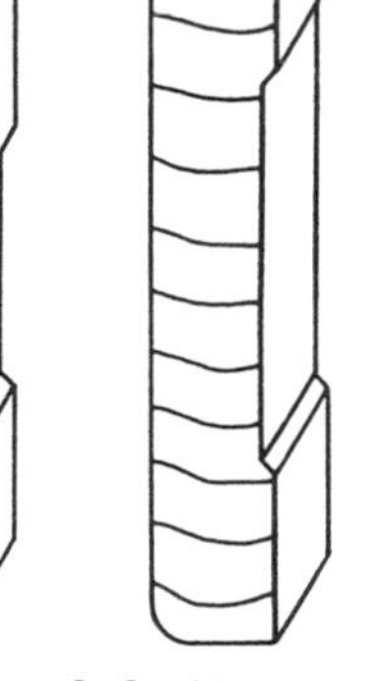

Figure 86: Casings.

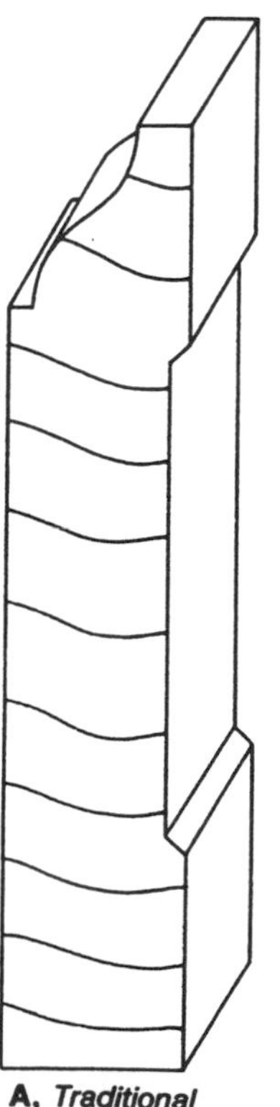

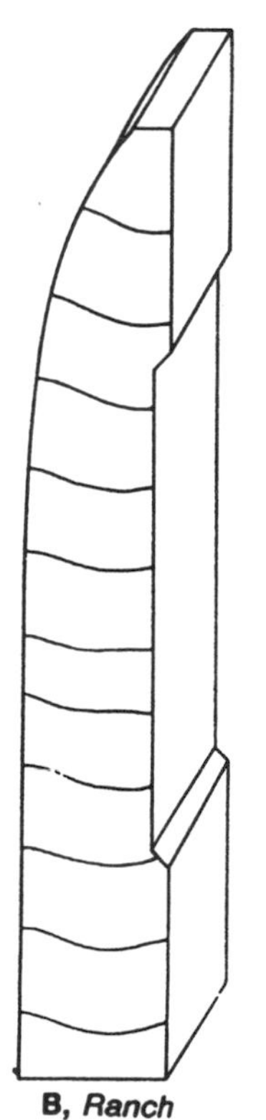

Figure 87: Base moldings.

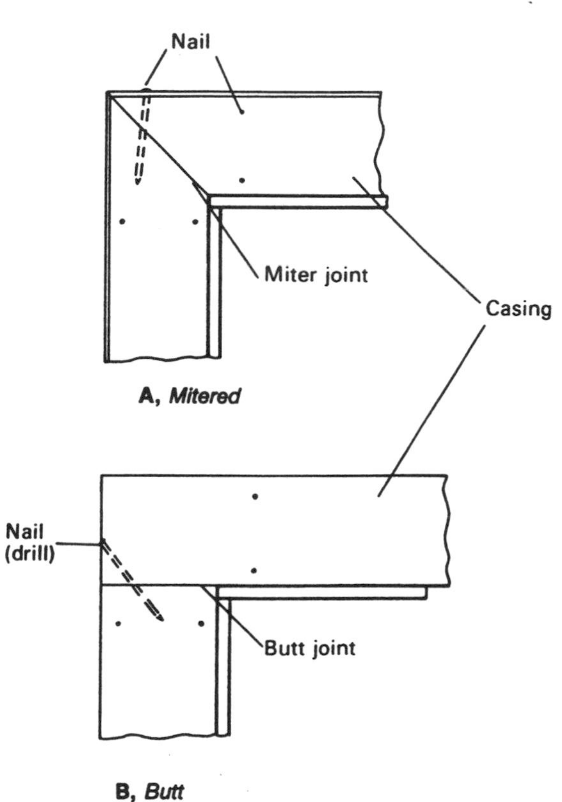

Figure 88: Casing joints.

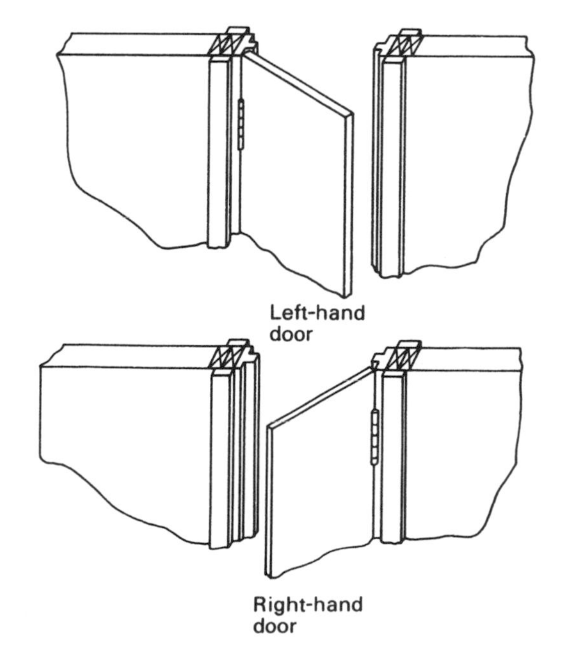

Figure 89: Direction of swing of right- and left-hand doors.

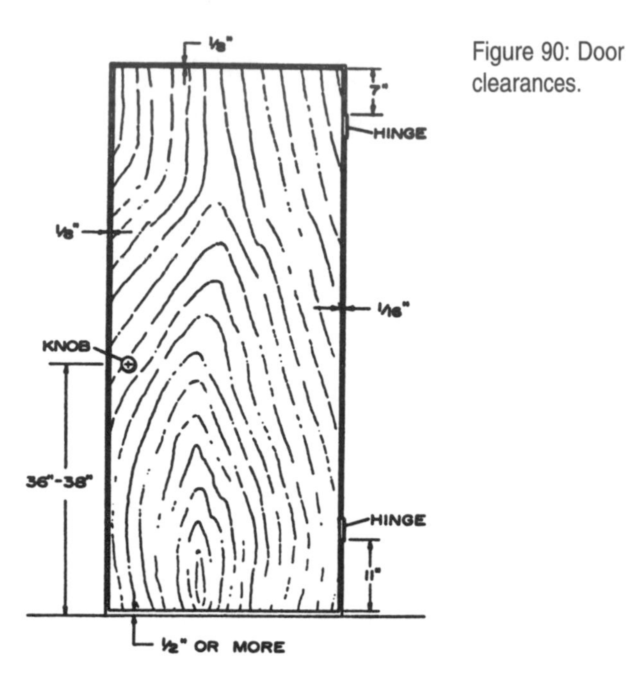

Figure 90: Door clearances.

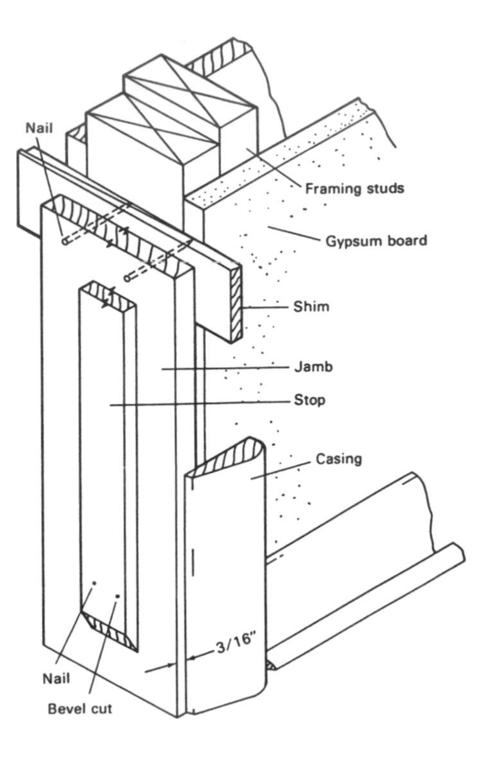

Figure 91: Installation of door frame and trim in rough opening.

TR19 CORRECT any imperfect trim and staining.

TR20 PAY trim sub.

TRIM—Sample Specifications

- All materials except tools are to be furnished by builder.
- Interior doors will be prehung.
- Finish nails are to be used exclusively and set below the surface, puttied and sanded over smooth.

Bid Is to Include:

- Hang all interior doors. All doors are to be right- and left-handed, as indicated on attached drawings.
- Install all door and window trim according to attached schedule.
- Install base molding and base cap molding as specified. Raise base molding in areas of hardwood flooring.
- Install apron and crown molding in living room, dining room and foyer area.
- Install two-piece chair rail molding in dining room as specified.
- Install 1″ shoe molding in all rooms not carpeted.
- Install den bookshelves as indicated on attached drawings.
- Install staircase trim as indicated below:
  - 12 oak treads
  - 12 pine risers
  - solid oak railing with involute
  - gooseneck
  - 36 spindles
- Build fireplace mantel as indicated on attached drawings.
- Install door locks, doorknobs, window hardware and dead bolts.
- All trim will be paint grade unless otherwise stated.
- Install all closet trim, shelves and closet rods.

TRIM—Inspection

Doors

☐ Proper doors are installed—correct style, size, type, etc.

☐ Doors open and close smoothly and quietly.

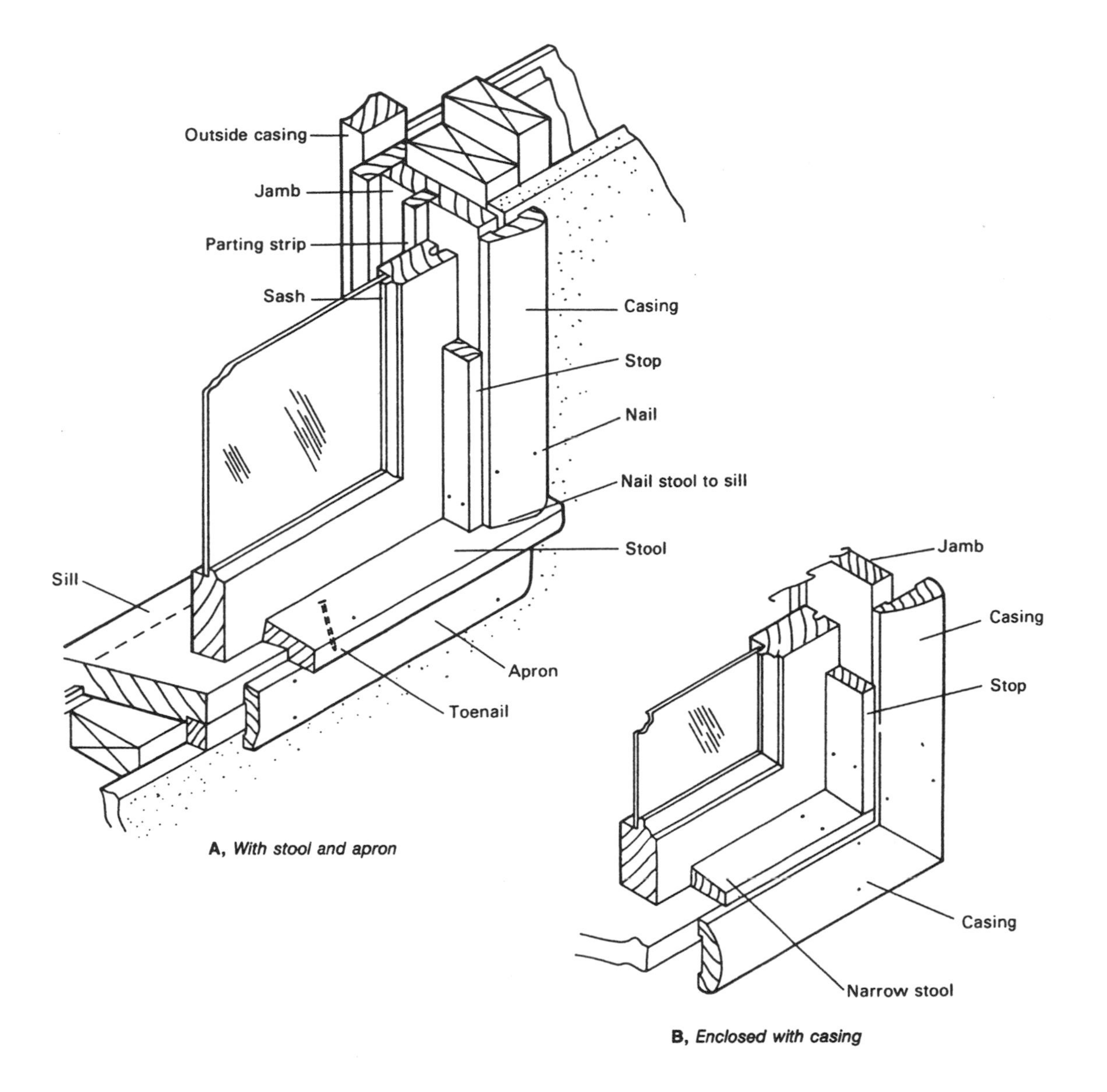

Figure 92: Installation of window trim.

Hinges do not bind or squeak. Doors should swing freely with no noise or friction against adjoining surfaces. Open doors at 30°, 45° and 60° angles. Doors should remain where positioned. If not, remove center pin, bend it slightly with a hammer and replace it.

☐ Doorknobs and latches align with latch insets. All dead bolt locks should align properly. Privacy locks are installed on proper sides of doors. Passage locks are on proper doors.

☐ All exterior doors lock and unlock properly. Locks should function freely.

☐ All keys are available for every door. Have all the locks keyed the same for ease of use.

☐ All doors open in the proper directions. Door latches face the proper directions.

☐ All doors are plumb against door jambs. With door slightly open, check for alignment and evenness of opening, two screws in each strike plate and a pin in each door hinge.

☐ All door casing nails are set below surface and filled with putty.

☐ Doorknobs, door locks and dead bolts work properly without sticking.

☐ No hammer dents are in door or door casing.

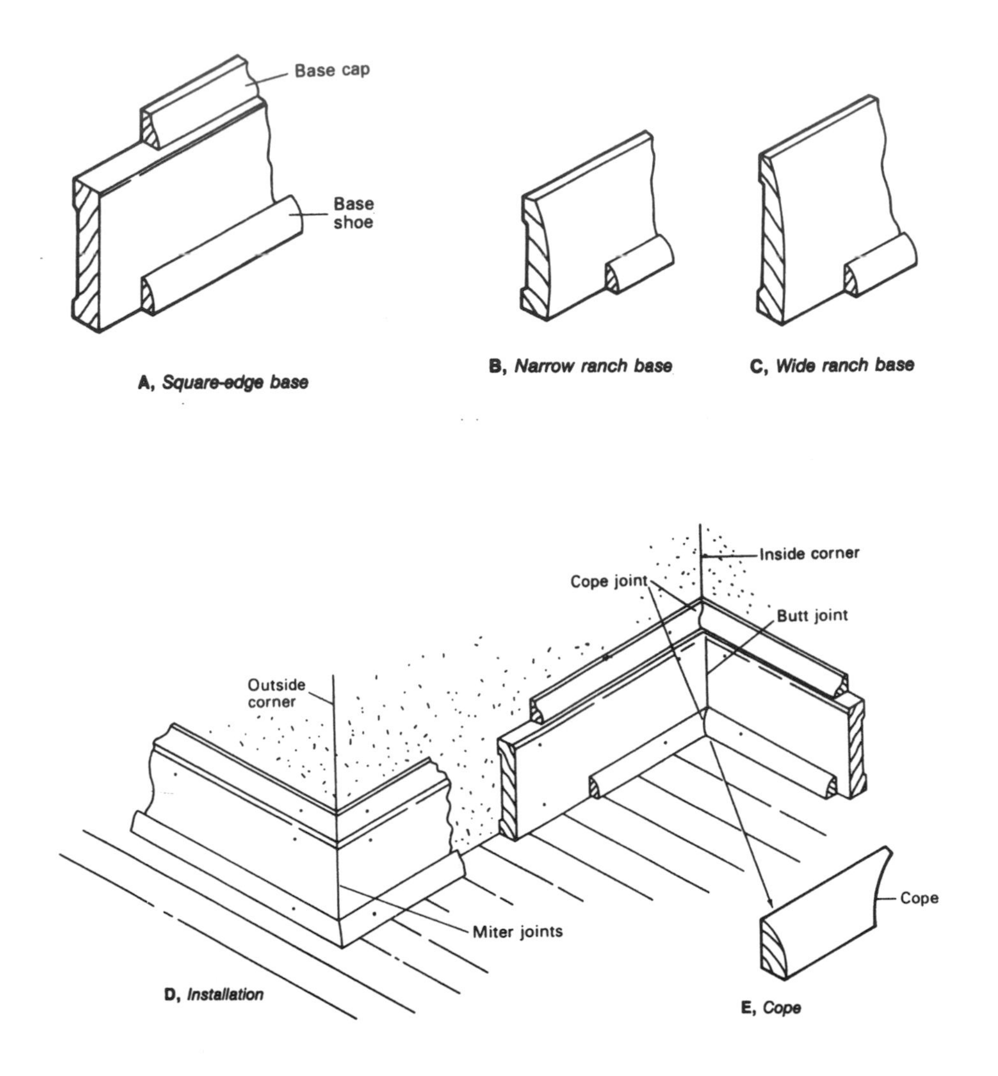

Figure 93: Base molding at floor.

- ☐ Thresholds are in place and properly adjusted.
- ☐ Weather stripping is in place, if specified.
- ☐ Proper clearance exists from floor (about ½″ above carpet level). Consider height of carpet and pad, tile, wood floor, etc.
- ☐ Door stops are in proper places.

Windows

- ☐ Window frames are secured in place with leveling wedges or chips.
- ☐ Windows are installed with less than ½″ gap between wall and frame.
- ☐ Windows open and close smoothly and easily. Windows should glide easily along tracks. Windows should close evenly and completely.
- ☐ Window sash locks are installed and operating properly.
- ☐ No hammer dents are in window or window casing.
- ☐ All window casing nails are set below surface and sealed with putty.
- ☐ Weather stripping is in place, if specified.
- ☐ Screens are installed, if specified.
- ☐ All window latches function properly.
- ☐ All window pulls are in place and secure.
- ☐ No excessive damage has been done to millwork.

Trim and Paneling

- ☐ All trim is void of major material defects.

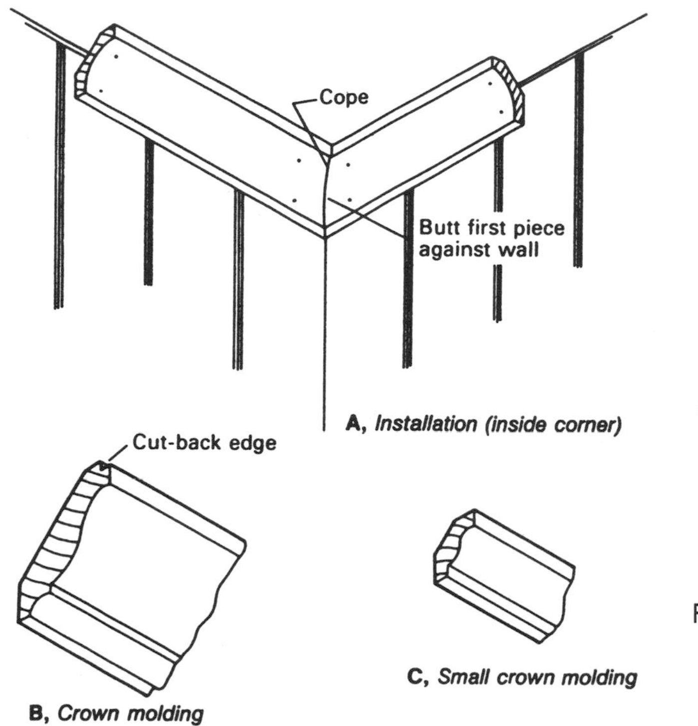

Figure 94: Ceiling moldings.

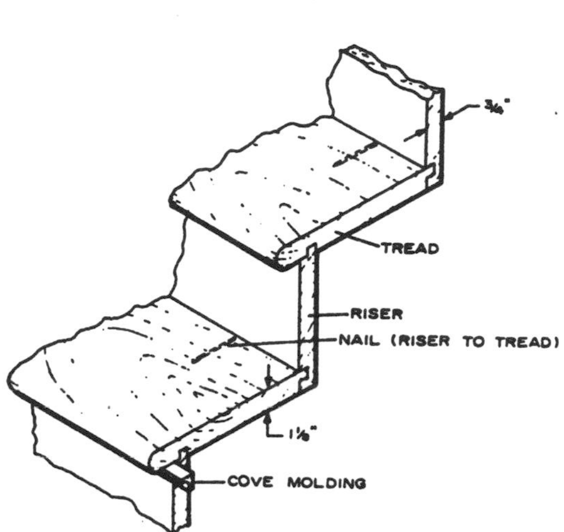

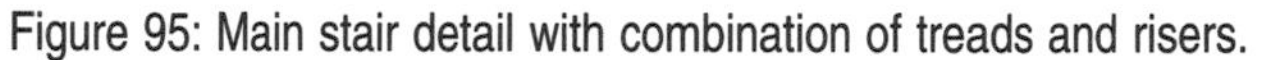

Figure 95: Main stair detail with combination of treads and risers.

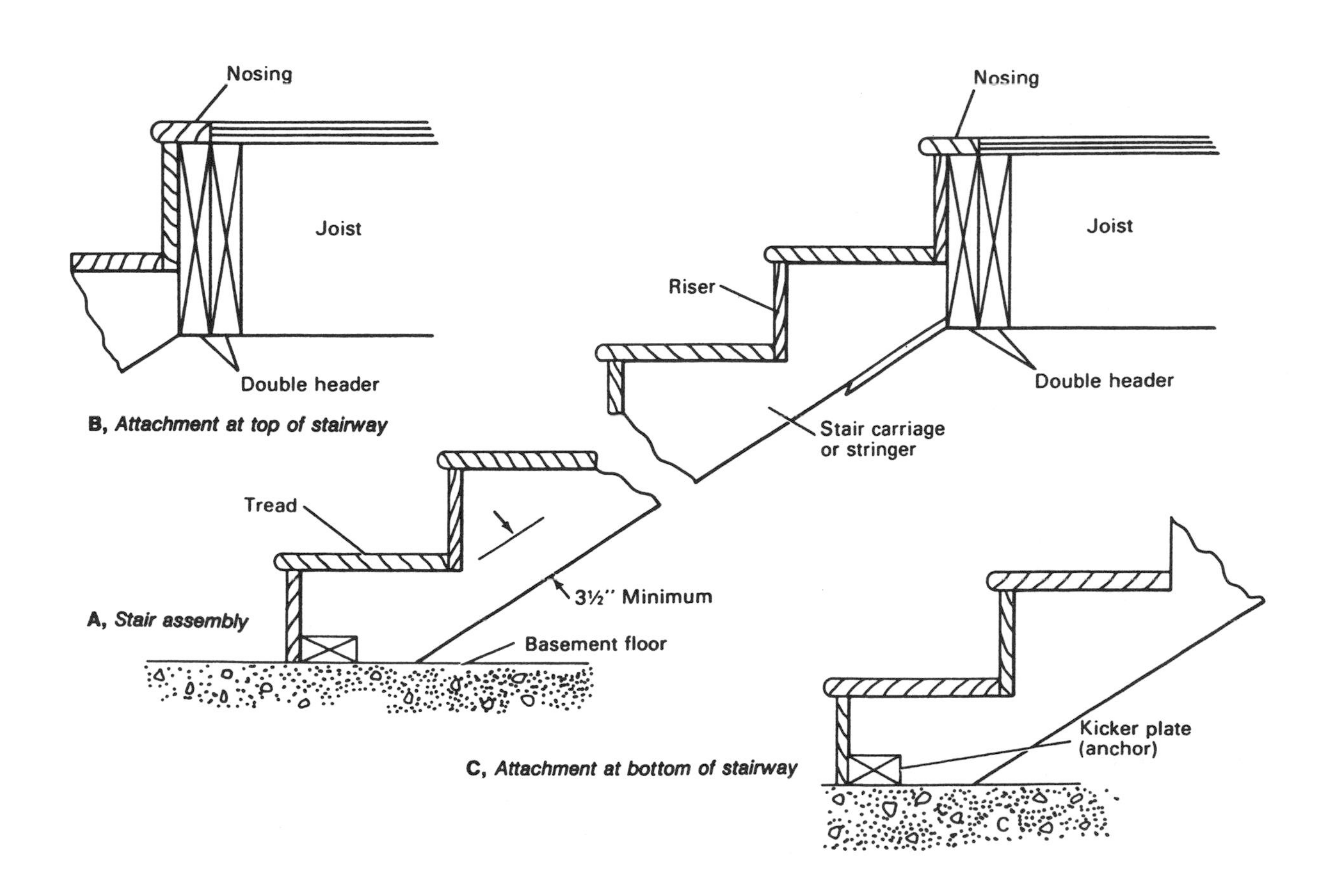

Figure 96: Stair details.

☐ All crown molding is installed and finished as specified.

☐ All base molding is installed and finished as specified.

☐ All chair rail molding is installed and finished as specified.

☐ All wainscoting is installed and finished as specified.

☐ Trim joints are caulked, sanded smooth and undetectable, both by sight and feel.

☐ Trim intersects with walls, ceilings and floors evenly, with no gaps or other irregularities.

☐ All finishing nails are set below surface and sealed with wood putty.

☐ All den paneling and shelving is installed and finished as specified.

☐ Fireplace mantel is installed and finished as specified and will support heavy load.

☐ Fireplace surround is installed and finished as specified.

☐ All closet shelving and coat racks are installed at proper height and level.

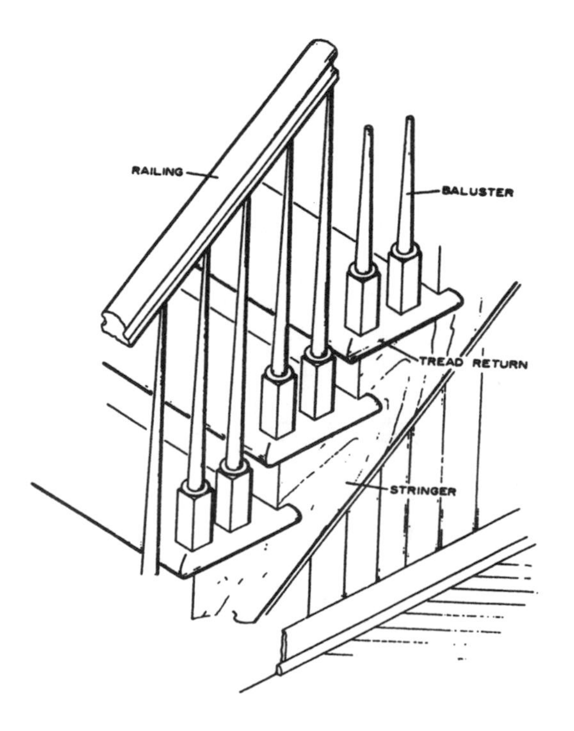

Figure 97: Details of open main stairway.

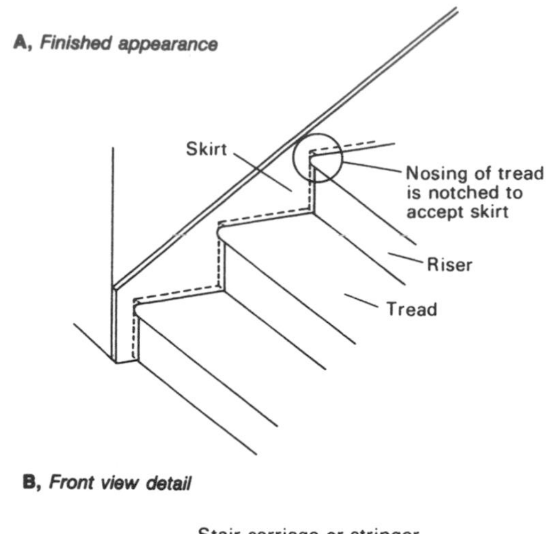

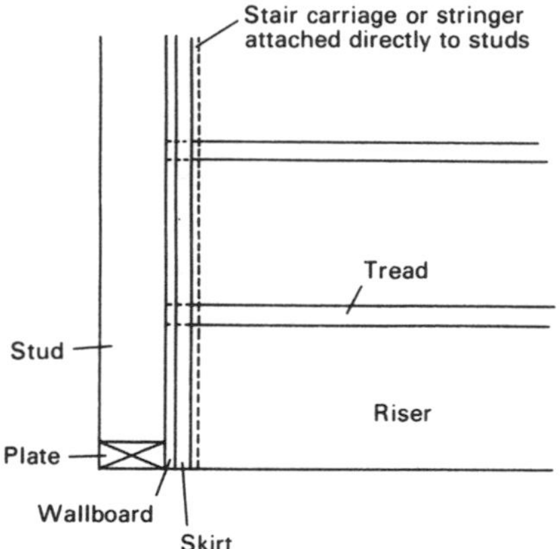

Figure 98: Stairway skirt board.

Staircase

☐ Proper treads are used.

☐ Finished and curved treads are installed correctly.

☐ All railings are in place, secure, and angled parallel to stairs.

☐ Spindles and newel posts are securely in place and properly finished.

☐ Balusters are securely in place and properly finished.

☐ No squeaks have developed in staircase steps or elsewhere on staircase.

Painting and Wallcovering

PAINT

If there are any jobs you can feel safe about doing yourself, they are painting and wallpapering. These may be worth several thousand dollars to you. Although painting can be easy, lots of preparation is essential for a professional look. Even the smallest defect will show up on a painted wall or piece of millwork. The goal of preparation is to make every surface smooth—getting rid of all cracks, bumps, nicks, scratches and rough areas.

Paints can be purchased in standard colors or custom mixed. Custom-mixed paints can't be returned, so be careful with color schemes and order quantities. Try to select a minimum of different colors, to avoid having to store a lot of different colors. If you decide to do the painting yourself, purchase the highest quality paints you can find—but make sure to insist on the builder price. One-coat paints are a good choice. New drywall should always receive two coats, but the one-coat paints guarantee the best coverage. Colors on small samples in a store will appear darker when they are applied to a room. If you plan to stipple ceilings, this should be done before any painting has begun. Although roller application is quick and yields good results, consider renting an electric paint sprayer. You may need to thin the paint if you go this route, and you will definitely need some form of eye and nose protection. Keep paint off areas to be stained, or it will show up in the final work. When enamel thickens, thin it with paint thinner.

HIRING A PROFESSIONAL

Painting your home can be a very rewarding experience, but it is time-consuming and tedious. Before you decide to tackle the painting, ask yourself the following questions:

(1) Do you have the time and patience for the job? If you are working full time, your time is worth money. Is your time worth more than the painting contractor?

(2) Do you have the skills to do the high-quality job you demand? Just because you know a good job when you see one doesn't mean you have the talent to do the job yourself.

(3) Do you have the physical stamina? Painting itself is not hard, but setting up scaffolding, climbing ladders and scraping paint can be backbreaking work.

If you hire professional painters, make sure they carry worker's compensation and liability insurance. Since painting can be more of an amateur trade than most, anyone can call himself a painter. These subs are less likely to carry the professional liability coverages that most other tradesmen carry. Ask for references from other jobs they have done and specify the masking and drop cloth requirements to protect your property.

TYPES OF PAINT

Paint formulations and technology have changed tremendously in the past few years. Before choosing the correct paint for your project, you need to know the nature of the material underneath the paint, the composition of the existing paint finish and the properties of the paint you plan to use. The many types of paint fall into three main categories:

(1) Oil-based pigments use an oil—such as linseed oil—as the carrier. This class of paint adheres well to most surfaces and provides a tough, durable, water-repellent surface.

(2) Latex paints use water as the carrier for the pigment. These paints produce little or no fumes, dry fast, are easy to use and produce a breathable finish that allows water vapor to escape. Latex

is not quite as durable as oil-based paints, but recent improvements have narrowed the gap considerably.

(3) Varnishes and other solvent-based finishes use solvents such as mineral spirits, alcohol and acetone as the carrier. These finishes provide myriad uses as penetrating finishes, floor finishes, epoxies, elastomer finishes, stains and other finishes.

All paints are made up of similar compounds that determine the quality of the paint. The carrier in paint is the liquid that suspends the pigments and additives and allows these materials to penetrate the painted surface. As the carrier evaporates, the pigments harden into a solid film that protects the surface against moisture and wear. Additional solvents also serve as carriers and are added to thin the paint to the proper consistency.

The pigments in the paint supply the "holdout," or covering ability of the paint. Pigments provide both opacity and tint to the paint. The most common pigments found in almost all paints today are the oxides: zinc oxide and titanium oxide. Lead oxide used to be quite popular twenty years ago, until it was discovered to be a health hazard, especially to children. Removal of old paint that may contain lead oxide is a serious consideration when stripping old finishes. You need to take precautions to avoid inhaling the paint dust or fumes. The amount and quality of pigments in your paint will determine the total quality of the paint grade. Certain types of clay are used as "budget" pigments in less expensive paints to reduce the need for the oxide pigments.

Additives make up a diverse selection of compounds that serve many purposes in special paint formulations. Some examples are: rust inhibitors, drying agents, fungicides, bonding agents, antifreeze, emulsifiers and thickening agents. The specific use of the paint will determine which ingredients are present and their quantities.

Most paint is sold in at least three quality grades: premium, budget and professional. For maximum quality, durability and coverage, always choose the premium grades of paint. These may cost a little more, but they last much longer. Surprisingly, the "professional" grades of paint are actually the cheapest grades. These paints are sold in large quantities to contractors at low prices. Many contractors are mostly concerned with the price and coverage of the paint, not the durability. These grades usually contain more clays and chalk pigment, which cover very well but don't last as long as the oxide pigments. Don't assumc that your professional contractor is using the best quality paint. Specify in your contract the grade of paint that you want for your project.

Oil-Based Paint

Until recently, the most common paints were oil-based. These paints are still popular, but have slowly been edged out in popularity by latex paints. Oil paints are still the most durable paints and are most appropriate for exterior finishes and trim. Oil paint bonds strongly to wood and metals, and also bonds better on substandard finishes, such as chalky, dirty or oily surfaces. Because oil repels water, however, it does *not* adhere to damp surfaces at all. Oil paints *must* be applied to a dry surface. Cleanup also requires the use of smelly solvents, such as mineral spirits. This makes oil paints less attractive for inside projects.

Oil-Based Stains

Oil-based stains are similar to the paints but contain much less pigment. Since they do not cover well, they are designed to be used on new, unfinished wood as a protectant and waterproofer, and allow the natural beauty of the wood to show through. They come in transparent, semitransparent and opaque stains. Penetrating stains are effective and economical finishes for all kinds of lumber and plywood surfaces. They are especially well suited for rough-sawn, weathered and textured wood and plywood. Knotty wood boards and other lower quality grades of wood that would be difficult to paint can be finished successfully with penetrating stains.

These stains penetrate into the wood without forming a continuous film on the surface. Because there is no film or coating, there can be no failure by cracking, peeling and blistering. Stain finishes are easily prepared for refinishing and are easily maintained.

Penetrating semitransparent stains form a flat and semitransparent finish, which allows only part of the wood grain pattern to show through. A variety of colors is available, including shades of brown, green, red and gray. The only color which is *not* available is white—it can be provided only through the use of white paint. The opaque stains contain enough pigment to behave much like paint and can be used over previous stain colors.

Stains are quite inexpensive and easy to apply. To avoid the formation of lap marks, the entire length of a course of siding should be finished without stopping. Only one coat is recommended on smoothly planed surfaces—it will last two to three years. After refinishing, however, the second coat will last six to seven years, because the weathered surface has absorbed more of the stain than the smoothly planed surface did.

Water-Repellent Finishes

These finishes contain little or no pigment and are designed only to penetrate and seal the wood grain. These are used mainly on patios and decks where foot traffic would damage a surface finish. A simple treatment of an exterior wood surface with a water-repellent finish greatly slows the natural weathering process. Staining seals the wood and promotes uniform natural tan color in the early stages of weathering and a reduction of uneven graying, which is produced by the growth of mildew on the surface.

Water-repellent finishes generally contain a preservative, a small amount of resin and a very small amount of a water-repellent that is frequently wax-like in nature. The water-repellency greatly reduces warping, excessive shrinking and swelling, which can lead to splitting. It also retards the leaching of chemicals from the wood and the staining from water at the ends of boards.

This type of finish is quite inexpensive, easily applied and very easily refinished. Water-repellent finishes can be applied by brushing, dipping and spraying. Rough surfaces will absorb more solution than smoothly planed surfaces—the treatment will also be more durable. It is important to thoroughly treat all lap and butt joints and the ends of all boards.

Two-coat application is possible on rough-sawn or weathered surfaces, but both coats should be applied within a few hours of each other. When using a two-coat system, the first coat should never be allowed to dry before the second is applied. If it does, the surface will be sealed, preventing the second coat from penetrating.

Initial applications may be short-lived (one year), especially in humid climates and on woods that are susceptible to mildew, such as sapwood and certain hardwoods. Under more favorable conditions, such as on rough cedar surfaces that will absorb large quantities of the solution, the finish will last more than two years.

Latex

Latex has become the paint of choice in recent years. It is durable, safe and easy to clean up with water. It dries quickly and produces no fumes, making it ideal for interior application in living areas. Latex can also be used for exterior painting with some precautions.

Latex paints are called "breather" paints and are more porous than conventional oil paints. If these are used on new wood without a good oil primer, or if any paint is applied too thinly on new wood (a skimpy two-coat paint job, for example), rain or even heavy dew can penetrate the coating and reach the wood. When the water dries from the wood, the wood chemicals leach to the surface of the paint. This happens most often with red cedar and redwood. Consider using an oil primer coat first and then finishing with latex.

Latex also works well on stucco and masonry finishes. It is more resistant to the alkali by-products that can leach to the surface of masonry. Latex is thick and easily applied to stucco with a paint roller.

Varnishes

Clear finishes based on varnish, which forms a transparent coating or film on the surface, are used mainly for coating interior finishes such as floors and trim. The polyurethane-type varnishes are very durable and resistant to water, alcohol and oil.

Regular varnishes should not be used on exterior wood exposed fully to the sun. These finishes deteriorate and often begin to disintegrate within one year. A special exterior varnish called "spar varnish" is appropriate for outside use. It contains "UV blockers" that filter out the ultraviolet rays from the sun that damage regular varnish (and give you a sunburn!).

Special-Purpose Paints

Several other types of paints are available for special painting needs. New water-based varnishes are available for interior finishing. These unique formulations combine the hardness and durability of polyurethane with the easy-to-clean properties of latex. They also create no fumes, which can make an interior paint job much more pleasant.

Porches and decks receive too much foot traffic to use standard exterior paints. Porch and deck paint dries to a much harder finish than standard paint. It comes in a limited number of colors and should be applied in two coats with no primer.

Steel or wrought iron fixtures are very prone to rusting. If painted with standard paint, any moisture that seeps through can cause a rust spot to develop. This rust spot will continue to grow and will eventually push the paint off the surface. Rust inhibitor paints and primers are available that contain the rust inhibiting chemicals. They bond tightly to the metal and create a chemical bond that prevents rust from starting. Since they come in a limited number of colors, they are better used as primer coats instead of final finishes.

Concrete sealants are highly elastic paints designed to maintain a waterproof seal even if cracking of masonry develops. They are so elastic that the surface will stretch and span any small cracks. They usually come in white only, and are not very durable since the finish is soft. They will not hold back a large amount of moisture and can blister badly in extremely wet conditions. You can paint over them with latex, however, if you want a broader color selection.

Primers

Primers are designed to seal the surface so that subsequent layers of paint will be absorbed evenly. Primers are specially formulated to provide an optimum surface for final coats, and come in most of the same formulations as finish coats. The difference is in the bonding and sealing agents. Primers are usually thinner so they can soak in and seal the surface better. They usually dry very fast. Oil-based primers can be used under latex paints. A combination of oil primer and latex finish coat can provide the best of both worlds. The big disadvantage, however, is that two different cleanups are needed—one for the primer and one for the latex. Use primer coats under your final finish whenever possible. The final finish will usually be superior to just two coats of regular paint.

MATERIALS AND EQUIPMENT

Before starting your painting project, you will need to collect all the necessary equipment. Depending on whether you are painting indoors or outdoors, you may need some or all of the following equipment:

Brushes

You will need a variety of types and sizes of brushes—small brushes for trim and touch-up and 4″ brushes for covering large areas. Natural bristle or "China bristle" brushes are the highest quality brushes for oil paints. Do not use these for latex, however; the water will cause the bristles to curl like hair on a humid day!

Synthetic bristle brushes made of nylon work well for both oil and latex paint. Look for the highest quality rating. The bristle ends should be frayed (similar to split ends) and the bristles soft but not limp. Flat wall brushes for painting large areas should be cut straight across. Smaller brushes for trim work should have the bristles cut at an angle. These are called "sash" brushes because the angled cut allows you to work paint into corners and maintain a straight paint line around trim. Don't try to save money on cheap brushes—they will always do a bad job. Properly maintained high-quality brushes will produce a superb finish for years.

Rollers

Paint rollers come in several varieties for different jobs. The rollers are made up of the roller handle, the roll itself, and some type of pan to hold the paint. Most handles come with a threaded hole that will accept an extension handle, allowing you to reach high areas without a ladder. The most common roller size is 9″. The main differences between rollers are the size and composition of the roller covers. Rolls with a close nap look like velvet, apply the smoothest finish and should be used on the smoothest surfaces only. They don't hold much paint, however, so frequent trips to the paint are necessary. The largest nap looks like wool fleece and will hold a great deal of paint. This thick nap allows the roller to deliver a lot

Wood	Ease of keeping well painted (I = easiest, V = most exacting[a])	Weathering: Resistance to cupping (1 = best, 4 = worst)	Weathering: Conspicuousness of checking (1 = least, 2 = most)	Appearance: Color of heartwood[b]	Appearance: Degree of figure on flat-grained surface
Softwoods					
Cedars					
Alaska-cedar	I	1	1	Yellow	Faint
(California) incense-cedar	I	–	–	Brown	Faint
Port-Orford-cedar	I	–	1	Cream	Faint
Western redcedar	I	1	1	Brown	Distinct
White-cedar	I		–	Light brown	Distinct
Cypress	I	1	1	Light brown	Strong
Redwood	I	1	1	Dark brown	Distinct
Products[c] overlaid with resin-treated paper	I	–	1	–	–
Pine					
Eastern white	II	2	2	Cream	Faint
Sugar	II	2	2	Cream	Faint
Western white	II	2	2	Cream	Faint
Ponderosa	III	2	2	Cream	Distinct
Fir, commercial white	III	2	2	White	Faint
Hemlock	III	2	2	Pale brown	Faint
Spruce	III	2	2	White	Faint
Douglas-fir (lumber and plywood)	IV	2	2	Pale red	Strong
Larch	IV	2	2	Brown	Strong
Lauan (plywood)	IV	2	2	Brown	Faint
Pine					
Norway	IV	2	2	Light brown	Distinct
Southern (lumber and plywood)	IV	2	2	Light brown	Strong
Tamarack	IV	2	2	Brown	Strong
Hardwoods					
Alder	III	–	–	Pale brown	Faint
Aspen	III	2	1	Pale brown	Faint
Basswood	III	2	2	Cream	Faint
Cottonwood	III	4	2	White	Faint
Magnolia	III	2	–	Pale brown	Faint
Yellow-poplar	III	2	1	Pale brown	Faint
Beech	IV	4	2	Pale brown	Faint
Birch	IV	4	2	Light brown	Faint
Cherry	IV	–	–	Brown	Faint
Gum	IV	4	2	Brown	Faint
Maple	IV	4	2	Light brown	Faint
Sycamore	IV	–	–	Pale brown	Faint
Ash	V/III	4	2	Light brown	Distinct
Butternut	V/III	–	–	Light brown	Faint
Chestnut	V/III	3	2	Light brown	Distinct
Walnut	V/III	3	2	Dark brown	Distinct
Elm	V/IV	4	2	Brown	Distinct
Hickory	V/IV	4	2	Light brown	Distinct
Oak, white	V/IV	4	2	Brown	Distinct
Oak, red	V/IV	4	2	Brown	Distinct

[a]Woods ranked in group V for *ease of keeping well painted* are hardwoods with large pores that must be filled with wood filler for durable painting. When so filled before painting, the second classification in the table applies.
[b]Sapwood is always light.
[c]Plywood, lumber and fiberboard with overlay or low-density surface.

Table 6: Characteristics of woods for painting and weathering (omissions in the table indicate inadequate data for classification).

Finish	Initial treatment	Appearance of wood	Cost of initial treatment	Maintenance procedure	Maintenance period of surface finish	Maintenance cost
Preservative oils (creosotes)	Pressure, hot and cold tank steeping	Grain visible. Brown to black in color, fading slightly with age	Medium	Brush down to remove surface dirt	5-10 yr only if original color is to be renewed; otherwise no maintenance is required	Nil to low
	Brushing	Brown to black in color, fading slightly with age	Low	Brush down to remove surface dirt	3-5 yr	Low
Waterborne preservatives	Pressure	Grain visible. Greenish in color, fading with age	Medium	Brush down to remove surface dirt	None, unless stained, painted or varnished as below	Nil, unless stains, varnishes or paints are used. See below
	Diffusion plus paint	Grain and natural color obsured	Low to medium	Clean and repaint	7-10 yr	Medium
Organic solvents preservatives*	Pressure, steeping, dipping, brushing	Grain visible. Colored as desired	Low to medium	Brush down and reapply	2-3 yr or when preferred	Medium
Water repellent†	One or two brush coats of clear material or, preferably, dip applied	Grain and natural color visible, becoming darker and rougher textured	Low	Clean and apply sufficient material	1-3 yr or when preferred	Low to medium
Stains	One or two brush coats	Grain visible. Color as desired	Low to medium	Clean and apply sufficient material	3-6 yr or when preferred	Low to medium
Clear varnish	Four coats (minimum)	Grain and natural color unchanged if adequately maintained	High	Clean and stain bleached areas, and apply two more coats	2 yr or when breakdown begins	High
Paint	Water repellent, prime and two topcoats	Grain and natural color obscured	Medium to high	Clean and apply topcoat; or remove and repeat initial treatment if damaged	7-10 yr‡	Medium to high

Source: This table is a compilation of data from the observations of many researchers.
*Pentachlorophenol, bis (tri-n-butyltin oxide), copper naphthenate, copper-8-quinolinolate and similar materials.
†With or without added preservatives. Addition of perservative helps control mildew growth and gives better performance.
‡Using top-quality acrylic latex topcoats.

Table 7: Exterior wood finishes: types, treatment and maintenance.

of paint to rough surfaces, such as stucco, masonry and rough siding. These "wooly mammoth" style rollers are great for outside jobs—but they create a lot of overspray. Make sure you have an ample covering of drop cloths. The most popular rollers fall somewhere between these two extremes.

Power rollers are available that force paint up a hose to the roller, eliminating the need to go back to the paint can. You can paint continuously. This makes the smaller nap rollers more practical, so your paint job will usually be smoother, with less overspray. Power rollers are available at most hardware stores and are well worth the investment for large paint jobs.

A large, flat pan is available to apply paint to the roller, and allows you to roll out the excess paint on a ramp built into the pan. If you are starting a large paint job, you can forgo the pan and use a roller *screen* instead. This is a piece of wire mesh designed to fit into 5-gallon professional paint cans. You can dip your paint roller right into the 5-gallon can, which is wide enough to accept a standard roller. The wire screen allows you to roll out the excess paint from the roller. If you expect to use more than three or four gallons of paint for the job, these large paints cans are more economical and convenient.

Sprayers

Paint sprayers come in four types: the hand-held airless sprayer, the high-pressure compressed-air sprayer, the professional airless pump sprayer and the new low-pressure air sprayers. The hand-held airless sprayers are the common sprayers many consumers buy for small paint jobs. They don't hold much paint, they are loud and they emit paint in a diffuse cloud of spray that is difficult to direct to the desired surface. You are likely to get as much paint on yourself as on the target area. These are not appropriate for large jobs.

Type of exterior wood surfaces	Water-repellent preservative		Stains		Paints	
	Suitability	Expected life* (yr)	Suitability	Expected life† (yr)	Suitability	Expected life‡ (yr)
Siding						
Cedar and redwood						
Smooth (vertical grain)	High	1-2	Moderate	2-4	High	4-6
Rough sawn or weathered	High	2-3	Excellent	5-8	Moderate	3-5
Pine, fir, spruce, etc.						
Smooth (flat grain)	High	1-2	Low	2-3	Moderate	3-5
Rough (flat grain)	High	2-3	High	4-7	Moderate	3-5
Shingles						
Sawn	High	2-3	Excellent	4-8	Moderate	3-5
Split	High	1-2	Excellent	4-8	—	—
Plywood (Douglas-fir and southern pine)						
Sanded	Low	1-2	Moderate	2-4	Moderate	3-5
Rough sawn	Low	2-3	High	4-8	Moderate	3-5
Medium-density overlay §	—	—	—	—	Excellent	6-8
Plywood (cedar and redwood)						
Sanded	Low	1-2	Moderate	2-4	Moderate	3-5
Rough sawn	Low	2-3	Excellent	5-8	Moderate	3-5
Hardboard, medium density ‖						
Smooth						
Unfinished	—	—	—	—	High	4-6
Preprimed	—	—	—	—	High	4-6
Textured						
Unfinished	—	—	—	—	High	4-6
Preprimed	—	—	—	—	High	4-6
Millwork (usually pine)						
Windows, shutters, doors, exterior trim	High ¶	—	Moderate	2-3	High	3-6
Decking						
New (smooth)	High	1-2	Moderate	2-3	Low	2-3
Weathered (rough)	High	2-3	High	3-6	Low	2-3
Glued-laminated members						
Smooth	High	1-2	Moderate	3-4	Moderate	3-4
Rough	High	2-3	High	6-8	Moderate	3-4
Waferboard	—	—	Low	1-3	Moderate	2-4

Source: This table is a compilation of data from the observations of many researchers. Expected life predictions are for an average continental U.S location; expected life will vary in extreme climates or exposure (desert, seashore, deep woods, etc.).

*Development of mildew on the surface indicates a need for refinishing.

†Smooth, unweathered surfaces are generally finished with only one coat of stain, but rough-sawn or weathered surfaces, being more absorptive, can be finished with two coats, with the second coat applied while the first coat is still wet.

‡Expected life of two coats, one primer and one topcoat. Applying a second topcoat (three-coat job) will approximately double the life. Top-quality acrylic latex paints will have best durability.

§ Medium-density overlay is generally painted.

‖ Semitransparent stains are not suitable for hardboard. Solid-color stains (acrylic latex) will perform like paints. Paints are preferred.

¶Exterior millwork, such as windows, should be factory treated according to Industry Standard IS4-81. Other trim should be liberally treated by brushing before painting.

Table 8: Suitability of finishing methods for exterior wood surfaces.

Traditional air-compressor sprayers will work for larger jobs, but they require that the paint be thinned out so that it will atomize properly. The sprayer has several adjustments to change the pattern, volume and pressure of the paint. Thick paint will splatter out the nozzle in lumps. This type of spray gun works best for the thinner stains and varnishes, and less well for thicker latex and oil paints. It also requires an air compressor. Most paint stores rent these machines.

A new type of sprayer has taken the market by storm in recent years—the high-volume, low-pressure sprayer. These devices come with their own high-volume turbine compressors that look and act like vacuum cleaners in reverse. The high volume of compressed air drives the paint toward the target, so more paint reaches its destination with less overspray. The compressed air is hot and helps to speed adhesion and drying of the paint. The adjustments on the sprayer provide a variety of paint patterns and volumes. This sprayer can produce a very professional job and can be adjusted to a fine spray pattern for working around trim areas. Like the traditional air gun, this type works best with thinner paints.

The airless diaphragm sprayer is the type you will most likely see professional painters using. These contain a high-volume airless compressor that pumps paint through a long hose to the spray nozzle. Most are designed to fit right on the edge of a 5-gallon paint can. They deliver a high-volume stream of paint with low overspray and can be used for inside and outside painting. It is amazing to watch a professional painter use one of these machines. They work well with the thicker latex paints and can paint an entire house in one day. You can rent these machines from paint stores as well.

CHOOSING THE PAINTING METHOD

If you decide to paint the project yourself, you will need to determine the most effective painting method to use—brushing, rolling or spraying. Each has its advantages and disadvantages. Generally, cleanup and preparation time is inversely proportional to the speed of painting. Brushing is the slowest method but requires the least masking and preparation. Spraying is extremely fast but requires careful preparation to protect surrounding areas from overspray. Using a roller falls somewhere in between.

If your project is extensive, such as exterior painting, and everything needs to be painted, spraying is probably worth the extra preparation time. Once you prepare the site, you can often finish the job in one or two days. This helps to avoid interruptions for bad weather. Proper spraying requires a certain level of skill, however, so make sure you are up to the task.

Saving Time

You can save a significant amount of time if you use a power roller. These units keep adequate paint on the roller with a push of a button. Avoid using paint sprayers for interior walls, except for the primer coat. Sprayers do not hide minor blemishes and drywall sanding marks like paint rollers do and will give disappointing results.

You can save a tremendous amount of time painting if you coordinate the paint, drywall and trim processes between subs. Apply both coats of wall paint and the first coat of trim paint before the doors and trim are installed. This will allow you to roll paint the walls right up to the door openings without worrying about getting paint on the trim. Put one coat on the doors and trim before installation. These two steps will cut interior painting time in half. Make sure to coordinate this with your trim carpenter first. Many trim subs do not like to work with prepainted trim. It creates an irritating dust when sawed. Test masking tape to be sure that it does not pull off paint when removed.

Painting Tips

- If you plan to do all or most of this yourself, don't forget to wear your worst clothing and bring your portable radio.
- Punch holes in the inside rim of opened paint cans with a nail or ice pick. This will allow the extra paint to drip back in the can.
- Tapered, angle-cut brushes are recommended for a smoother finish. Avoid cheap brushes that shed bristles into your freshly painted surfaces.
- It is not necessary to use a water-based primer for a water-based final coat and an oil-based primer for an oil-based final coat. But the base coat must be dry and stable. Consult with your local paint supplier.
- Don't worry about what the surface looks like. If it feels smooth, it's ready to paint. When in doubt, sand more.
- Always use nose protection or a respirator when sanding.
- Wrap oil-based rollers and brushes in plastic wrap (for short periods of time) or place them in a jar of paint thinner.
- Rinse water-based rollers and brushes in water when stopping for several hours.

- Try to paint in daylight. If you must paint at night, use powerful halogen lamps.
- Allow coats of paint to dry before recoating.
- Apply paint evenly. Applying too heavily will leave trails of paint on the wall that must be sanded later.
- Be careful around flammable paints and solvents: no smoking, no open flames and no matches.
- Be careful of harmful vapors in paints and solvents. Obey all cautions printed by manufacturers.
- Stain-grade moldings and handrails should be sanded with a very fine grit sandpaper before staining. Fill holes with a wood filler that accepts stain or color match filler to stain used. Use a good quality wiping stain and don't leave the stain on too long.

Wallcoverings

Wallpaper is an excellent do-it-yourself project. Most wallpaper outlet stores will be glad to show you the ins and outs of quality wallpaper selection and installation. Most wallpaper comes in double rolls of 32 feet per roll. Single rolls can be purchased at a premium. Wallpaper has a standard width of 24″.

When purchasing wallpaper, make sure all rolls have the same run number. The same number tells you that wallpaper of a certain pattern was all made at the same time, and the colors will be exactly the same from roll to roll. Repeat pattern is a big factor in estimating the number of rolls that you will need. Repeat pattern is the width of the pattern before it starts over. On large repeat patterns, you will waste a lot of paper getting patterns to match up.

Wallpaper comes in unpasted and prepasted forms. Many wallpaper hangers still use a paint roller to roll on wallpaper glue even on prepasted wallpaper. Wallpaper that will be used for baths should be either foil or vinyl, due to the humid environment.

PAINTING—Steps

PT1 SELECT paint schemes. Minimize number of colors. Try not to rely only on paint samples, since final colors may vary. Try a few test samples on your wall. Check them in daylight and with a portable lamp at night. Colors often have different "looks" between day and night. If you need to mix custom colors, one quart is normally the minimum. Some custom colors can only be mixed in gallon sizes due to the minute amounts of some pigments required. Custom colors are not returnable.

PT2 PERFORM standard bidding process. You may want to get this bid broken out between materials and labor. You may wish to use the painter as labor and furnish your own paint. If you do, have the painter help you select your paint. The painter might be able to get you a better price. It is advisable to buy the best paint you can afford. More expensive paint has more pigment and its durability can withstand more scrubbing. *Consumer Reports* can help you identify which brands of paint have been tested for best quality, durability and value. If you are planning on doing the painting yourself, this will help you determine how much you will save. If you are not up to doing the entire job, consider hiring the painter for the exterior work only and doing the interior work yourself.

PT3 PURCHASE all painting materials if you plan to do the painting yourself. Normally, you'll want flat water-based latex for walls and ceilings. Semigloss is normally recommended for baths and children's rooms. Materials needed will include:

- 1 gallon or quart plastic/cardboard buckets for carrying around small amounts of paint for small work
- Bucket of drywall mud (for fixing walls and ceilings)
- Can opener
- Cans of interior and exterior wood filler (for fixing millwork)
- Caulk gun
- Disposable wooden paint stir sticks (normally free where you purchase your paint).
- Drop cloths; rosin paper, which comes in large roles, can be used as a substitute.

Tape it down flat and it will stay out of your way. It is good at absorbing small spills and drips.

- Extendable sanding poles with swivel heads—Fiberglass composition is recommended.
- Extension cords for power sprayers, sanders, vacuum cleaners, lamps and portable radios
- Halogen lamps, for painting at night
- Liquid soaps and hand cleaners, sponges and buckets for cleanup
- Medium- and fine-grained sandpaper
- Paintbrushes (1″, 2″ and 3″ for painting millwork)
- Paintbrushes (4″—for doing inside wall corners and around trim)
- Paint rags
- Paint roller (fine nap) or power roller (self-feeding). Consider getting a professional-type roller that has a 20″ coverage width to save time.
- Paint sprayer
- Paint stir sticks
- Paint thinner (Penetrol for oil-based paints, Flotrol for water-based paints)
- Paint tray and paint tray liners. Liners are inexpensive and eliminate the need to clean the pan out each day.
- Rolls of masking tape. 1″ and 2″. If tape stays down too long, it can damage the wall surface. You may wish to tape windows to minimize clean-up later.
- Single-edged razor blades and handle
- Six-foot step ladder
- Trim guard
- Tubes of trim caulk. A large home may require several dozen tubes.
- Vacuum cleaner to suck up paint sanding dust.

Exterior

PT4 PRIME and caulk all exterior surfaces: related trimwork, windows, doors, exterior corners and cornice. Exterior millwork should be primed immediately after installation. An all-weather exterior primer should be used. Priming protects this raw millwork from moisture damage. In this step, all finish nails should be set below the surface with a nail punch. Holes should be filled with exterior water-based wood filler or oil-based wood dough.

PT5 PAINT exterior siding, trim, shutters and wrought iron railing. Iron railing should be painted with a rust-retarding paint. Painting two-story fixtures can be simplified by coordinating with stucco, siding, or masonry subs. They will have scaffolding on-site that can be used for painting.

PT6 PAINT all cornice work. For two-story homes, this usually requires very long, adjustable ladders. A professional is recommended.

PT7 PAINT gutters if needed. It is often handy to paint metal gutter sections on the ground prior to installation. If you do so, don't paint the joints. These must be kept clean so that they can be jointed properly. Gutters may still get scratched up during installation, but this gives you a head start on a difficult painting process. You should use either a latex- or alkyd-based metal primer for metal gutters. *NOTE:* If the gutter seams require soldering, don't paint them until the soldering is finished.

Interior

PT8 PREPARE painting surfaces. This is one of the most important steps in this book. Nothing is more important in the final appearance of your home than good paint preparation. Sand all trim and repair with wood filler as needed. Ceilings and walls should be sanded smooth with a pole sander. If you intend to use an oil-based enamel finish coat, your millwork should be as smooth as a baby's behind; imperfections are magnified greatly once a gloss or high-gloss finish coat is applied. Apply trim caulk between base moldings and wall, between crown moldings and wall, between crown moldings and ceiling and between adjoining pieces of crown molding, base molding and wainscoting. Water-based caulk is normally recommended. This involves performing the following tasks:

- DUST off all drywall with a dry rag.
- APPLY trim caulk to joint between trim and wall. Smooth it with your finger. Wipe off excess. When dry, sand to a smooth joint. This is an important step that will help to yield professional-quality results.
- REPAIR any dents in drywall or moldings with spackle or wood filler.
- SAND all repaired areas to a smooth finish.
- *NOTE:* If ceiling is to be stippled, you don't need to do a thing to it.
- *NOTE:* If you plan to paint stair spindles, do them before installing them. The painting process will be much easier.

PT9 PAINT prime coat on ceilings, walls and trim. Water-based primer takes about four hours to dry—less in hot, dry weather. The prime coat provides a good bonding surface for the finish coat. Just about any good painting job involves two coats of paint. Before you even start, make sure you have excellent lighting available in every room to be painted. Even if you intend to wallpaper, paint a primer coat so that it will be easier to remove the wallpaper if you ever want to later. If you have Masonite interior doors, prime them also. Ceiling white makes a good neutral primer. Since there is nothing in the house that you can harm, you need not worry about drop cloths. You can either rent a commercial paint sprayer, purchase your own or use a roller. For a professional job, lightly sand the walls before the final coat after the primer has dried. Sand trim also.

PT10 PAINT or stipple ceilings. A paint roller with an extension arm makes this job easier. Wear goggles to keep paint splatters out of your eyes. If you have stippled ceilings, you won't be painting them.

PT11 PAINT walls. Even if you want switch and outlet plates to match your paint, paint them separately. Don't install them first. Start with a small brush and paint around the ceiling, all windows and doors. Then cover the large areas with a roller. Do one wall at a time.

PT12 PAINT OR STAIN trim. This also includes kitchen cabinets, window frames, doors and door jambs, handrails, treads and risers as needed. If you plan to stain these, a wiping stain is recommended. Test all materials on a sample piece of like material before applying to finished product. All areas to be stained should be prepared using stain-grade putties. If you will have exposed, stained stair treads and painted stair rail pickets, it is easier to paint stair pickets prior to installing them, to avoid getting paint on the rail and treads. Once dry, stained surfaces should be coated with BIN—an alcohol-shellac-based primer. Start with the highest trim and work down. Allow the walls to dry first. It will be easier if you use a 1½″ sash brush for doing windows and a 2″ brush for all other trim. Some painters don't bother to mask off all the little trim in multipaned windows; they just scrape all the dried paint off with a razor blade. Stained trim should be coated with a varnish or other sealer. Hardwood floors are stained and sealed after sanding. This should normally be done by a professional.

PT13 REMOVE paint from windows with one-sided utility razor blades. Do this only when the paint has completely dried or the soft paint shavings will stick to the surface and make a mess.

PT14 INSPECT paint job for spots requiring touch-up. Make sure you are looking at the paint job either in good natural light or under a bright light.

PT15 TOUCH UP paint job.

PT16 CLEAN UP. Latex paints will come off with soap and warm water. Enamel, which you may have used on your trim, will require mineral spirits or Varsol for removal. Dry out the brushes and store them wrapped up in aluminum foil with the bristles smoothed out. Place plastic wrap on the paint cans prior to storing them.

PT17 PAY painter (if applicable) and have him sign an affidavit.

PT18 PAY retainage after final inspection.

PAINTING—Specifications

Interior Painting

- Bid is to include all material, labor and tools.
- Contractor is to ensure that all cans of custom-mixed paint match. (This is important!)
- Primer coat and one finish coat are to be applied to all walls, including closet interiors.
- Walls are to be touch-sanded after primer has dried.
- Flat latex is to be used on drywall; gloss enamel to be used on all trim work.
- Ceiling white is to be used on all ceilings.
- All trim joints are to be caulked and sanded before painting.
- All paint is to be applied evenly on all areas.
- Window panes are to be cleaned by painter.
- Painting contractor is to clean up after job.
- Excess paint is to be labeled and to remain on-site when job is completed.

Exterior Painting

- Five-year warranty exterior-grade latex paint is to be used on exterior walls and shutters.
- All exterior trim and shutters are to be primed with an exterior-grade primer.
- All trim areas around windows, doors and corners are to be caulked with exterior-grade caulk before final coat. Caulk color is to match paint color.

PAINTING—Inspection

- ☐ Proper paint colors are used.
- ☐ Paint application appears even, with no variation between cans of custom-mixed paint.
- ☐ All ceilings and walls appear uniform in color, with no visible brush strokes.
- ☐ Trim is painted with a smooth appearance. Gloss or semigloss enamel is used as specified.
- ☐ All intersections (ceiling-wall, trim-wall, wall-floor) are sharp and clean. A clean, straight line separates wall and trim colors.
- ☐ Window panes are free of paint inside and out.
- ☐ Extra touch-up paint is left at site.
- ☐ All exterior areas are painted smoothly and evenly.
- ☐ No dried paint drops mar the job.

Cabinetry and Countertops

CABINETS

Because of the cabinetry, the kitchen invariably turns out to be the most expensive room in the house. Cabinetry and countertops are used in both the kitchen and baths. Money spent on high-quality cabinets can really add value to your home. Cabinetry can be prefabricated or custom-built for your installation. Although custom cabinets are nice, the extra expense and time involved are rarely worth it. There is an abundance of high-quality prefab cabinets available today.

When shopping for cabinets, pay attention to space-saving devices, such as revolving corner units, drop-down shelves and slide-out shelves. These options can significantly increase the usable space in your cabinets.

Kitchen Cabinets

Visit a few kitchen cabinet stores or kitchen design stores to become acquainted with the many functional types of cabinets that are available. Look at sample kitchen layouts to get ideas. Determine your material (wood, Formica, etc.), hinge and knob styles and finish. Also determine how much cabinet space you need. The amount of cabinet space is not as important as how well the space is used. Many American craftsmen are turning out impressive reproductions of fine, expensive European models at a fraction of the price. The cabinetry should be an integral part of your kitchen design, complementing appliances, lighting and space. Remember, kitchens and baths sell houses.

Bathroom Cabinets and Vanities

Purchase all your cabinets from the same place if you can. Not only does this make your cabinet selections more uniform, but you can probably get a better price from your cabinet sub. Always look for bathroom cabinets with drawers for storing toiletries.

COUNTERTOPS AND VANITY TOPS

Countertops are normally made of Formica. But other materials have become popular, including tile, Corian, granite, marble and other surfaces. Selection of material largely depends upon your budget and the look you want. Granite is more durable than marble. Tile is good but some people tire of cleaning the grouted surfaces. But Formica is still by far the most popular and least expensive. Some companies sell bathroom vanities and countertops with the sink built into the countertop.

Countertops typically should have a standard depth of 24″ and may have a 3″ or 4″ backsplash against

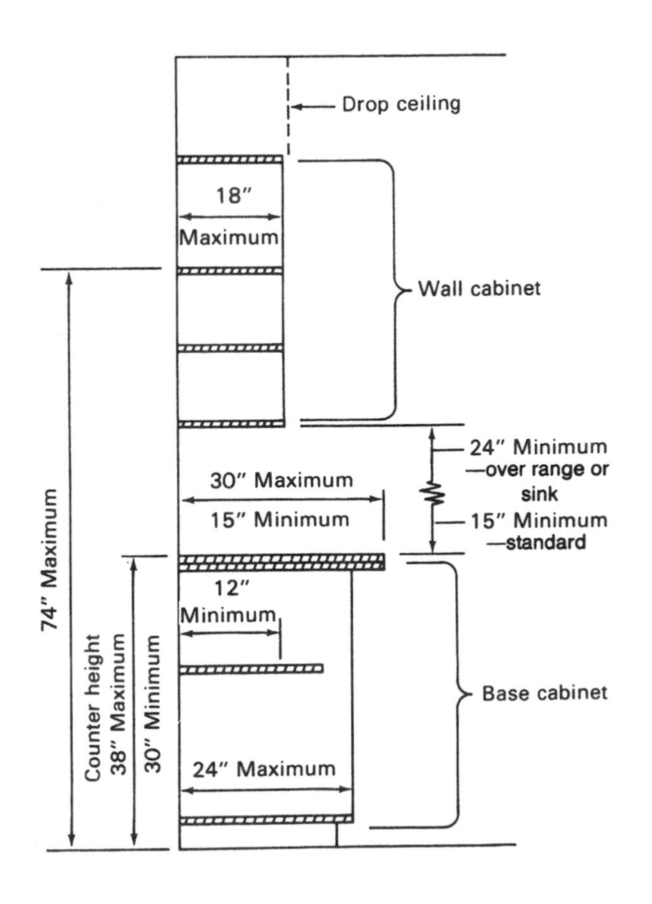

Figure 99: Kitchen cabinet dimensions.

the wall. If you desire, add a few inches of additional depth to the countertop. This will leave more room to clean behind the sink. Countertops are priced and installed by the foot, and special charges may be incurred to cut openings for stoves, sinks and fancy edgework. If you get a separate company to do the counter and vanity tops, which is most likely when cultured marble is used, installation is normally included in the price—but get this confirmed. Advantages to cultured marble tops are the lack of a seam between the top and the sink, no sink bowl to purchase and an easier job for the plumber.

Cabinet Sub

This sub will normally produce or supply all cabinets and install them along with the countertops. This includes making all the cuts for sinks, stoves, etc. Cabinet subs will normally bid the job at one price for material and labor.

CABINETRY—Steps

CB1 CONFIRM kitchen design. Make sure you have enough cabinets to store all pots, pans, utensils and food items.

CB2 SELECT cabinetry/countertop styles/colors.

CB3 COMPLETE diagram of cabinet layout for kitchen and baths. This is a critical step to perform prior to electrical and plumbing installation. Show locations of all plumbing connections on diagram: ice maker, dishwasher and sink.

CB4 PERFORM standard bidding process. There are lots of independent cabinetmakers who can copy expensive European cabinet designs.

CB5 PURCHASE or have constructed all cabinetry and countertop surfaces. If your countertops come from a separate sub, coordinate the cabinetry and countertop subs. You can't install tops until vanities are in. Unless you know what you are doing, wait for the drywall to be hung before having the cabinet sub measure. This gives him more exact figures to work from and a place to mark on the wall. But do order your cabinets as soon as possible, because they can be a source of big delays later.

Give your cabinet sub specs for all appliance dimensions and locations of all doors and windows. This will ensure that the cabinets will install properly under the windowsill. At a minimum, wait for framing to be finished. Walk through the structure with the cabinet sub, covering kitchen, baths and utility room. Countertops are typically 36″ above the floor, kitchen desks 32″ high with a minimum 24″ knee space. When measuring, remember to account for ½″ drywall on walls, and widths of adjacent trim. Mark dimensions on the floor with builder's chalk or crayon.

CB6 STAIN and seal, paint or otherwise finish all cabinet woodwork. Stains will stain tile grout and other nearby items, so cabinets must be stained before they are installed.

CB7 INSTALL bathroom vanities. These must be in place before your plumber can connect the sinks, drainpipes and faucets.

CB8 INSTALL kitchen wall cabinets. If you install base cabinets first, they get in your way while installing wall units.

CB9 INSTALL kitchen base cabinets. Also make necessary cutouts for sinks and cook tops. Show the cabinet sub where cutouts are to be made. Don't make any cutouts until the units have arrived.

CB10 INSTALL kitchen and bath countertops, aprons and backsplashes.

CB11 INSTALL and adjust cabinetry hardware—pulls, hinges, etc.

CB12 INSTALL utility room cabinetry.

CB13 CAULK all wall/cabinetry joints as needed.

CB14 INSPECT all cabinetry and countertops.

CB15 TOUCH up and repair any scratches and other marks on cabinetry and countertops.

CB16 PAY cabinet sub. Have him sign an affidavit.

CABINETRY—Specifications

- Bid is to provide materials and complete cabinetry and countertop installation as indicated on attached drawings.

- All cabinets are to be installed by a qualified craftsman.
- All cabinets are to be installed plumb and square and as indicated on attached drawings.
- All hardware such as hinges, pulls, bracing and supports is to be included in bid.
- All necessary countertop cutouts and edging are to be included in bid.
- All exposed corners on countertops are to be rounded.
- Install one double stainless-steel sink (model/size).
- Special cabinet interior hardware includes:
  - Lazy Susan
  - Broom rack
  - Pot racks
  - Sliding lid racks

VANITY TOPS—Specifications

- Bid is to manufacture and install the following cultured marble tops in ¾″ solid-color material. Sink bowls to be an integral part of top.
- Vanity A is to be 24″ deep, 55″ wide, finished on left with backsplash; 19″ oval sink cut with rounded edge, with center of sink 22″ from right side. Color is solid white. Spread faucet is to be used.
- Vanity B is to be 24″ deep, 45″ wide, finished on right with backsplash; 19″ oval sink cut with rounded edge, with center of sink 18″ from left side. Color is solid almond. Standard faucet is to be used.

CABINETRY—Inspection

- ☐ All cabinet doors open and close properly with no binding or squeaking. Doors open completely and remain in place half-opened.
- ☐ All pulls and other hardware are securely fastened. Check for excessive play in hinges.
- ☐ All specified shelving is in place and level.
- ☐ All exposed cabinetry has an even and smooth finish.
- ☐ No nicks, scratches, scars or other damage/irregularity are on any cabinetry and countertops.
- ☐ Countertops are level, checked by placing water or marbles on surface. All joints are securely glued with no buckling or delamination.
- ☐ All drawers line up properly.
- ☐ All mitered and flush joints are tight.
- ☐ Cutouts for sinks and cook tops are done properly, and units are fitted into place.

Flooring, Tile and Glazing

The most widely used floorings include:

- Carpet
- Hardwood (oak strip, plank and parquet)
- Vinyl floor covering
- Ceramic tile

CARPET

Carpet is measured, purchased and installed by the square yard and comes in rolls 12′ wide. Do not try to install carpet yourself unless you have experience; it is not as easy as it appears. Carpet should be installed with a foam pad, which adds to the life of the carpet and makes the carpet feel more plush. A stretcher strip will be nailed around the perimeter of the room, and the carpet will be stretched flush with the wall. Stretching the carpet prevents it from developing wrinkles. After two to three months, the carpet may need to be stretched again. Make sure your carpet company will agree to do this at no cost.

When pricing carpet, always get the installed price with the cost of the foam pad included. A thicker foam pad will make your carpet feel more luxurious, but may cost extra. Do not install carpet until all worker traffic has ceased. Make sure that all bumps and trash have been cleaned from the floor before installation. Even the smallest bump will be noticeable through the carpet after installation.

WOOD

Hardwood flooring is measured by the board foot (oak or other plank flooring) or by the square yard (parquet). Oak is laid in tongue-and-groove strips nailed to the subfloor, stained and sealed. Parquet is normally installed as 6″ squares with an asphalt adhesive. Parquet squares are available in stained and unstained styles. Avoid laying hardwood floors in humid weather because, when the air dries out, gaps will appear between the boards. Buy the flooring a week or two before it is needed and store it in the rooms where it will be laid to let the moisture content stabilize. "Select and better" is the normal grade used in homes. It has a few knots in it. "Character grade" has more knots but is less expensive.

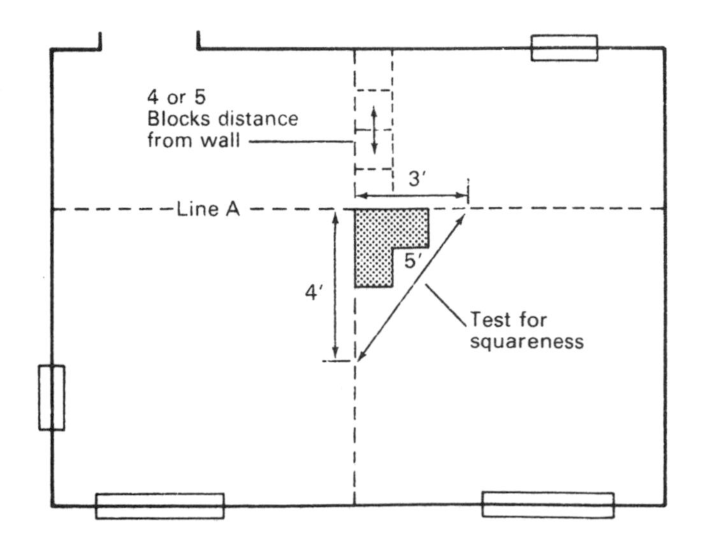

Figure 100: Working lines for laying block in a square pattern.

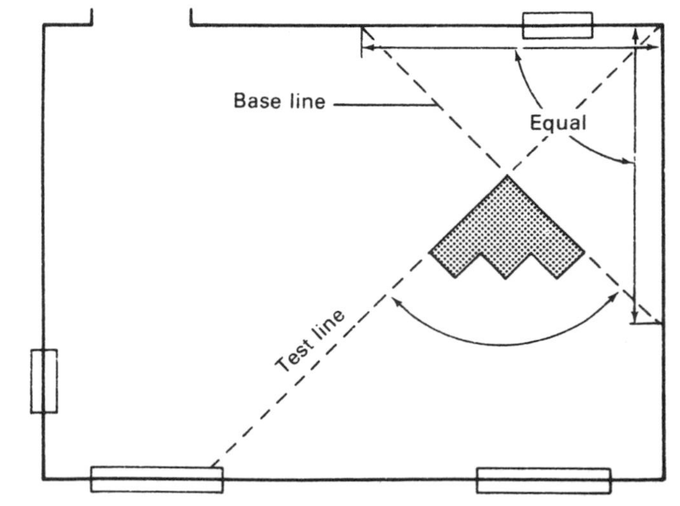

Figure 101: Working lines for laying block in a diagonal pattern.

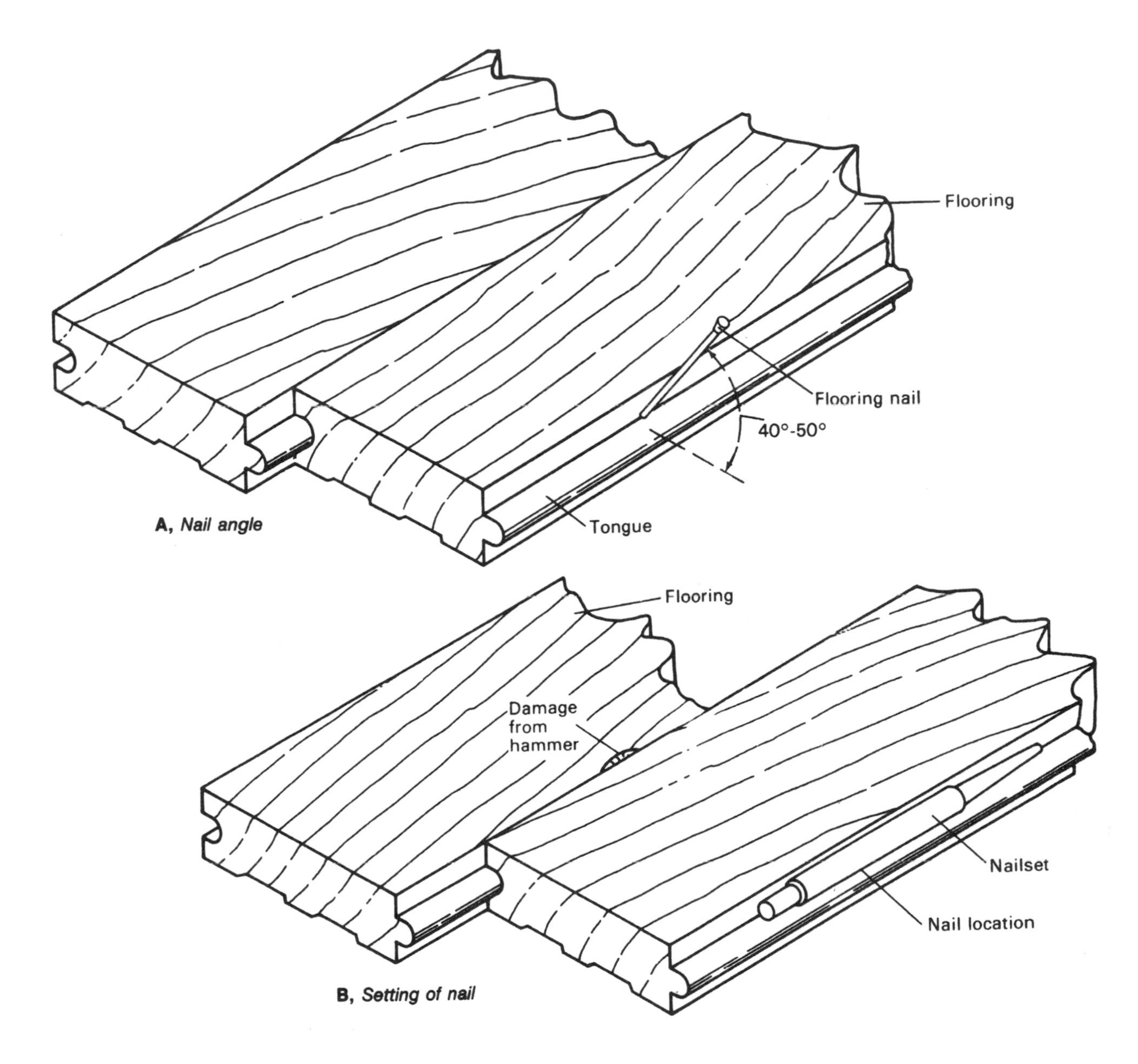

Figure 102: Nailing of flooring.

After installation, you may want to sand the floor for an extra smooth finish. Pine flooring normally isn't sanded after installation. On all other wood floors, sanding should only be done with the grain, or sanding marks will show up. Use a professional-size drum sander. An "edger" should be used to sand the edges. If you lay hardwood floors on a concrete slab, you need to seal the concrete and use a concrete adhesive. A better method is to lay down runner strips and nail the floor to the strips. Wood strip floor installers usually work as a pair, one nailing strips in place while the other saws pieces to the proper length and sets them in place. If you can avoid it, do not lay wood floors on slabs.

VINYL

Vinyl floor covering is measured by the square yard and comes in a variety of styles and types. Padded vinyl rolls have become very popular because of their ease of installation and maintenance. Installation procedures vary depending on the material and manufacturer, so take care to follow the manufacturer's recommendations when installing. Warranties can be voided if the material is not installed properly.

When installing vinyl floors, consider laying an additional subfloor of plywood over the area to be covered. This additional layer provides a smoother

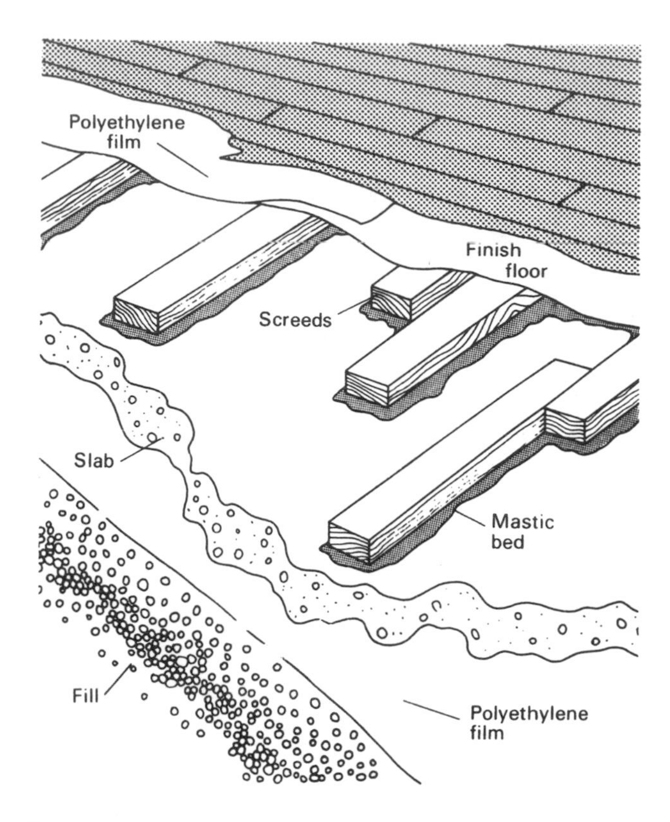

Figure 103: Base for wood flooring on concrete slab.

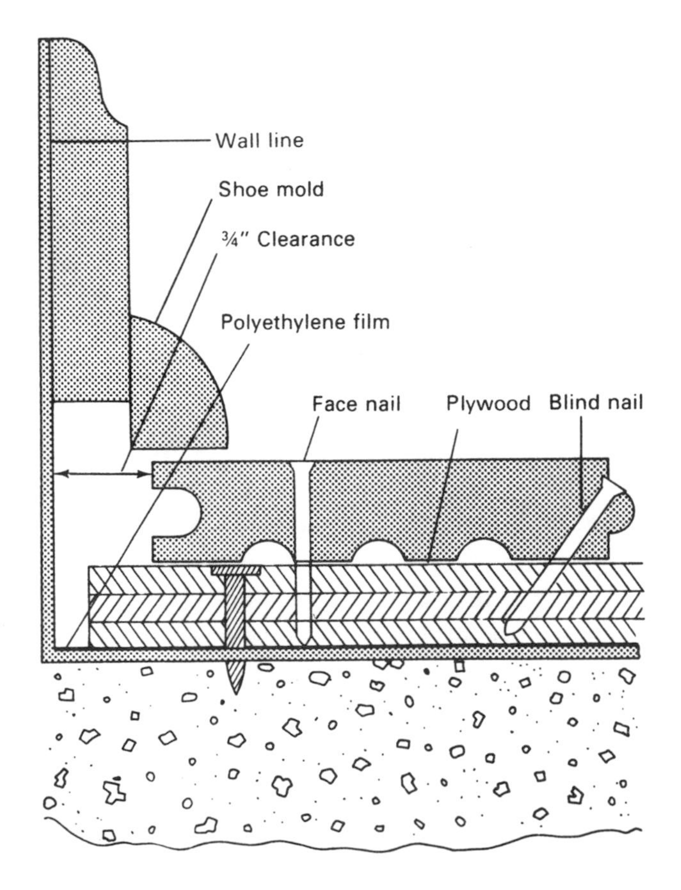

Figure 104: Plywood-on-slab method of installing strip oak flooring.

surface for the floor and brings the level of the floor up to the same level as any surrounding carpet.

TILE

Tile is measured and sold by the square foot, although tile units can come in any size. Tile comes either glazed or unglazed and can be installed over a wood subfloor, although a special adhesive cement and latex grout must be used. Consider using a colored grout, which can hide dirt. If you coat the tile grout with a protective coating, you may void the tile warranty.

Tile installation is a tedious and critical procedure. Make sure your tile sub is experienced and quality conscious. Improper installation can result in buckling, cracking and delamination. To ensure that tiles adhere properly, avoid installation in cold weather or make sure that the floor is heated for 24 hours after installation.

Colored grouts add a special touch without adding to costs. Don't bother with a colored grout unless the joint is at least ⅛" wide. You may wish to use a grout additive instead of water. This will strengthen the grout and bring out the color. To save time, also consider using a special protective coating applied to tile prior to grouting. This will allow you to wipe the grout film off the tile easily when the job is done.

If you plan to install tile in kitchens or other large areas, check the height of the installed floor (tile and mortar bed) to ensure that you aren't higher than the adjoining carpet or hardwood floors.

Remember that you have a ½" fudge factor around floor perimeters if you use shoe molding. If you install the tile yourself, ask your supplier to loan or rent you a tile cutter with a new blade, or buy a new blade—this will make a world of difference.

Tiling shower and bath stalls is trickier than tiling floors. You may want to do the floors but hire a sub for the rest. Use spacers when tiling the floor to keep the tiles straight.

GLAZING

Glazing, while not part of flooring, normally occurs during the flooring phase. Glazing refers to any mirrors and plate glass windows that must be custom cut from sheets of plate glass. No special specifications are needed.

GLAZING—Steps

GL1 DETERMINE mirror and special glazing requirements. This means all of the following items:

- Mirrors
- Closet door mirrors
- Shower doors
- Fixed picture windows
- Medicine cabinets

GL2 SELECT mirrors and other items. Here you must decide such trivia as beveled-edge or nonbeveled-edge mirror glass, chrome or gold frame for shower doors, etc. Glazing subs will have catalogs for you to look through.

GL3 CONDUCT standard bidding process for glazing sub.

GL4 INSTALL fixed-pane picture windows. Check that window has no scratches and is adequately sealed.

GL5 INSTALL shower doors and medicine cabinets. If you can, turn on shower head and point toward the shower door to check for leaks. Door should glide freely along track.

GL6 INSTALL mirrors and closet door mirrors. Mirrors should be installed with a nonabrasive mastic. Check for hairline scratches in the silver coating; this cannot be fixed aside from replacing the mirror.

GL7 INSPECT all glazing installation.

GL8 CORRECT all glazing problems.

GL9 PAY glazier and have him sign an affidavit.

FLOORING AND TILE—Steps

FL1 SELECT carpet brands, styles, colors and coverage. Also select your padding. Keep styles and colors to a minimum. This helps your buying power, minimizes scrap and provides for easier resale.

FL2 SELECT hardwood floor type, brand, style, color and coverage.

FL3 SELECT vinyl floor covering brand, style, color and coverage.

FL4 SELECT tile brands, styles, colors and coverage. This includes tile for bath walls, bath floors, shower stalls, patios and kitchen countertops. Once your subfloor is in, borrow about a square yard of your favorite tile from your supplier. Lay it down where it will be used and see how it looks.

FL5 CONDUCT standard bidding process. For best prices on carpet, go to a carpet specialist or carpet mill. When possible, have the supplier bid for material and labor. If your supplier does not install, ask for installer references.

Tile

FL6 ORDER tile and grout. If you do the job yourself, borrow or rent a tile cutter from the tile supplier. Look for good bargains on "seconds"—tile with slight imperfections—to use as cut pieces. Don't use sanded grout with marble tiles. The sand will scratch the marble surface. Make sure that all tile comes from the same batch. Tile colors can vary from batch to batch.

FL7 PREPARE area to be tiled. Sweep floor clean and nail down all squeaky floor areas with ridged nails.

FL8 INSTALL tile base in shower stalls. This normally involves applying concrete over a wire mesh. In shower stalls, this is done on top of the shower pan. If you have tile in the kitchen, do not tile under the cabinets or the island. Prior to tiling around a whirlpool tub, make sure the tub is secure and has been fully wired, grounded and plumbed.

FL9 APPLY tile adhesive. Use a trowel with grooves suitable for your tile and grout. Use large, curved, sweeping motions.

FL10 INSTALL tile and marble thresholds. Don't forget to use tile spacers to ensure even tile spacing. Make sure tubs are secure in place before tiling. Make sure to save a few extra tiles in case of subsequent damage to floor tiles. This will ensure a perfect color match if repairs are needed.

FL11 APPLY grout over tile. Apply silicone sealant between the tub and tile.

FL12 INSPECT tile. Refer to specifications and inspection criteria.

FL13 CORRECT any problems that need fixing.

FL14 SEAL grout. This should only be done after grout has been in place for about three weeks. Use a penetrating sealer to protect the grout from stains.

FL15 PAY tile sub. Get him to sign an affidavit.

FL16 PAY tile sub retainage after floor has proven to be stable and well glued.

Hardwood

FL17 ORDER hardwood flooring. Hardwood flooring should remain in the room in which it will be installed for several weeks in order for it to expand and contract based on local humidity conditions. You may want to order a few extra bundles of shorter, cheaper lengths to do closets and other small areas.

FL18 PREPARE subfloor for hardwood flooring. If you have a wood subfloor, you should put down at least two layers of heavy-duty building paper. Make sure that all dirt and drywall compound are thoroughly cleaned off the floor. The surface must be totally clean and level. If you are laying hardwood on a slab, you have two choices:

- Seal the slab with a liquid sealer and install flooring with an adhesive.
- Prepare a subfloor.
- Sweep slab.
- Apply 1″×2″ treated wood strips called bottom sleepers with adhesive and 1½″ concrete nails, 24″ apart, perpendicular to the oak strips.
- Lay 6-mil poly vapor barrier over strips.
- Lay second layer of 1″×2″ wood strips over vapor barrier.
- Nail wood flooring strips to the sleepers.

FL19 INSTALL hardwood flooring. You may want to wait until all drywall and painting work is complete before installing hardwood, to protect the finish. Ask installers to leave some scraps. Store them away for repair work or other use.

FL20 SAND hardwood flooring. Although most floors are presanded, you may still need to have them sanded for the smoothest possible finish. Sand only with the grain. Cross-grain sanding will not look bad until the stain is applied. Once floors are sanded, there should be no traffic until after the sealer is dry.

FL21 INSPECT hardwood flooring. Refer to specifications and inspection guidelines.

FL22 CORRECT any problems with hardwood flooring. Common problems are split strips, hammer dents in strip edges, uneven spacing, gaps, squeaks, nonstaggered joints and last strip not parallel with wall.

FL23 STAIN hardwood flooring. The floors must be swept very well before this step is performed. Close windows if it is dusty outside.

FL24 SEAL hardwood flooring. Polyurethane is the most popular sealant. You normally have an option of either gloss or satin finish. Place construction paper over dry floors to protect them from construction traffic.

FL25 INSPECT hardwood flooring again, this time primarily for finish work.

FL26 PAY hardwood flooring sub. Have him sign an affidavit.

FL27 PAY hardwood flooring sub retainage.

Vinyl Floor Covering

FL28 PREPARE subfloor for vinyl floor covering. Sweep floor clean and nail down any last squeaks with ringed nails. Plane down any high spots.

FL29 INSTALL vinyl floor covering and thresholds. Ask installers to leave behind any large scraps that can be used for repair work.

FL30 INSPECT vinyl floor covering. Refer to specifications and inspection guidelines.

FL31 CORRECT any problems with vinyl floor.

FL32 PAY vinyl floor covering sub. Have him sign an affidavit.

FL33 PAY vinyl floor covering sub retainage.

Carpet

FL34 PREPARE subfloor for carpeting. Sweep floor clean and nail or plane down down any high spots. This is your last chance to fix squeaky floors easily, and to install hidden speaker wires.

FL35 INSTALL carpet stretcher strips.

FL36 INSTALL carpet padding. You should use an upgraded pad if you use a tongue-and-groove subfloor. If you don't use a carpet stretcher on this step, you are doing the job wrong.

FL37 INSTALL carpet. This should be done with a carpet stretcher to get a professional job. Ask installer to leave larger scraps for future repair work.

FL38 INSPECT carpet. Refer to specifications and inspection guidelines.

FL39 CORRECT any problems with carpeting installation.

FL40 PAY carpet sub. Have him sign an affidavit.

FL41 PAY carpet retainage.

FLOORING AND TILE—Specifications

Carpet

- Bid is to furnish and install wall-to-wall carpet in the following rooms:
  - Living room
  - Dining room
  - Master bedroom and closets
  - Second bedroom and closets
  - Third bedroom and closets
  - Second-floor stairway
  - Second-floor hallway
  - 8″ accent border to be installed in master bathroom, color to be of builder's choice
- Full-size upgraded pad is to be provided and installed under all carpet.
- Stretcher strips are to be installed around carpet perimeters.
- All seams are to be invisible. (This is critical to a good carpet job.) Seams should be located in inconspicuous and low-traffic areas.
- All major carpet and padding scraps are to remain on-site.
- All necessary thresholds are to be installed.
- All work is to be performed in a professional manner.

Oak Flooring

- Bid is to provide and install "select and better" grade flooring in the following areas:
  - Foyer and first-floor hallway
  - Den
- Oak strips are to be single width, 3¼″ strips.
- Two perpendicular layers of heavy-gauge paper are to be installed between subfloor and oak flooring.
- All oak strips are to be nailed in place with flooring nails at a 50° angle. Nails are to be installed only with a nailing machine, to prevent hammer head damage to floor.
- Shorter lengths are to be used in closets.
- All oak strips are to be staggered randomly so that no two adjacent joint ends are within 6″ of each other.
- Oak flooring and oak staircase treads are to be sanded to a smooth surface.
- Oak flooring and oak staircase treads are to be swept, stained, sealed and coated with satin finish polyurethane.
- Stain is to be a medium brown.
- Ridged or spiral flooring nails are to be used exclusively.
- Only first and last strips are to be surface-nailed, concealable by shoe molding.

Parquet Flooring

- Bid is to provide and install parquet flooring in the following areas:
  - Sunroom
  - Sunroom closet
- Parquet flooring is to be stained and waxed.
- Install marble thresholds in the following areas:
  - Between hallway and living room
  - Between hallway and dining room
  - Between hallway and den
  - Between kitchen and dining room

Vinyl Floor Covering

- Bid is to provide and install vinyl floor covering in the following areas:
  - Kitchen and kitchen closet
  - Utility room
- All seams are to be invisible.
- All major scraps are to remain on-site.
- All necessary thresholds are to be installed.

Tile

- Bid is to provide and install specified tile flooring in the following areas:
  - Master bath shower stall walls: one course of curved base tile around perimeter, tile to go all the way to ceiling
  - Master bath shower floor
  - Surround and one step for whirlpool bath
  - Second bath shower stall walls, all the way to ceiling
  - Second bath floor per attached diagram
  - Half-bath floor per attached diagram; one course of base tile around perimeter
  - Kitchen floor per diagram: tan grout sealed, one course of base tile around perimeter
  - Kitchen wall backsplash; see drawings attached
  - Kitchen island top
  - Hobby room and closet
- Grout colors are to be as specified by builder.
- All necessary soap dishes, tissue holders, towel racks and toothbrush holders are to be supplied and installed in white unless otherwise specified by builder in writing.
- Matching formed base and cap tiles are to be used as needed to finish edging and corners.
- Install with ¼″-per-foot slope toward drain in shower stalls.
- Install all necessary tub and shower drains.
- All grout is to be sealed three weeks after grout is installed.
- Extra tile and grout are to remain on-site.

FLOORING AND TILE—Inspection

Carpet

- ☐ Proper carpet is installed.
- ☐ No seams are visible.
- ☐ Carpet is stretched tightly and secured by carpet strips.
- ☐ Colors match as intended. Watch for color variations.
- ☐ Carpet installed on stairs fits tight on each riser.

Oak Flooring and Parquet

- ☐ Proper oak strips and/or parquet is installed (make, color, etc.).
- ☐ No flooring shifts, creaks or squeaks occur under pressure.
- ☐ Flooring has even staining and smooth surface.
- ☐ Flooring has no scratches, cracks or uneven surfaces. Scratches normally indicate sanding across the grain or using too coarse a sandpaper for final pass.
- ☐ Smaller lengths are used in closets, to reduce scrap and waste.
- ☐ All necessary thresholds are installed and are parallel to floor joints.
- ☐ Only edge boards have visible nails. No other boards have visible nails.
- ☐ Edge boards are parallel to walls.

☐ Floor has no irregular or excessive spacing between boards.

☐ No floor registers are covered up by flooring.

Vinyl Floor Covering

☐ Proper vinyl floor covering is installed (make, color, pattern, etc.).

☐ Vinyl floor covering has smooth adhesion to floor, with no bubbles.

☐ No movement of flooring occurs when it is placed under pressure.

☐ Floor has no visible or excessive seams.

☐ Floor has no scratches or irregularities.

☐ All necessary thresholds have been installed and are parallel to floor joints.

☐ All thresholds are of proper height for effective door operation and have no more than ⅛″ gap under interior doors.

☐ Pattern is parallel or perpendicular to walls as specified.

Tile

☐ Proper tiles and grouts are installed (make, style, dimensions, color, pattern, etc.).

☐ Tile joints are smooth, evenly spaced, parallel and perpendicular to each other and walls, where applicable.

☐ Tiles are secure and do not move when placed under pressure.

☐ Tiles have no scratches, cracks, chips or other irregularities.

☐ Grout is sealed as specified.

☐ Grout lines are parallel and perpendicular to walls.

☐ Tile pattern matches artistically.

☐ Tiles are laid symmetrically from the center of the room out. Cut pieces on opposite sides of the room are of the same size.

☐ All tile cuts are smooth and even around perimeter and floor register openings.

☐ Marble thresholds are installed level, with no cracks or chips.

☐ Marble thresholds are parallel with door when closed.

☐ Metal thresholds are installed between tile and wood joints.

☐ Marble thresholds are at proper height.

☐ Matching formed tiles are used at base of tile walls as specified.

☐ Cap tiles are used at top of tile walls.

☐ Proper tile edging is used for tubs, sinks and shower stalls.

☐ No tiles are cracked or damaged.

☐ Tile floor is level.

☐ Extra tile and grout are available on-site.

MISCELLANEOUS FINISH

Weather stripping

☐ All exterior doors have weather stripping and threshold installed.

☐ Doors open and close securely with a tight seal.

☐ No leaks appear after a heavy rain.

Mirrors and Shower Doors

☐ Door fixtures match faucets.

☐ Doors slide properly with no friction or binding.

☐ Mirrors have no scratches.

Landscaping

When you consider how much resale value and curb appeal a good landscaping job can add to a home, it is surprising that so many builders scrimp on this important effort. Many builders just throw up a few bushes and then seed or sod. You should be willing to spend just a bit more time and effort to do the job right. If you don't want to spend all the money at once, you can devise a two- or three-year implementation plan, beginning with the critical items such as ground coverage.

Important items to keep in mind are:

- Plan for curb appeal. Your lawn should be a showplace.
- Plan for low maintenance. If you don't want to trim hedges and rake leaves forever, be smart about the trees and bushes you install or keep.
- Work with the sun, not against it. Maximize sunlight into the home by careful planning of trees.
- Plan for proper drainage. This is *critical*! Water must drain away from the dwelling and not collect in any low spots.
- Consider doing much of this work yourself.
- Consider assistance from a landscape architect.

LANDSCAPE ARCHITECT

Every lot has one best use and many poor ones. A good landscape architect can help you do more than select grass. Landscape architects can be helpful in:

- Determining the best position of the home on the lot.
- Determining attractive driveway patterns.
- Determining which trees should remain and what

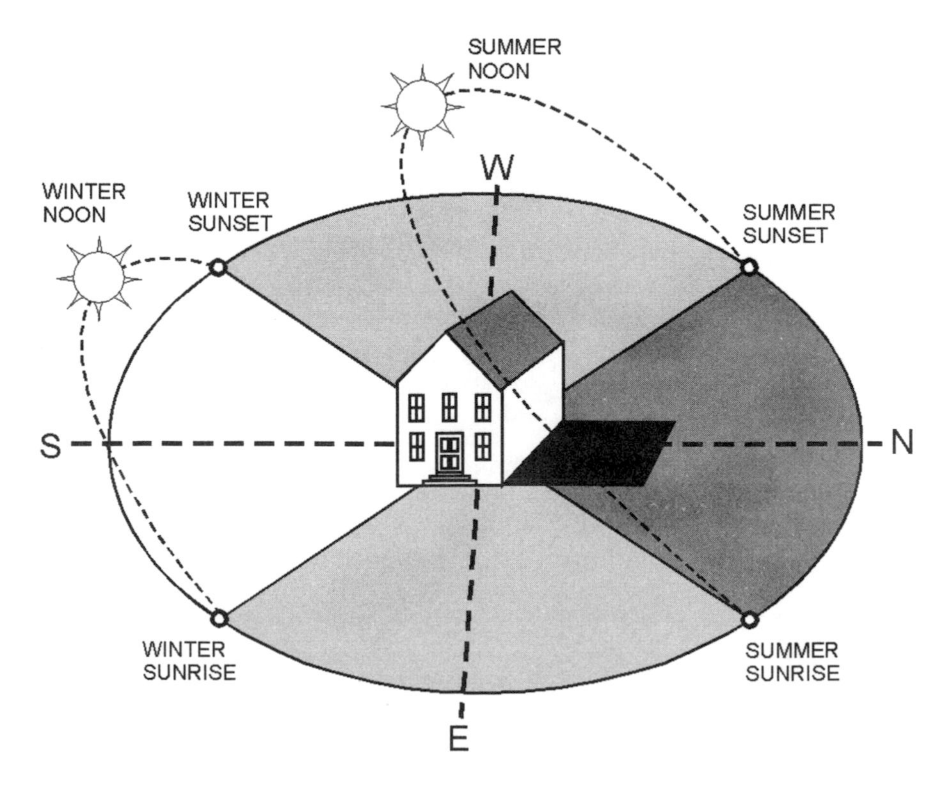

Figure 105: Landscaping orientation affects the placement of home and trees.

additional trees and/or bushes to plant for privacy and lowest maintenance.

- Determining soil condition and helpful additives.
- Determining proper drainage and ground coverings.

You can normally pay an independent landscape architect by the hour, but if you do a lot of work with him, work up a fixed price. Meet him at your lot with the lot survey. Walk the lot with him, getting his initial impressions and ideas. From there, you may request a site plan. The site plan will normally be a large drawing showing exact home position, driveway layout, existing trees and bushes, sculptured islands and other topographical features.

SITE PLAN AND DESIGN

Before you even lay out the foundation of your house, you should examine the site and plan the placement with landscaping and sun orientation in mind. A well-oriented, well-designed house will maximize energy savings. When purchasing the lot, consider its orientation and the exact site where you may want to situate your home. Look at the lay of the land and the way the lot drains. Examine your homesite's exposure to sun, wind, shade and water. Also note the location and proximity of any nearby buildings, fences, bodies of water, trees and pavement that might have climatic effects. Nearby buildings provide shade and a windbreak. Nearby fences and walls can block or channel wind. Bodies of water can moderate temperature but can increase humidity and produce glare. Trees will provide shade, windbreaks or wind channels. Dark pavement absorbs heat; light pavement reflects heat. Take a soil sample and have the county extension office do a complete soil analysis. This will tell you if soil supplements are needed.

In northern climates, place the home's windows, decks and sunrooms so that they face south or southeast. The home's north- and west-facing walls should have less window area because heat will radiate out the windows. North-facing windows receive minimal direct sunlight.

Before you start landscaping, you need to draw a landscape site plan. The landscaping plan should include all the trees, shrubs, flowers, berms, walls, fences, trellises, sheds, garages and drainage paths. This will help you refine the landscape plan before you begin construction. To develop a landscape plan:

- Use plan paper with ¼″ or ⅛″ grids.
- Sketch a simple, scaled drawing of your yard's basic features, showing buildings, walks, driveways and utilities.
- Show the location of streets, driveways, patios and sidewalks.
- Identify potential uses for different zones of the yard: vegetable gardens, flower beds, patios and play areas. Use arrows to represent sun angles and prevailing winds for both summer and winter.
- Circle the areas of your yard needing shade or wind protection.
- Mark areas that may have poor drainage or standing water. You will need to plant moisture-tolerant plants here, or even install a drainpipe.
- Pencil in the name and location of each plant. Draw the plant to scale.

LANDSCAPING ISN'T JUST FOR LOOKS

A well-designed landscaping job serves many valuable purposes in addition to improving the appearance of the yard. These other advantages can actually add up to more benefits than the cost of the landscaping.

Landscaping Can Save Energy

By placing trees in strategic locations, you can reduce heating and cooling costs by up to 25 percent. How? In the summer, deciduous trees can shade the house from the sun, reducing cooling costs. In the winter, that same tree will drop its leaves, allowing the low winter sun to enter windows and warm the house. The U.S. Department of Energy estimates that three properly placed trees can save $100 to $250 in annual energy costs—enough savings to return your initial investment in less than eight years. Shade from trees can also reduce the temperature in the yard by 5° to 10°.

The sun's heat, passing through windows and being absorbed through the roof, is the main cause of heat buildup in the summer. Shading rooftops will reduce

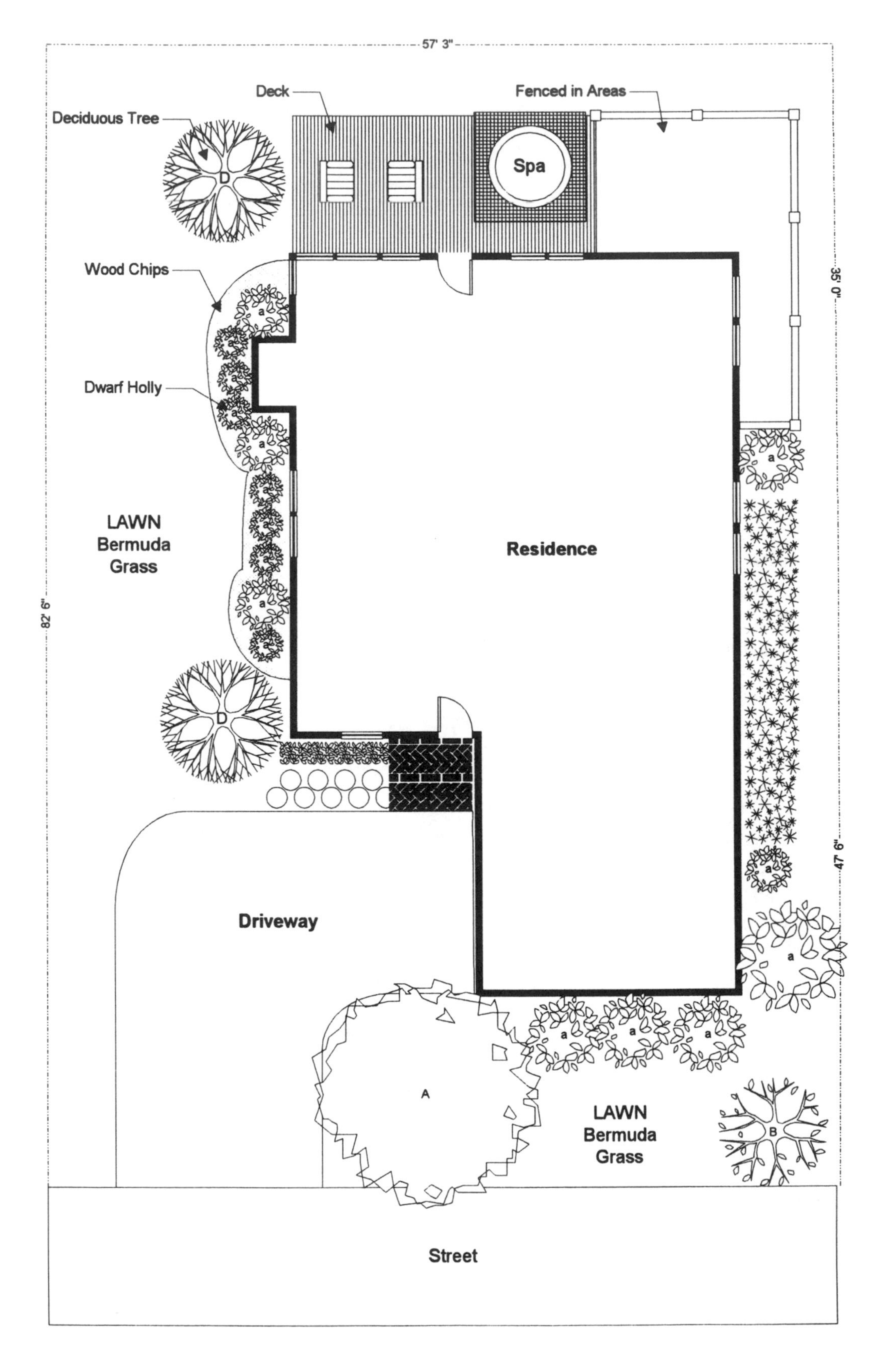

Figure 106: A complete landscaping site plan. This can be produced by you or a professional landscape designer. The master plan will help you visualize the landscaping and produce an order list of materials.

solar heat gain and air-conditioning costs. However, using shade effectively requires knowledge of the position of the sun and the size, shape and location of the object casting the shadow.

The Sun's Changing Angle Is the Key

As the seasons change, the sun's angle in the sky changes, affecting your placement of plants, trees and windows. At the winter solstice, the sun rises 30° south of east and stays low in the sky even at noon. The sun then begins to rise in the sky, tracing a more northerly path, as the season progresses and approaches summer. At the summer solstice, the noon sun is almost straight overhead and the sun sets 30° north of west. The exact angle of the sun will vary, depending on whether you live in a northern or southern latitude. The sun's angle is higher in southern latitudes, but the general effect is the same. These changing angles will affect the sun's ability to cast shadows. When the sun is low in the sky, objects in its path cast long shadows horizontally. Horizontal overhangs or tall trees cast short vertical shadows because the sun shines underneath them. However, short shrubbery can block the low sun from windows on the east and west sides of the house. In the summer, on the other hand, overhangs and tall trees can block much of the high, direct sunlight from entering south-facing windows, keeping the house much cooler. Knowing these angles will help you to place plants properly.

Trees should be selected for appropriate size, density and shape for the specific shading application. To block solar heat in the summer and allow it in the winter, use hardwoods (deciduous trees). To provide continuous shade or to block heavy winds, use evergreen trees or shrubs that don't drop leaves in the winter. Tall deciduous trees should be planted south of the house to provide maximum summertime roof shading. Trees with lower branches are more appropriate to the west, where shade is needed from lower afternoon sun angles. Plant trees or shrubs so that they shade the air-conditioning unit. This can increase its efficiency by as much as 10 percent. Trees, shrubs and ground cover can also shade the ground and pavement around the home, reducing heat radiation and cooling the air before it reaches your home's walls and windows.

Plants Also Provide Wind Protection

Properly placed landscaping can provide excellent wind protection, reducing winter energy costs. You should place the windbreaks on the north and west sides of the house to provide shelter from cold winter winds. Plant trees and shrubs together, to block wind from the ground to the treetops.

Evergreen shrubs and small trees can be planted as a solid wall at least 4′ to 5′ away from the north side, providing a windbreak. However, it is better to have dense growth further away so that air movement can occur during the summer.

If snow tends to drift and collect in your area, plant low shrubs on the windward side of your windbreak in order to trap snow before it blows next to your home. In addition to more distant windbreaks, consider planting shrubs, bushes and vines adjacent to the dwelling in order to create dead-air spaces that insulate the dwelling in both winter and summer. Allow at least one foot of space between full-grown plants and the dwelling's wall.

Landscaping Can Improve the Environment

Well-placed trees, shrubs and flowers can have many health benefits:

- Trees and ground cover can control erosion in areas where grass will not grow, such as on sloped or rocky areas.
- Large-scale tree and shrubbery planting will clean the air by absorbing carbon dioxide and releasing oxygen.
- Flowers can provide a home for beneficial insects and birds that can control harmful pest infestations, such as Japanese beetles, mosquitoes and leaf-eating caterpillars.

CHOOSING PLANTS FOR YOUR AREA

By choosing plants properly, you can reduce the amount of care and feeding needed to keep your yard in prime condition. Some species of trees, bushes and grasses require less water than others. Other species are naturally more resistant to certain pests. Using these species can reduce energy, water and time associated with lawn care, watering and trimming.

Xeriscaping

Xeriscaping is a landscaping technique that uses drought-resistant plants able to survive with very little water. These are usually plants from arid regions and various types of cactus. When planted properly, these sites are attractive, grow slowly and require virtually no watering. Xeriscaping saves energy and reduces water consumption. Make sure these plants will grow in your area. Excessively damp climates or poorly drained soil will encourage diseases.

Select Plants to Match Your Climate

Your goals and plant selection will differ depending on the climate in your area. The plants will serve different purposes in different climates. Climates fall into four basic categories:

Cool Climate

- Use evergreen shrubbery as yard borders to block cold winter wind.
- Leave south-facing areas clear of trees to allow winter sun to enter southern windows.
- Use deciduous trees that will drop leaves in the winter.
- Place evergreen shrubbery to shelter north-facing windows and reduce radiant heat loss.

Hot and Dry Climate

- Use dry-loving trees and bushes.
- Plant tall evergreens to create maximum cooling shade for roof, walls and windows.
- Funnel summer winds toward house to cool it naturally.

Hot and Humid Climate

- Channel cool summer breezes toward the dwelling.
- Maximize summer shade with deciduous trees that allow penetration of low-angle sun in winter.
- Avoid placing plants in low-lying areas that drain poorly.
- Choose disease-resistant plant varieties.

Temperate Climate

- Use deciduous trees to maximize the cooling effects of sun in the summer and warming effects of bare trees in the winter.
- Deflect winter winds away from dwelling.
- Funnel summer breezes toward dwelling.

Growth Rate and Root Systems

Slow-growing trees require many more years of growth before shading a roof, but will generally live longer than a fast-growing tree. Also, slow-growing trees often have deeper roots and stronger branches, making them less prone to breakage by windstorms or snow loads, and more drought resistant than fast-growing trees. Avoid placing fast-growing trees with large root systems near the house or walkways. For example, never place a weeping willow tree near the house. Its massive root system can literally uproot foundation footings. Magnolia trees have roots that like to grow right on the surface, making lawn mowing difficult.

GRADING AND DRAINAGE

Before your planting begins, the landscaper or grader will need to complete the final grade. This will be your last chance to make sure the lot drains well. Lay out a grid of stakes and strings that have been leveled with a string level. Crisscross the lot and examine the lay of the land. It is almost impossible to judge drainage sloping if the lot is relatively flat. The string grid will help you to follow the lay of the land and calculate the path of water runoff. This is your final opportunity to get the drainage just right.

If the grading contractor has to move a lot of dirt or install a retaining wall, he will probably use a loader, switching to a tractor for the finish grading. He should rake the yard to remove large- and medium-size rocks and to mix in any topsoil that you have saved for this purpose. The string grid will have to be removed during grading, so check the drainage again with the string once the grader is finished, but before he leaves. Have him make any final adjustments. Make sure the drainage is perfect now. It is much harder to correct drainage problems later. If your yard has a low-lying area that is impossible to build up, dig a trench and install drainage pipe to provide an underground drainage path.

RETAINING WALLS

Retaining walls are used to alter topography or to provide improved drainage. In some local jurisdic-

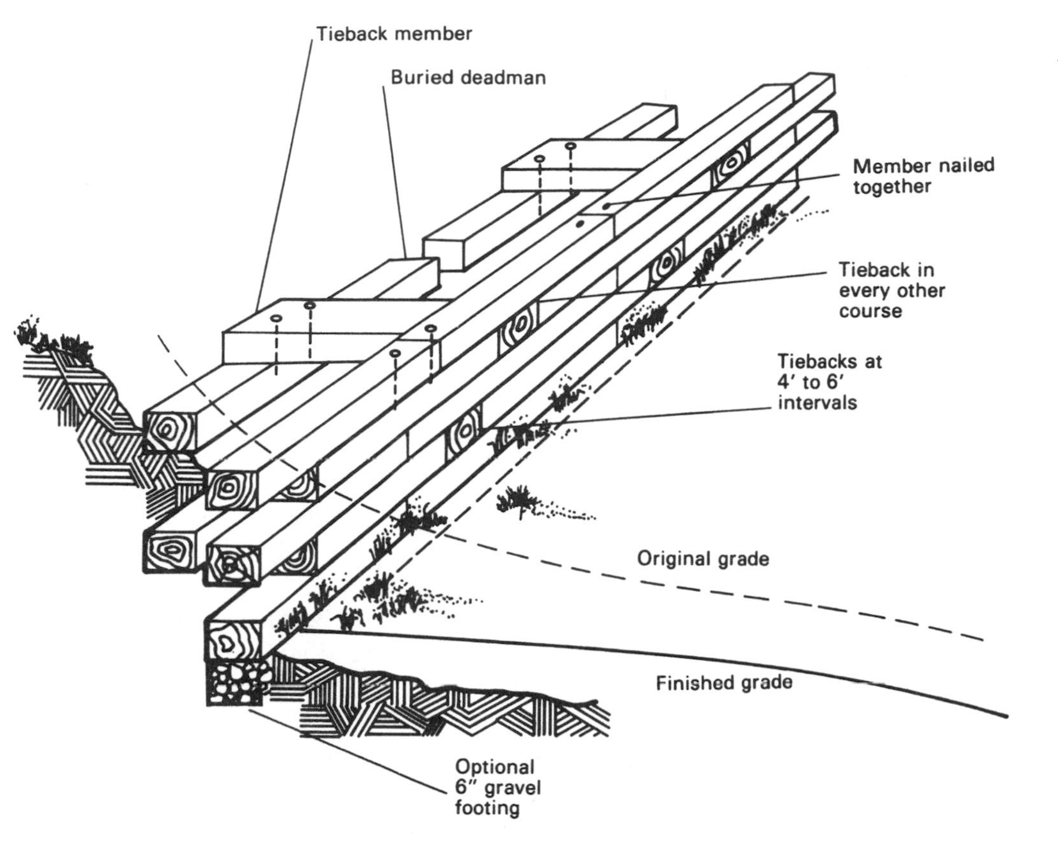

Figure 107: Pressure-treated timber retaining wall.

tions, a special permit is required to erect a retaining wall in excess of a given height, such as 36″. Materials used for constructing retaining walls include pressure-treated wood, masonry and poured concrete. If your site requires a large concrete retaining wall, chances are you will construct it while the concrete subcontractor is pouring the foundation. This will save money and help prevent mudslides and flooding during construction. Refer to the foundation section for more information on poured-concrete walls. If you are installing wood or masonry retaining walls, you might consider doing these yourself.

Wood Retaining Walls

Pressure-treated rectangular wood timbers or railroad ties may be used to construct retaining walls. The timbers are stacked so that the butted ends of the members in one course are offset from the butted ends of the members in the courses above and below. The bottom course should be placed at the base of a level trench. In well-drained, sandy soil, the footing is unnecessary. In less well-drained soils, apply 12″ to 24″ of gravel backfill

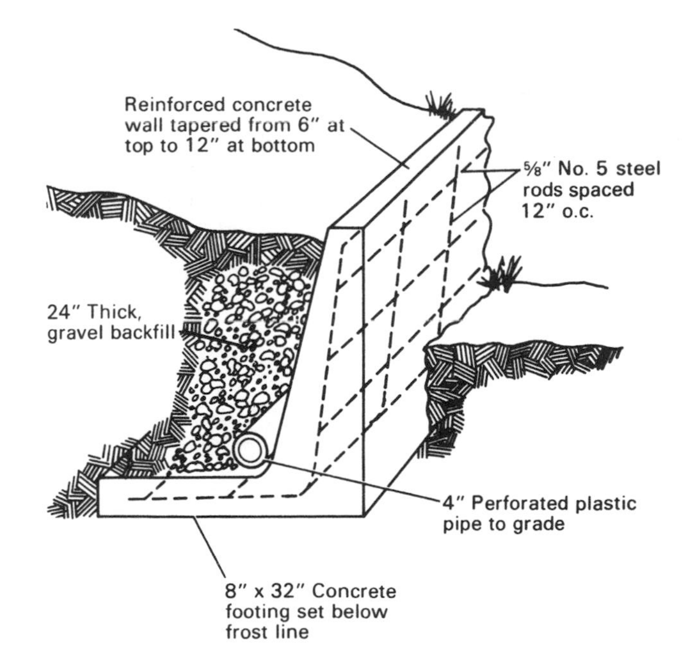

Figure 108: Reinforced concrete retaining wall (4′ high above grade).

behind the wall and a 6″-deep gravel footing below the bottom course. Nail the courses together with galvanized spikes. Every second course of timbers should include members inserted into the face of the wall that extend horizontally into the soil behind the wall. This "tieback" timber will anchor the wall to the soil and prevent the weight of the soil from collapsing the retaining wall. The tieback timbers should extend into the soil for the same distance as their distance above the base of the wall. The end of this tieback member should have a "deadman" timber 24″ long, attached to it in the shape of a *T*. This works like the anchor of a boat. These tiebacks and deadmen should be installed every 4′ to 6′ along the retaining wall.

An alternative retaining wall design consists of pressure-treated rectangular timbers or railroad ties set in holes spaced 4′ apart. Rough-sawn, pressure-treated 2″ lumber is then placed behind the vertical members. The 2″ crosspieces are held in place by backfilling as they are inserted. In poorly drained soils, the backfill should consist of 12″ to 24″ of gravel. In this design, the vertical members should be set in post holes to a depth of 4′ or to frost line depth, whichever is greater, in order to resist tipping from the pressure of the retained soil.

Block Retaining Walls

A reinforced-concrete block retaining wall can be constructed from several different types and styles of block specially designed for this purpose. Check with your local landscaping supplier. Several new types of block don't require mortar or footings as long as the wall is not more than 36″ tall. Some blocks are designed with a cavity for planting cover plants or vines.

For taller walls, use a mortared concrete block wall. Pour an extra wide footing to a depth below the frost line. After the footing concrete has hardened, the retaining wall blocks are laid. After the wall has set, install 12″ to 24″ of gravel behind the wall as backfill to provide drainage and to minimize the pressure from the soil freezing behind the wall.

Poured-Concrete Wall

The retaining wall can be constructed solely of poured concrete instead of concrete block. This wall will be identical to poured walls used for the foundation. Because of the cost of setup, fabrication and pouring, try to install these retaining walls at the same time as poured foundation walls. The same subcontractor can do both. The form for the face of the wall should be vertical, but the back of the wall should be built at an angle to provide a wall that is thicker at the base. Reinforcing rods should be placed in the form and wired together. The concrete should be poured in the form to the depth of the footing and allowed to partially set before the remaining concrete is poured. Backfilling the wall with 12″ to 24″ of gravel is recommended.

LAWNS

It is hard to imagine a suburban landscape without its well-manicured lawns. Regardless of your landscaping design, most of the square footage of your yard will be covered with grass, unless you plan an "all natural" landscape. The trick is to get this lawn established before rain and wind carry off the valuable topsoil.

Installing Sod vs. Seed

Sodding a lawn provides instant beauty and maturity. Turf is attractive and, in most cases, relatively easy to maintain. It's the perfect way to quickly stop erosion and runoff. Sod is also good for yards with little or no topsoil, since the sod brings its own topsoil along. Sod may be too expensive to use on the entire yard. You can compromise by using it in the front yard and using seed in the backyard. Sod can be successfully installed in every month of the year. It can be harvested and laid almost anytime during the winter, provided the ground is not frozen. There is not a day in the summer that is too hot for sodding—provided the sod is watered immediately after laying. Freshly laid sod requires frequent watering until the roots have grown into the soil.

Seeding the Yard

Sowing grass is not simply a matter of scattering seed. Soil preparation and composting can really make a difference in the amount of water your lawn needs and uses. This is especially important in yards with little or no topsoil. Properly prepared soil allows moisture to be absorbed and retained. The end result is less surface runoff and less overall water consumption.

Make sure to choose the right type of grass for your climate and maintenance requirements. Do you need heat tolerance or cold tolerance? Does the area re-

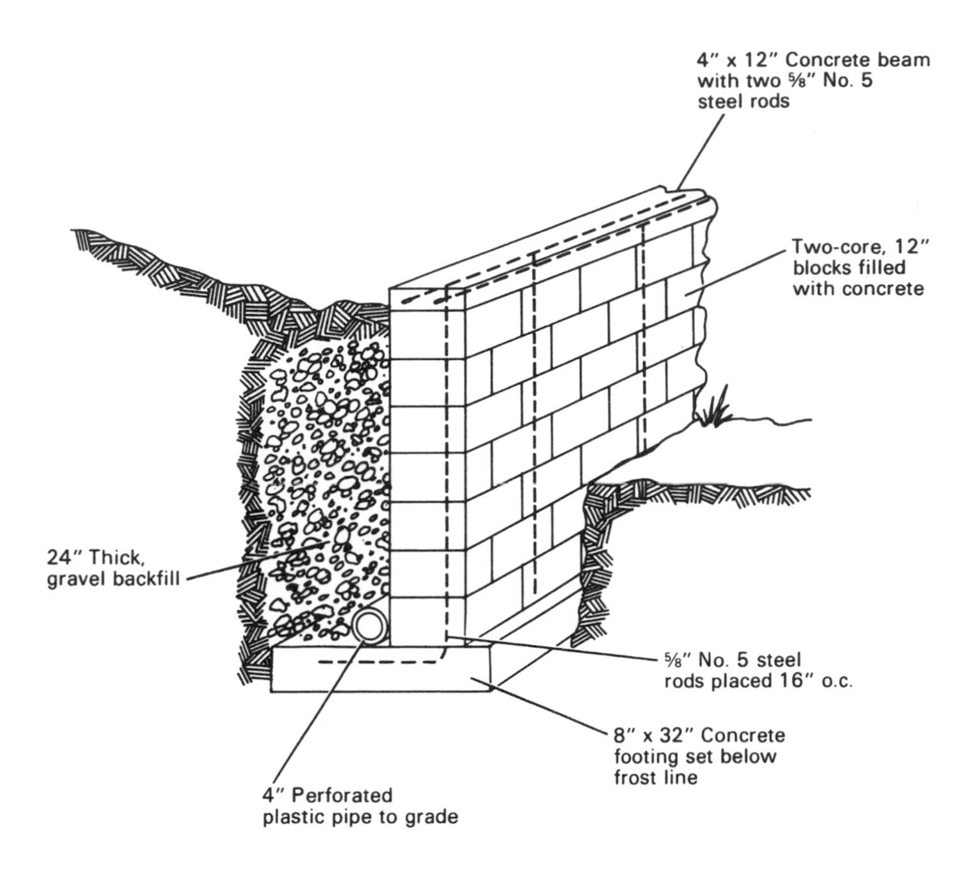

Figure 109: Reinforced concrete block retaining wall (maximum 4′ high above grade).

ceive a lot of traffic? Is it in full sun or shade? Will it be irrigated regularly or is it subject to drought conditions? Here is a brief summary of the major grass types:

- **Bluegrass**—cool-weather grass with fine texture, ability to withstand wear, and rich green color. Shade tolerant. Needs good soil, fertilizer and regular watering.
- **Fescue**—moderate-weather grass with tough, coarse texture. Tolerates wear, drought, disease and hot temperatures. Grows easily from seed.
- **Ryegrass**—cold-weather grass with fine blades. Comes in annual and perennial varieties. Annual variety used as a cover crop to allow other grasses to establish themselves. Will die in warm weather. Grows easily from seed.
- **Bermuda grass**—hot-weather grass with thick, fine blades. Tolerates heat, drought, poor soil, salt and heavy traffic. Makes an excellent turf. Turns white in winter. Requires regular watering and fertilizer. Needs full sun and will spread into nearby planting beds. Best grown from turf or sprigs.
- **Zoysia Grass**—excellent smooth, even-textured grass that grows low to the ground and creates a smooth thick turf. Tolerant of poor soil, drought and heat. Turns white in the winter. Slow growing and takes one or two years to cover yard. Must be grown from sprigs.
- **Centipede grass**—hot-weather grass that grows low to the ground. Very low maintenance with tolerance for poor soil and heat. Tolerates shade and chokes out weeds. Disease- and insect-resistant. Can be grown from seed or sprigs.

To establish a seeded yard, make sure the soil has been smoothed completely and has a good, soft texture. The challenge will be to get the grass established quickly. Otherwise, rain will wash away the seeds and soil. Use a mixture of your final grass selection and annual ryegrass. Rye germinates and grows rapidly, establishing a root system quickly. This will hold the soil in place while the final grass gets established. Rye has the additional advantage of fixing nitrogen into the soil, actually improving it for the other grass. The rye will eventually die off, allowing the final grass selection to take over.

Use a commercial seed spreader to spread an even layer of seeds. Cover the seeds with some type of mulch, such as wheat straw or a commercial spray-on mulch. Don't apply it too heavily, or you will block out the sun and smother the newborn grass seedlings. Keep the ground moist until the grass is fully established. After about two weeks, you should have a pretty green carpet of rye. Go back and fill in any bare spots with additional seed.

SPRINKLER SYSTEMS

Sprinkler systems are great for saving time on watering. Underground sprinkler systems should be installed after the final grading and before sodding or seeding. Sprinkler systems can actually save water—and money. Difficult soils absorb water very slowly, and sprinkler systems can help. The timer can be set to water during the night, for short intervals or repeatedly. This method will give your lawn an adequate soaking, without leaving gallons of water running down the street.

A sprinkler system is composed of:

- An electronic control system
- Antisiphon valves, where required
- PVC pipe
- Valve stems and sprinkler heads

Landscape supply companies stock all types of sprinkler systems. Because of the complexity of sprinkler systems, it is best to let the suppliers or subcontractors install them.

LANDSCAPING—Steps

Before Construction

LD1 EVALUATE your lot in terms of trees to keep, drainage, privacy, slopes, high and low spots and exposure to morning and afternoon sun. A landscape architect can help you greatly with this step. He can also help you determine whether you will have too little or too much soil after excavation and what to do about it. Excavated soil may be used to fill in low spots or otherwise assist in developing a drainage pattern or landscaping feature.

LD2 DETERMINE best possible position for home on lot. Considerations should include:

- Build with good southern exposure for parts of the house with a lot of glass.
- Reserve the best view for most important rooms when at all possible—don't give your best view to a garage or windowless wall.
- Situate your house with the best angle for curb appeal. Consider angling your home if on a corner lot.

LD3 DEVELOP site plan. This is a bird's-eye view of what you want the finished site to look like, including exact position of home, driveway, walks, patios and other paved surfaces, major trees, islands, grass areas, flower beds, bushes and even an underground watering system if you intend to have one. The site plan should show all baselines and easements. If any portion of your home falls outside a baseline, you will need a land variance. When determining grass areas, consider shade. Most grasses do not grow well or do not grow at all in shade. No amount of fertilizer or lawn care can compensate for a lack of sun.

LD4 CONTACT a landscape architect to critique and/or improve on your design. They normally have more experience and can surprise you with new and interesting ideas. For example, instead of having a driveway go straight from the street to the garage, consider putting a lazy curve in the drive and a slight mound on the street side. This helps hide the expanse of concrete, which is not normally a focal point. Landscape architects can recommend specific types of bushes to suit special privacy needs.

LD5 FINALIZE site plan. Make several copies in blueprint form. You will need at least one for yourself, one for a landscaper if you hire one, and one for the county.

LD6 SUBMIT site plan to building officials and bank along with your blueprints.

LD7 DELIVER fill dirt, if needed, to site. Choose a spot to pile topsoil from the foundation grading. Do not mix this dirt with fill dirt unless both types are topsoil. Make sure fill dirt is clean and dry. Wet dirt will form lumps and clods as it dries.

LD8 PAY landscape architect (if you used one).

After Construction

LD9 CONDUCT soil tests. Soil samples should be made for every several thousand feet of yard. Soil to be tested should be taken from 6″ down. Soil testing will help to determine pH (level of acidity) and the lack of essential nutrients. Most counties have an extension service that will test the samples for you at little or no charge. Good general-purpose soil conditioners are peat moss, manure and lime.

LD10 TILL first 4″ to 6″ of soil, adding soil conditioners as needed. Cut any unwanted trees down and grind stumps before sod is laid down. Rake surface smooth first one way and then another.

LD11 APPLY soil treatments as needed. You may also wish to add a slow release, nonburning fertilizer (no stronger than 6-6-6) to help your young grass grow. Apply gypsum pellets to break down clay.

LD12 INSTALL underground sprinkler system as needed. This normally consists of a network of ½″ PVC pipe purchased from a garden supply store.

LD13 PLANT flower bulbs. This is the easiest and best time to plant daffodils, crocus, hyacinths, gladiolus, etc.

LD14 APPLY seed or sod, depending upon your choice. Seed is much cheaper, but sod means an instant thriving lawn and elimination of topsoil runoff. Seed is hard to start on a hill or in other adverse conditions. If you do choose seed, use a broadcast spreader or a special seed dispensing device. Cover the seed with hay or straw to protect seeds from washing away with rain. If you sod, this will be a tougher job, but doing it yourself could really save you some money. Sod always has at least three prices: direct from a sod farm, delivered by a sod broker or delivered and installed by a sod broker. If you can, buy direct from a sod farm. Visit the farm if you have time to see before you buy. Have the sod dropped in shady areas all around the grounds to be covered, to reduce carrying. Keep sod slightly moist until installed. Have a good crew to help you lay sod because it will only live on pallets for a few days.

LD15 SOAK lawn with water if sod is applied. Sprinkle lightly for long periods of time if a seeded lawn is used. Roll lawn with a heavy metal cylinder.

LD16 PLANT bushes and trees. If you plan to put in trees, keep them at least 4′ from the driveway area. Trees are notorious for breaking up a driveway when the roots begin to spread.

LD17 PREPARE landscaped islands. Place pine bark, pine straw, mulch or gravel in designated areas. Over time, pine straw will kill some grasses, particularly Bermuda. Keep pine straw in islands only.

LD18 INSTALL mailbox if you have the kind that sits on a post. If you have a masonry or large frame mailbox, it will be installed during the framing or masonry stages of construction.

LD19 INSPECT landscaping job. Refer to specifications and checklist.

LD20 CORRECT any problems noted.

LD21 PAY landscaping sub for job. Have him sign an affidavit.

LD22 PAY landscaping sub retainage after grass is fully established without bare areas.

LANDSCAPING—Specifications

- Bid is to landscape as described below:
- Loosen soil in yard to a depth of 6″ with no dirt clumps larger than 6″.
- Finely rake all area to be seeded or sodded.
- Install the following soil treatments:

Peat moss	21 cubic feet per 1000 square feet
Sand	500 lbs. per 1000 square feet
Lime	200 lbs. per 1000 feet
6-6-6	as prescribed by mfr. for new lawns

- All grass is to be lightly fertilized with 6-6-6 and watered heavily immediately to prevent drying.
- Sod is to be Bermuda 419, laid within 48 hours of delivery.
- Sod pieces are to butt each other with no gaps and joints staggered.
- Islands are to be naturally shaped by trimming sod.
- Sod is to be rolled after watering.
- Prepare three pine straw islands as described in attached site plan.
- Put in trees and bushes according to attached planting schedule and site plan.
- Cover all seeded areas with straw to prevent runoff.
- Install splash blocks for downspouts.
- Burlap is to be removed from root balls prior to installation.
- Trees and shrubs are to be watered after installation.

LANDSCAPING—Inspection

- ☐ All areas are sodded/seeded as specified.
- ☐ All trees and bushes are planted as specified.
- ☐ Sod and/or seed is alive and growing.
- ☐ Sod is smooth and even with no gaps.
- ☐ Straw covers seed evenly with no bare spots.
- ☐ All specified bushes and trees are alive and in upright position.
- ☐ Splash blocks are installed.
- ☐ Grass is cut, if specified.
- ☐ No weeds are in yard.
- ☐ No exposed dirt is within 16″ of siding or brick veneer.
- ☐ Pine bark or pine straw islands are installed as specified.

Decks and Sunrooms

These structures may or may not be considered part of the house's heated living space. Sunrooms usually require a full foundation and roof and should usually be constructed along with the rest of the house. An exception is the prefabricated sunroom offered by many service companies. Decks are considered non-integral structures and will be the last structures to be added to your home. Sunrooms should usually be built by an experienced framer, since they require the same structural integrity as the rest of the house.

Porches are usually integral to the design of the house and will be framed along with the rest of the main structure. Their construction is the same as the house framing and is not covered in the steps here.

SUNROOMS

A sunroom is very popular because it adds light and conditioned space to the house. The least expensive installation will consist of fixed panes of glass, but without ventilation, this can cause the room to heat up unbearably in the summer. Sliding glass doors or screened windows will provide additional ventilation, but will also drive up the cost. Try to settle on a combination of fixed panes and screened openings that are strategically placed to provide cross ventilation. Ceiling fans will also help keep the air moving. Since heat buildup is a strong consideration, think about using energy efficient double-pane glass. This is commonly called Low-E glass because of its "low energy" requirement. This glass is usually tinted to keep out excessive heat in the summer, and the double panes help hold heat in during the winter. The additional cost will usually justify itself by creating a much more comfortable environment.

Solar heating in the winter is usually sufficient to heat a sunroom without additional heating systems. If the sunspace is open to the rest of the house, the heat can also be distributed to reduce conventional heating needs. The summer is another story, however. In southern climates, heat buildup can be considerable. Make sure to provide for adequate cross ventilation. Also consider installing several deciduous trees around the sunroom (but not too close!) to provide additional shade. They will drop their leaves in the winter and allow the solar energy in just when it is needed.

Many contractors specialize in installing new and retrofit sunrooms. They have all the specialized tools and knowledge to complete the job effectively. Many of these new prefab sunrooms are well-designed and cost-efficient. They usually consist of premanufactured metal channels, special "sunroom" windows and foam-core roofs. This job, unlike a deck, is best left to the professionals. Since most of the cost of a sunroom lies in the materials, you will save little by attempting the installation yourself.

DECKS

An outdoor deck is a standard do-it-yourself project for beginning and advanced remodelers. It provides maximum value for the amount invested and requires a moderate level of expertise to construct. Modern fasteners, custom lumber sizes, installation manuals and videos make the construction of a world-class deck an easy weekend project. Many building supply stores provide a computerized deck design program to help in determining your design and material costs. If you purchase the materials there, the service is usually free. If you decide to let the professionals do it, talk with the framer or trim carpenter who did the work on the main structure.

Decks are not considered part of the living space and can be added or expanded after initial construction, but the building inspector will require a deck with a handrail on any back entranceway taller than 30″. If you plan to add a more substantial deck later, plan a small section to meet code that can be easily expanded into the full-size deck at a later date. You should wait until all the grading and landscaping is complete before framing the deck. The landscaping grader will need access to the area close to the house for smoothing and backfilling.

Even though it is not part of the heated structure, decks can add considerably to the value and traffic flow of the house. Decks can also extend the perceived space of the house by providing alternate areas for relaxing and socializing. Decks are very inexpensive additions that provide an escape from the confined space of the house.

Plan for the future with the deck. Are there any new trees close to the deck area? These can grow large in a few years and extend limbs over the deck area. Are you planning to convert the deck into a sunroom at some future date? If so, build your deck with standard footings and foundation. This will allow you to incorporate the existing deck structure into the sunroom. Otherwise, you may have to rip up the entire deck and start over. Think about the placement of steps leading down from the deck. Will they work structurally if the sunroom is added later? Is there room and an attachment point for a roof over the deck? Have you planned for the flashing that will be required between the house and the sunroom's roof?

Types of Decks

Wood Deck

The most common decks are low- or high-level decks that are attached to the back of the house. Detached low-level decks are also popular on one-story houses and around pool or patio areas.

Most decks are constructed from planking nailed down with a space between each board for ventilation and drainage. Less common, but useful occasionally, are solid wood decks. These decks are made of caulked planking or exterior plywood with a waterproof coating. This kind of deck can serve double duty as both an upper level deck and a roof for a carport or playroom. Sealed decks are hard to build properly and seal thoroughly, so plan carefully before constructing one of these types.

Decks constructed close to the ground are easy to lay out, requiring a minimum foundation, sometimes consisting of nothing more than concrete blocks set in the ground on solid dirt. This type of deck can be attached to the house at the foundation with nails or lag bolts.

Concrete Deck

This is the simplest and most cost-effective surface to construct. Concrete slabs are usually poured whenever the house foundation itself is a slab or

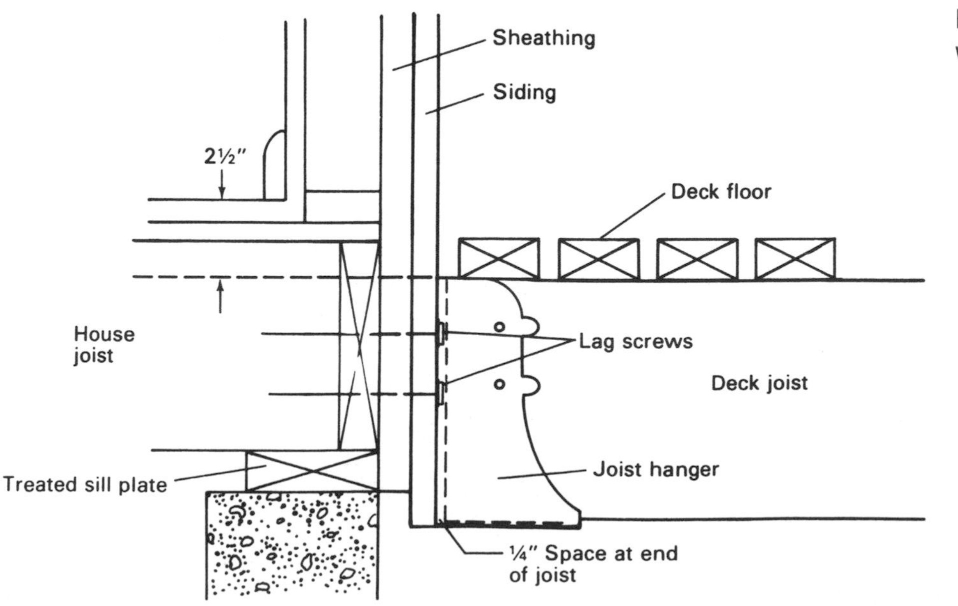

Figure 110: Deck attachment to house with joist hangers.

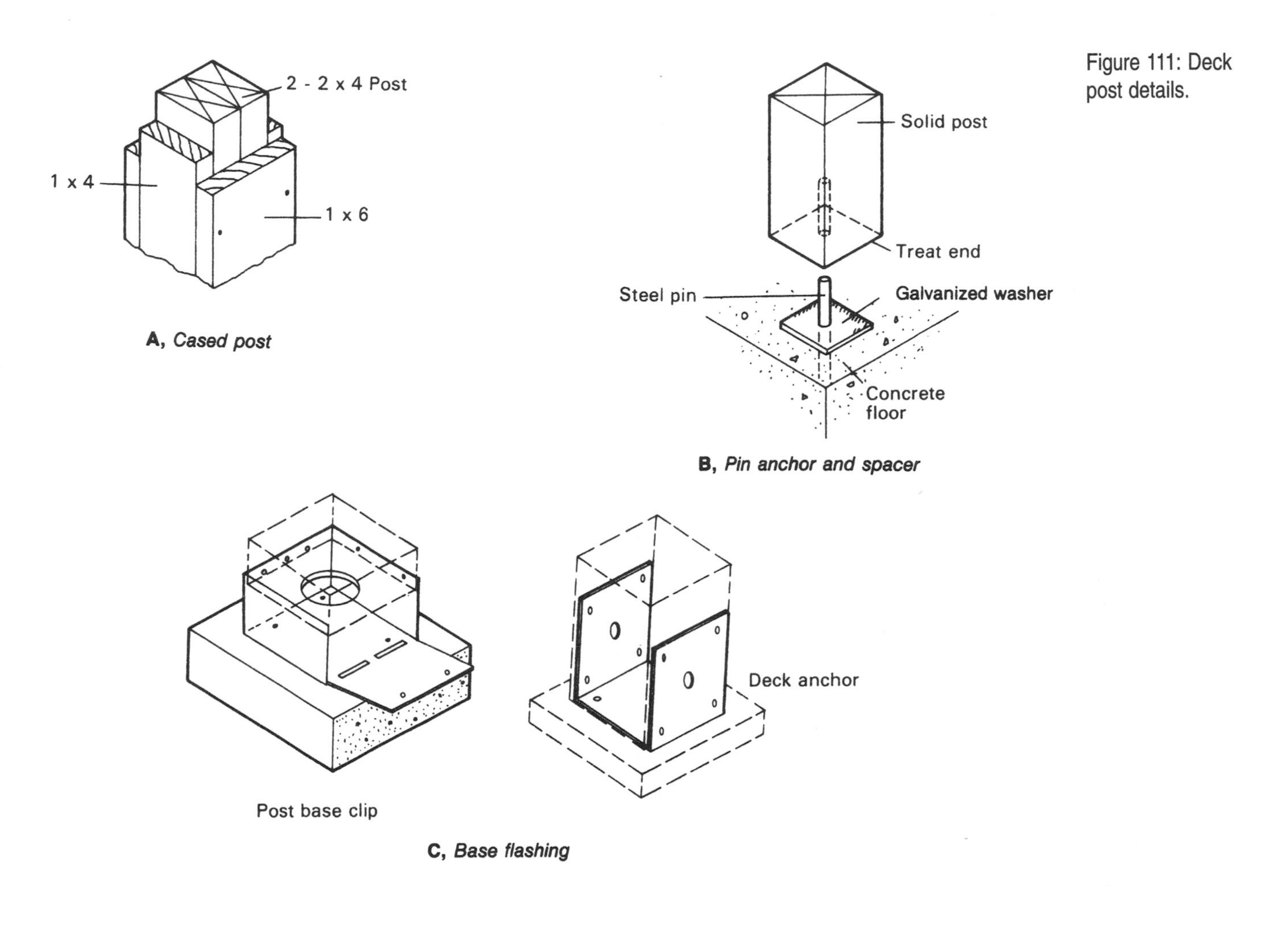

Figure 111: Deck post details.

close to the ground. If the house foundation is more than 30″ tall or the back yard area is uneven, slabs are impractical. Many people install a slab because it's inexpensive, planning to enhance it later by constructing a sunroom, porch or deck on it at some later date. In this case, you must use the standards for a regular foundation. Since porch slabs are not part of the structure, building codes don't usually require them to have reinforcement, edge footings, insect treatment or waterproofing. But if you plan to build a structure on it later, this standard will be inadequate. The slab will crack and settle under the weight of the new structure. Unless the concrete subcontractor knows your plans to build on the slab later, he will simply pour the least expensive slab possible. (This may become the most expensive slab you ever bought if you have to rip it up later!) If you plan to build a structure on it, ask the contractor to pour a full foundation slab as described in the Foundation section. It will save you headaches later!

Choosing Deck Materials

All deck materials are readily available at most building supply stores at reasonable cost. Just make sure that you buy material designed to hold up to the rigors of outside use and abuse, such as wind, rain, sunlight, snow and pests.

In the past, redwood was a popular deck material. It was very attractive and stood up well to wind and weather. Unfortunately, redwood has become extremely expensive, which may lower your enthusiasm for using this traditional material. Cypress is another very attractive but expensive material suitable for decks. Consider these materials if you are looking for a distinctive "cost-is-no-object" project.

Plywood is a common covering material for solid decks because it is easy to seal and make totally waterproof. A word of caution: Make sure that you use only exterior-grade plywood. Exterior-grade

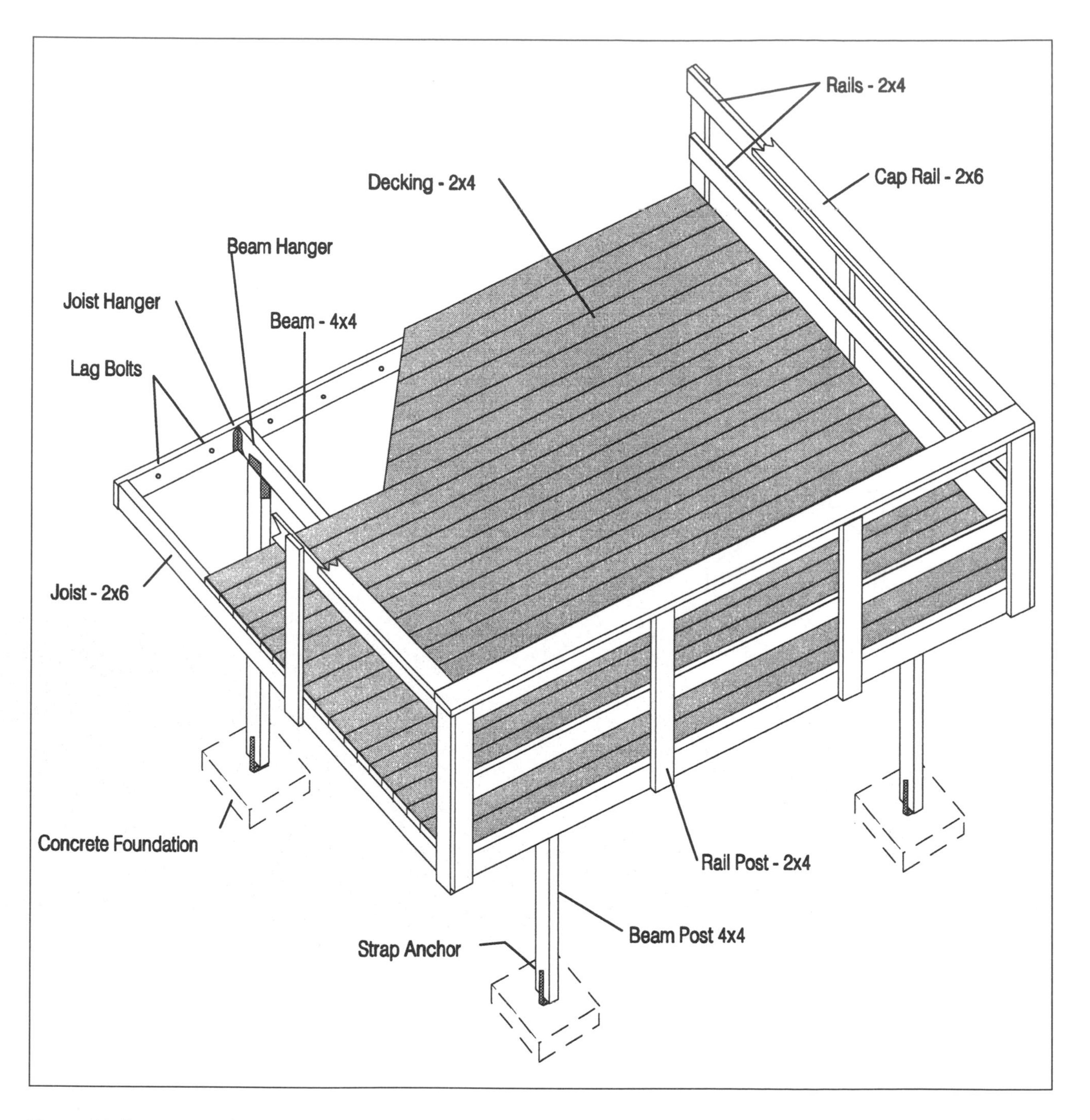

Figure 112: Components of a typical outdoor deck.

plywoods use special glues that are resistant to moisture. Plywood decks are notoriously difficult to drain properly, and will gladly succumb to rotting, splitting and termites if constructed out of inferior-grade materials or without proper drainage. For a premium-quality sealed deck, use marine-grade "high-density overlay" plywood with a skid-resistant surface.

Currently, the most common deck material is pressure-treated lumber. The wood has been impregnated under pressure with a copper arsenate solution that sinks into the wood, providing a pest- and rot-proof material that is more than skin deep. The copper solution provides an attractive color and surface and is the ultimate in simplicity for construction. Properly treated lumber can even be buried in moist soil with little danger of rotting or pests for 20 years or more.

With any decking material chosen, make sure to personally pick pieces for your project that are free from cracks, knots and warping. If a piece is warped before you use it, think what it will look like after a year in the sun!

Lumber Sizes

The most common lumber sizes for decks are the commonly used framing sizes; i.e., 2×4, 2×6 and 2×8. These sizes are so commonly used in house construction that they are very cost-effective. The 2″ thickness of the lumber allows for greater spans between supporting joists than standard 1″ flooring material. Use 2×4s for the floor decking whenever possible. Any stock wider than 4″ (such as 2×6) is more susceptible to warping and splitting, as well as being more expensive. A newer, more inexpensive deck board has become popular lately. It is a little over 1″ thick and has rounded edges for a pleasing appearance. If you use this deck board, make sure to decrease the span between joists to increase support.

Construction Hints

Proper drainage on a deck is all-important, so avoid construction techniques that allow water to build up and sit on parts of the deck. If the seam between the deck and the house is exposed to weather, cover it with flashing to prevent water from settling in the crack and rotting the siding of the house. On planked decks, make sure to leave a drainage gap of at least ⅛″ between each board. The easiest way to accomplish this is to use a 12d or 16d nail as a spacer when nailing down the boards. The surface of the deck should be slanted slightly away from the house to speed up drainage.

Galvanized joist hangers will greatly speed up the construction project and are much easier to use than toenailing the lumber into adjacent members. This is especially true when attaching the joists to a beam that is bolted to the house. Make sure the joist hangers are exterior-grade galvanized, to prevent rusting. You should use hot-dipped galvanized nails or screws if you want to prevent ugly rust spots around the nail heads. Smooth shank nails often lose their holding power when exposed to wetting and drying cycles. To prevent slippage, use a ring shank nail, spirally grooved nail or, better yet, the deck screw. The advent of modern battery-powered drills has made deck screws the fastener of choice. They go in easily (with no banged fingers), hold better than nails

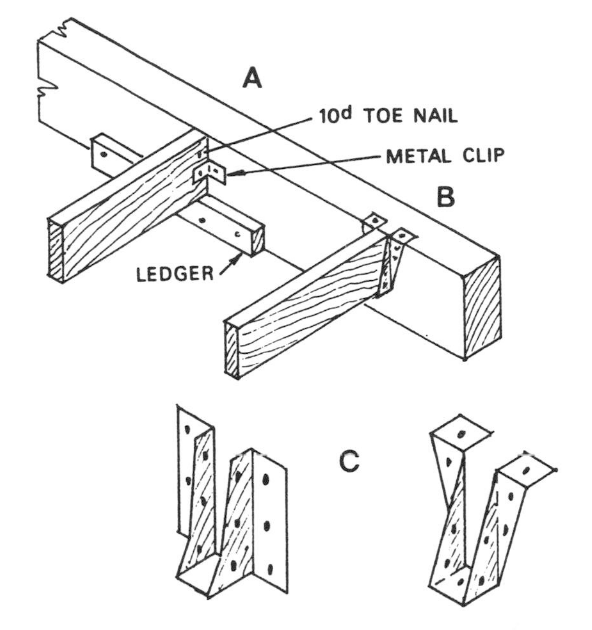

Figure 113: Joists between beams: A, ledger support; B, joist hanger support; C, joist hangers.

and can be removed if necessary. As with the nails, make sure to use galvanized deck screws.

Deck Foundation

In northern climes, you may need to pour a post foundation. This consists of a deep hole dug below the frost line, usually with a post hole digger, that is filled with concrete. The deck beam is then placed on top of the concrete column. In older construction, the beam is sometimes cast into the concrete. Avoid this technique if possible—the beam is exposed to a lot of moisture and will be difficult to replace if decayed.

The beams underneath the deck should rest securely on concrete foundation piers that have been poured. The piers should extend down to solid ground below the frost line for your area. Securely fasten the post to the foundation so that the post cannot slide laterally off the foundation. Galvanized strap anchors provide a convenient way to secure the deck beam to the foundation. These anchors will keep the beam from shifting or sliding and will keep the end of the beam from contacting the moist ground or foundation. This will prevent rotting of the end of the beam.

If the deck is close to the ground, you may have

trouble with grass and weeds growing up between the boards, with no way to cut or trim them. You can prevent this by covering the ground with a layer of sand or fine gravel.

Finishing

Deck surfaces are difficult to paint or stain because of the amount of abuse and foot traffic the surfaces will endure. Common paints and stains simply wear off the surface, leaving unattractive paths at points of heavy foot traffic. The best coating for a deck consists of transparent or semitransparent water-resistant compounds. These will seal the surface and reduce warping, splitting and fading without changing the natural coloring of the wood.

Pressure-treated wood has been saturated with a decay-resistant compound and may be very moist. Make sure the wood is completely dry before applying penetrating water-resistant compounds so that they can soak into the wood. This can take six to eight weeks or longer in cold weather. If you can purchase the wood ahead of time, find a dry area to stack the lumber, using spacers to allow ventilation. If the lumber has dried for at least four weeks, you can coat the underside of the lumber with penetrating finish, since this side will be difficult to reach after installation. Leave the other side uncoated so the lumber can continue to dry after installation. Apply the final coat to the exposed lumber after it has completely dried. See the Painting chapter for more information.

Safety

Even though constructing a deck is a simple and straightforward project, don't forget that this will become a structural part of your house. Think about the maximum number of guests that might crowd onto the deck. There have been many instances of decks collapsing under the weight of crowds of party-goers much greater than the deck was designed to handle. Better safe than sorry. Make sure to follow proper construction techniques to ensure that your deck structure is securely attached to the house and that proper bracing is applied to the deck posts where needed. If your deck rises more than three feet off the ground, you may need to reinforce the construction more carefully. As a rule, any deck over 6′ tall that is not securely fastened to the house should have diagonal braces attached between the deck posts.

Whenever possible, fasten the deck securely to the house with lag bolts screwed directly into the foundation or floor frame. Nails simply are not adequate. Use a lag bolt long enough to penetrate the deck header, house siding, and three-quarters the thickness of the house's framing.

DECK—Steps

DE1 EXAMINE the building site carefully, taking into consideration the height of deck in relation to the house and surrounding lot. Decks high above the ground will require larger supporting posts and perhaps cross bracing. Take this into consideration when ordering material.

DE2 DRAW a complete plan for the deck.

DE3 PURCHASE materials for deck, including nails, hangers, lumber and concrete for footings. Lay out lumber so it can balance its moisture content with the local humidity. Store ready-mix concrete inside or under a protective plastic cover to keep it from getting wet.

DE4 FINISH all grading and backfilling before starting deck construction.

DE5 LAY OUT the deck design with string and stakes to locate the positions of posts and concrete piers. Measure the diagonal distances from corner to corner of the layout to square up the design. The diagonal distances should be the same from corner to corner.

DE6 DIG footings and pour concrete piers. Place the tops of the piers slightly above the height of the surrounding terrain to prevent water from collecting around the piers. The hole itself can function as the form for the pier. Use 1×4s at the top of the hole as forms to extend the concrete pier above ground level and to provide a smoother surface on the exposed part of the pier. Insert any embedded-type post anchors into the concrete while it is still wet.

DE7 PULL a leveling line from corner to corner at the finish height of the deck and mark each corner post. The leveling line can be made with string and a string level or by using a garden hose. When using the garden hose,

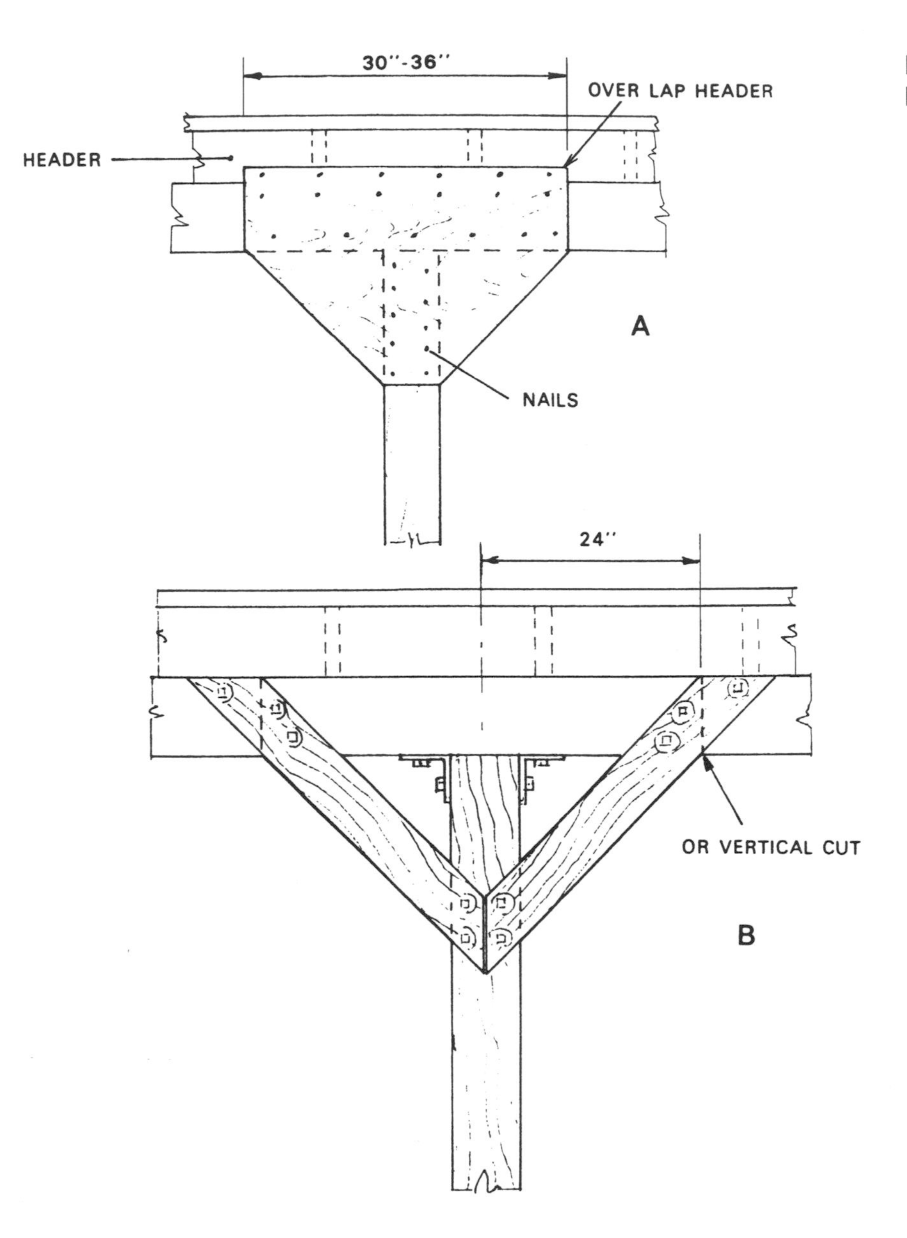

Figure 114: Partial braces: A, plywood gusset; B, lumber brace.

brace the posts in place and strap the ends of the garden hose to the posts. Fill the hose with water and adjust each end of the hose until each end is full of water all the way to the top. Keep adding water until this is accomplished. Since water always seeks its own level, this method produces very accurate results. Mark the posts at each end of the hose. When marking the posts, make sure to allow for the height of the anchor plate that will attach the post to the concrete pier.

DE8 INSTALL posts on pier anchors and make sure they align with each other and are plumb. Use a carpenter's level to adjust the posts. Attach temporary braces to the posts to hold them in the correct position.

DE9 ATTACH deck beams to the posts with lag bolts or galvanized beam hangers. This operation will take at least two people to position and attach the beams. Make sure the structure remains square as you put each corner of the deck together.

DE10 INSTALL flashing on the tops of the beams where the beam and post are joined. This will keep rain from soaking the end grain of the post.

DE11 MOUNT the joist header to the house by running lag or stove bolts through the header, siding and the header of the house. Make

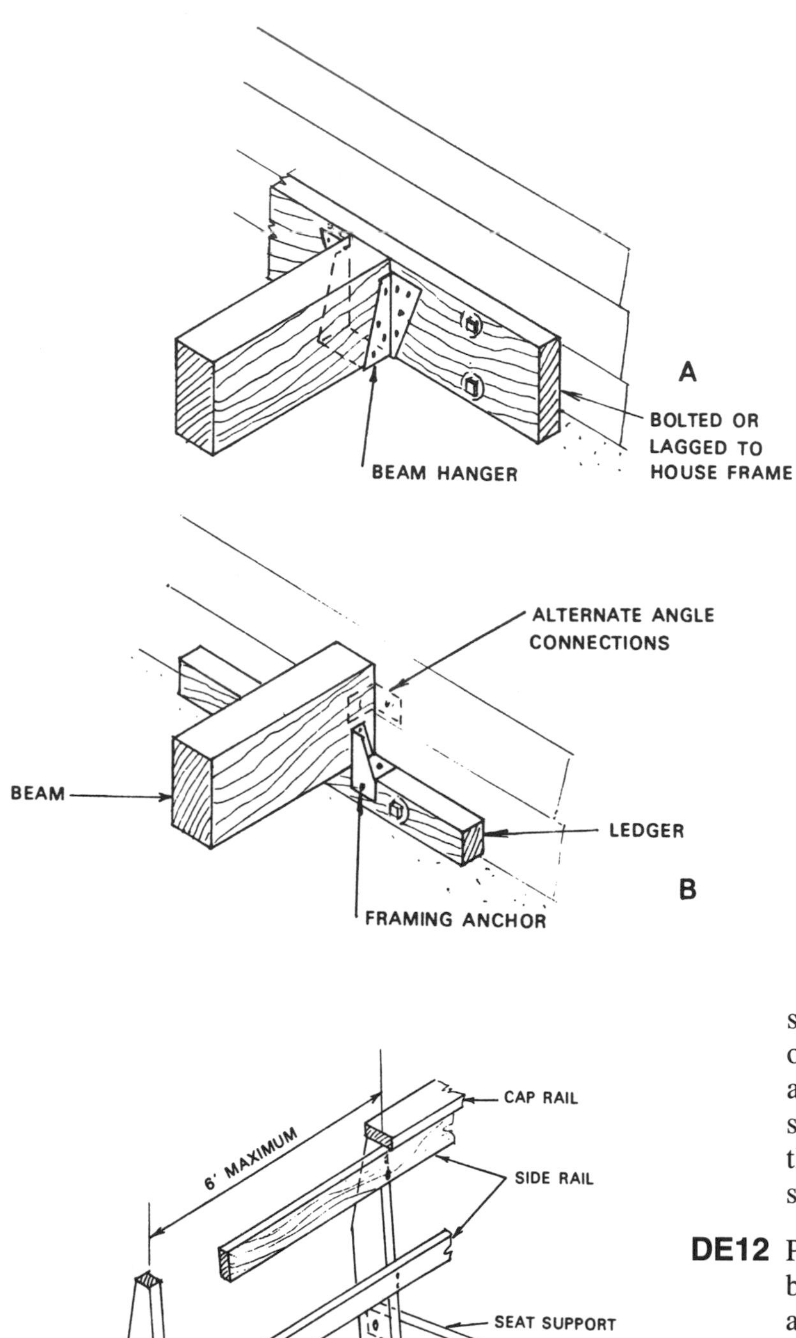

Figure 115: Beam to house: A, beam hanger; B, ledger support.

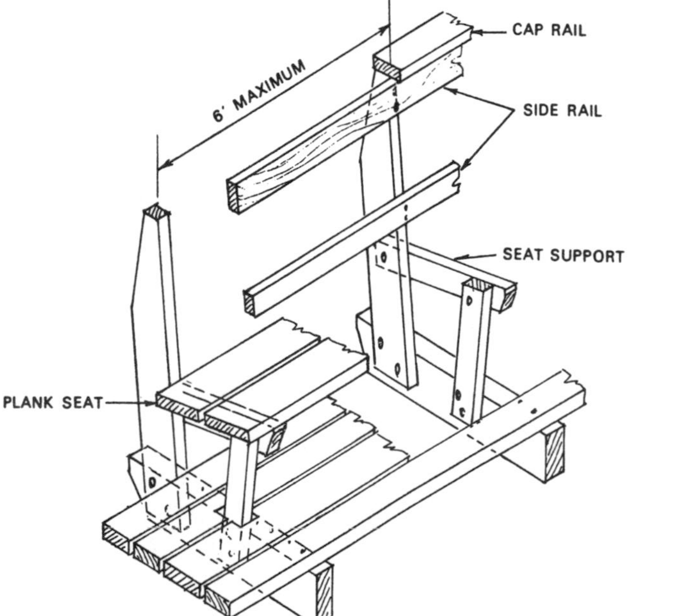

Figure 116: Deck bench.

sure to install a drip cap or flashing on top of the header to prevent standing water from accumulating between the header and the siding. The header should be placed so that the deck will slope away from the house slightly for better drainage.

DE12 PLACE the remaining headers on top of the beams and tie them together with bolts or angle irons.

DE13 INSTALL joists between the resulting frame members with joist hangers or by resting them on ledgers nailed to the joist headers.

DE14 MOUNT cross bracing between the posts if the height of the deck is greater than 5′. This will strengthen the deck against side-to-side movement. Use 2×4s for the bracing and attach them to the posts and joist header with galvanized stove bolts.

DE15 APPLY the deck boards to the joists with galvanized deck nails or screws. The best nails for the job are ring shank or spiral groove. Decking screws have become very popular recently for several reasons. They have greater gripping power and can pull a warped board into alignment during installation. They have the added benefit of being removed easily. If one of the deck boards warps or splits, simply unscrew and replace it with a new board. As the deck boards dry out, you can easily go back and retighten the deck screws to maintain a snug fit.

Leave gaps between boards for better drainage. The easiest way to do this is to use a 12d or 16d nail as a spacer. If the moisture content of the wood is high, leave a smaller gap—it will enlarge as the board dries and shrinks. Don't worry about cutting the exact length of each deck board. Leave a little extra hanging over the header joist.

DE16 SAW the deck boards flush with the header joist with a power saw after all boards are installed. This is much quicker, especially if you installed the deck boards diagonally. Pull a chalk line parallel to the header joist and follow the line with the power saw. This will give you a straight and accurate line.

DE17 ATTACH stairway stringers to header joist with galvanized straps and mount the other ends of the stringers to the concrete piers with post anchors.

DE18 INSTALL the treads between the stringers.

DE19 INSTALL post rails around the perimeter of the deck and down the stair stringers. Attach the cap rails to the tops of the posts.

DE20 TREAT the deck with two coats of water-resistant wood preservative. Apply the second coat after the first coat has soaked in but before it has dried. This will ensure that the second coat can soak into the wood completely.

Sources of Information

ASSOCIATIONS

APA—The Engineered Wood Association
7011 S. 19th St.
Tacoma, WA 98466-5399
(206) 565-6600

Air Conditioning & Refrigeration Institute
4301 N. Fairfax Dr., Suite 425
Arlington, VA 22203
(703) 524-8800

Aluminum Association (AA)
818 Connecticut Ave., N.W.
Washington, DC 20006

American Institute of Timber Construction
7012 S. Revere Pkwy., Suite 140
Englewood, CO 80112
(303) 792-9559

American Architectural Manufacturers Association
1540 E. Dundee Rd., Suite 310
Palatine, IL 60067-8316
(708) 202-1350

American Concrete Institute (ACI)
P.O. Box 19150
Detroit, MI 48219

American Gas Association
1515 Wilson Blvd.
Arlington, VA 22209
(703) 841-8589

American Hardboard Association
1210 W. Northwest Hwy.
Palatine, IL 60067-3607
(708) 934-8800

American Hardware Manufacturer's Association
801 N. Plaza Dr.
Schaumburg, IL 60173-4977
(847) 605-1025

American Institute of Architects
1735 New York Ave. N.W.
Washington, DC 20006
(202) 626-7300

American Institute of Steel Construction, Inc.
1221 Avenue of the Americas, Suite 1980
New York, NY 10020

American Institute of Timber Construction (AITC)
333 West Hampden Ave.
Englewood, CO 80110

American Iron and Steel Institute (AISI)
1000 16th St., N.W.
Washington, DC 20036

American Lighting Association
2050 Stemmons Fwy., Suite 10046
Dallas, TX 75258
(214) 698-9898, (800) 605-4448

American National Standards Institute (ANSI)
1430 Broadway
New York, NY 10018

American Plywood Association (APA)
1119 A St.
Tacoma, WA 98401

American Society for Testing Materials (ASTM)
1916 Race St.
Philadelphia, PA 19103

American Society of Heating, Refrigeration and Air Conditioning Engineers
United Engineering Center, 345 East 47th St.
New York, NY 10017

American Society of Interior Designers
1420 Broadway
New York, NY 10018
(212) 944-9220

American Welding Society, Inc. (AWS)
2501 N.W. 7th St.
Miami, FL 33125

American Wood Council American Forest & Paper Association
1111 19th St., N.W.
Washington, DC 20036
(202) 463-2769

American Wood Preservers Association (AWPA)
1625 I Street, N.W.
Washington, DC 20006

American Wood Preservers Bureau (AWPB)
P.O. Box 6085
Arlington, VA 22206

American Wood Preservers Institute (AWPI)
1651 Old Meadow Rd.
Arlington, VA 22101

Asphalt Roofing Industry Bureau (ARIB)
750 Third Ave.
New York, NY 10017

Association of Home Appliance Manufacturers
20 N. Wacker Dr.
Chicago, IL 60606-2806
(312) 984-5800

Association of Wall & Ceiling Industry International
307 E. Annandale Rd., Suite 200
Falls Church, VA 22042
(703) 534-8300

Better Heating & Cooling Council
P.O. Box 218

Berkeley Heights, NJ 07922
(908) 464-8200

Brick Institute of America
11490 Commerce Park Dr.
Reston, VA 22091-1525
(703) 620-0100

California Redwood Association
405 Enfrente Dr., Suite 200
Novato, CA 94949
(415) 382-0662

Canadian Wood Council
1730 St. Laurent Blvd., Suite 350
Ottawa, ON K1G 5L1 Canada
(613) 247-7077

Carpet and Rug Institute
P.O. Box 2048
Dalton, GA 30722-2048
(706) 278-3176

Cedar Shake & Shingle Bureau
515 116th Ave., N.W., Suite 275
Bellevue, WA 98004-5294
(206) 453-1323

Cellulose Insulation Manufacturer's Association
136 S. Keowee St.
Dayton, OH 45402
(937) 222-2462

Cellulose Marketing Council
P.O. Box 440
Meadow Vista, CA 95722
(916) 888-7456

Ceramic Tile Distributors Association
800 Roosevelt Rd., Bldg. C, Suite 20
Glen Ellyn, IL 60137
(708) 545-9415

Commercial Standards (U.S. Department of Commerce)
Superintendent of Documents, Government Printing Office
Washington, DC 20402

Concrete Reinforcing Steel Institute (CRSI)
180 North LaSalle St.
Chicago, IL 60601

Federal Specifications (FS)
Superintendent of Documents, Government Printing Office
Washington, DC 20234

Forest Products Laboratory (FPL)
United States Department of Agriculture
Madison, WI 53705

Forest Products Society
2801 Marshall Ct.
Madison, WI 53705
(608) 231-1361

Garage Door Hardware Association
2850 S. Ocean Blvd., Suite 311
Palm Beach, FL 33480-5535
(407) 533-0991

Gas Research Institute
8600 W. Bryn Mawr
Chicago, IL 60631
(312) 399-8249

Gypsum Association
810 First St., N.E., Suite 510
Washington, DC 20002
(202) 289-5440

Hardwood Plywood & Veneer Association
P.O. Box 2789
Reston, VA 22090-0789
(703) 435-2900

Home Automation Association
808 17th St., N.W., Suite 200
Washington, DC 20006-3910
(202) 223-9669

Home Ventilation Institute
30 West University Dr.
Arlington Heights, IL 60004-1806
(708) 394-0150

Hydronics Institute (HI)
35 Russo Place
Berkely Heights, NJ 07922

Insulated Steel Door Institute
30200 Detroit Rd.
Cleveland, OH 44145-1967
(216) 899-0010

Insulation Contractors Association
P.O. Box 26237
Alexandria, VA 22313
(703) 739-0356

International Masonry Industry All-Weather Council (IMIAWC)
208 South LaSalle St.
Chicago, IL 60604

International Society of Interior Designers
433 South Spring St., Suite 6-D
Los Angeles, CA 90013
(213) 680-4240

Italian Tile Association
305 Madison Ave., Suite 3120
New York, NY 10165-0111
(212) 661-0435

Laminating Materials Association
116 Lawrence St.
Hillsdale, NJ 07642-2730
(201) 664-2700

Manufacturing Chemists' Association, Inc.
1825 Connecticut Ave., N.W.
Washington, DC 20009

Marble Institute of America
33505 State St.
Farmington, MI 48335-3569
(810) 476-5558

Metal Building Manufacturers Association (MBMA)
2130 Keith Building
Cleveland, OH 44115

Metal Lath Association (MLA)
221 N. LaSalle St., Suite 1507
Chicago, IL 60601

Mobile Homes Manufacturers Association (MHMA)
14650 Lee Rd.
Chantilly, VA 22021

National Association of Brick Distributors
1600 Spring Hill Rd., Suite 305
Vienna, VA 22182
(703) 749-NABD

National Association of Building Manufacturers (NABM)
1619 Massachusetts Ave., N.W.
Washington, DC 20036

National Association of Floor Covering Distributors
401 N. Michigan Ave.
Chicago, IL 60611
(312) 644-6610

National Association of Home Builders (NAHB)

15th & M Streets, N.W.
Washington, DC 20005

National Association of Mirror Manufacturers
9005 Congressional Ct.
Potomac, MD 20854-4608
(301) 365-4080

National Association of the Remodeling Industry
1901 N. Moore St., Suite 808
Arlington, VA 22209
(703) 276-7600

National Automatic Sprinkler and Fire Control Association, Inc. (NASFCA)
45 Kensico Dr.
Mt. Kisco, NY 10549

National Bureau of Standards (NBS)
(U.S. Department of Commerce)
Superintendent of Documents,
Government Printing Office
Washington, DC 20402

National Clay Pipe Institute (NCPI)
1130 17th St., N.W.
Washington, DC 20036

National Concrete Masonry Association (NCMA)
2302 Horse Pen Rd.
Herndon, VA 22071-3499
(703) 713-1900

National Environment Systems Contractors Association (NESCA)
1501 Wilson Blvd.
Arlington, VA 22209

National Fenestration Rating Council
1300 Spring St., Suite 120
Silver Spring, MD 20910
(301) 589-6372

National Fire Protection Association (NFPA)
470 Atlantic Ave., N.W.
Washington, DC 20036

National Forest Products Association
1619 Massachusetts Ave., N.W.
Washington, DC 20036

National Kitchen & Bath Association
687 Willow Grove St.
Hackettstown, NJ 07840
(908) 852-0033

National Lime Association (NLA)
4000 Brandywine St., N.W.
Washington, DC 20016

National Mineral Wool Association, Inc. (NMWAI)
382 Springfield Ave., Suite 312
Summit, NJ 07901

National Particleboard Association (NPA)
18928 Premiere Ct.
Gaithersburg, MD 20879-1569
(301) 670-0604

National Pest Control Association (NPCA)
250 West Jersey St.
Elizabeth, NJ 07202

National Propane Gas Association
1600 Eisenhower Ln., Suite 100
Lisle, IL 60532
(708) 515-0600

National Spa and Pool Institute
2111 Eisenhower Ave.
Alexandria, VA 22314-4698
(703) 838-0083

National Tile Contractors Association
P.O. Box 13629
Jackson, MS 39236
(601) 939-2071

National Wood Window & Door Association
1400 E. Touhy Ave., Suite G-54
Des Plaines, IL 60018-3305
(708) 299-5200, (800) 223-2301

Northeastern Lumber Manufacturer's Association
P.O. Box 87A
Cumberland Center, ME 04021-0687
(207) 829-6901

Perlite Institute (PI)
45 West 45th St.
New York, NY 10036

Pittsburgh Testing Laboratory (PTL)
1330 Locust St.
Pittsburgh, PA 85013

Plumbing Manufacturers Institute
800 Roosevelt Rd., Bldg. C, Suite 20
Glen Ellyn, IL 60137-5833
(708) 858-9172

Portland Cement Association (PCA)
5420 Old Orchard Rd.
Skokie, IL 60076

Post-Tensioning Institute (PTI)
Suite 3500, 301 West Osborn
Phoenix, AZ 85013

Prestressed Concrete Institute (PCI)
20 N. Wacker Dr.
Chicago, IL 60606

Product Fabrication Service (PFS)
1618 West Beltline Highway
Madison, WI 53713

Red Cedar Shingle and Handsplit Shake Bureau (RDSHSB)
5510 White Building
Seattle, WA 98101

Sheet Metal & Air Conditioning Contractor's National Association (SMACCNA)
8224 Old Court House Rd.
Vienna, VA 22180

Society of Plastics Industry, Inc. (SPI)
355 Lexington Ave.
New York, NY 10017

Solar Energy Industries Association
122 C St., N.W., Fourth Fl.
Washington, DC 20001
(202) 383-2600

Southern Forest Products Association (SFPA)
P.O. Box 52468
New Orleans, LA 70152

Southern Gas Association (SGA)
4230 LBJ Freeway, Suite 414
Dallas, TX 75234

Southern Pine Council
P.O. Box 641700
Kenner, LA 70064-1700
(504) 443-4464

Southern Pine Inspection Bureau (SPIB)
P.O. Box 846
Pensacola, FL 32502

Southwest Research Institute (SRI)
8500 Culebra Rd.
San Antonio, TX 78228

Steel Door Institute
30200 Detroit Rd.

Cleveland, OH 44145-1967
(216) 899-0010

Steel Joist Institute (SJI)
703 Parham Rd.
Richmond, VA 23229

Steel Window Institute (SWI)
2130 Keith Building
Cleveland, OH 44115

Structural Board Association
45 Sheppard Ave. E., Suite 412
Willowdale, ON M2N 5W9 Canada
(416) 730-9090

Tile Council of America
P.O. Box 1787
Clemson, SC 29633-1787
(864) 646-TILE

Truss Plate Institute (TPI)
1800 Pickwick Ave.
Glenville, IL 60025

Underwriters' Laboratories, Inc. (ULI)
333 Pfingsten Rd.
Northbrook, IL 60062

Western Wood Products Association
522 S.W. Fifth Ave., Suite 400
Portland, OR 97204
(503) 224-3930

Wood Moulding & Millwork Producers Association
507 First St.
Woodland, CA 95695
(916) 661-9591

Woodwork Institute of California
P.O. Box 980247
Sacramento, CA 95798-0247
(916) 372-9943

World Floor Covering Association
2211 E. Howell Ave.
Anaheim, CA 92806
(714) 978-6440, (800) 624-6880

LOANS—FANNIE MAE REGIONAL OFFICES

Home Office
FNMA
3900 Wisconsin Ave. NW
Washington, DC 20016-2899
(202) 752-7000

Midwestern Regional Office
FNMA
One South Wacker Dr., Suite 3100
Chicago, IL 60606-4667
(312) 368-6200

Northeastern Regional Office
FNMA
1900 Market St., 8th Floor
Philadelphia, PA 19103
(215) 575-1400

Southeastern Regional Office
FNMA
950 East Paces Ferry Rd., Suite 1900
Atlanta, GA 30326-1161
(404) 365-6000

Southwestern Regional Office
FNMA
Two Galleria Tower, 13455 Noel Rd., Suite 600
Dallas, TX 75265-5003
(214) 991-7771

Western Regional Office
FNMA
135 North Los Robles Ave., Suite 300
Pasadena, CA 91101-1707
(818) 568-5000

LOANS—FREDDIE MAC REGIONAL OFFICES

Corporate Headquarters
Federal Home Loan Mortgage Corp.
1759 Business Center Dr.
Reston, VA 22090
(703) 759-8000

North Central Regional Office
Federal Home Loan Mortgage Corp.
333 West Wacker Dr., Suite 3100
Chicago, IL 60606-1287
(312) 407-4000

Northeast Regional Office
Federal Home Loan Mortgage Corp.
2231 Crystal Dr., Suite 900, P.O. Box 2408
Arlington, VA 22202-3798
(703) 685-4500

Seattle Branch
Federal Home Loan Mortgage Corp.
600 Stewart St., Suite 720
Seattle, WA 98101
(206) 622-9904

Southeast Regional Office
Federal Home Loan Mortgage Corp.
2839 Paces Ferry Rd., N.W., Suite 700, P.O. Box 723788
Atlanta, GA 30339
(404) 438-3800

Southwest Regional Office
Federal Home Loan Mortgage Corp.
12222 Merit Dr., Suite 700
Dallas, TX 75251
(214) 702-2000

Western Regional Office
Federal Home Loan Mortgage Corp.
15303 Ventura Blvd., Suite 500
Sherman Oaks, CA 91403
(818) 905-0070

LOANS—HUD OFFICES

Albuquerque Service Office
HUD
625 Truman St., N.E.
Albuquerque, NM 87110
(505) 766-3249

Anchorage Area Office
HUD
334 West Fifth Ave.
Anchorage, AK 99501
(907) 271-4175

Atlanta Area Office
HUD
Richard B. Russell Federal Building, 75 Spring St., S.W.
Atlanta, GA 30303
(404) 221-4017

Baltimore Area Office
HUD
Equitable Building, 10 N. Calvert St.
Baltimore, MD 21202
(310) 962-2144

Birmingham Area Office
HUD
Daniel Building, 15 S. 20th St.
Birmingham, AL 35322
(202) 254-1611

Boise Service Office
HUD
800 Park Blvd.

Boise, ID 83705
(208) 334-1338

Boston Area Office
HUD
Bulfinch Building, 15 New Chardon St.
Boston, MA 02114
(617) 223-4182

Buffalo Area Office
HUD
Statler Building Mezzanine, 107 Delaware Ave.
Buffalo, NY 11202
(716) 846-5710

Carribbean Area Office
HUD Federico Deoptau Federal Bldg.
U.S. Court House, Room 428, Carlos E. Chardon Ave.
Hato Rey, Puerto Rico 00918

Charleston MF Service Office
HUD
Kanawha Valley Building, Capito and Lee Streets
Charleston, WV 25301
(304) 347-7064

Chicago Area Office
HUD
One North Dearborn
Chicago, IL 60602
(312) 353-9174

Cincinnati MF Service Office
HUD
550 Main St.
Cincinnati, OH 45202
(513) 684-2884

Cleveland MF Service Office
HUD
770 Rockwell Ave., 2nd Floor
Cleveland, OH 44114
(216) 522-4032

Columbia Area Office
HUD
Strom Thurmond Federal Building
1835-45 Assembly St.
Columbia, SC 29201
(803) 765-5826

Columbus Area Office
HUD
200 N. High St.
Columbus, OH 43215
(614) 469-5704

Dallas Area Office
HUD
2001 Bryan Tower, 4th Floor
Dallas, TX 75201
(214) 767-8394

Denver Regional Area Office
HUD
Executive Tower Building, 1405 Curtis St.
Denver, CO 80202
(303) 837-4721

Des Moines MF Service Station
HUD
210 Walnut St., Room 259
Des Moines, IA 50309
(515) 284-4770

Detroit Area Office
HUD
McNamara Federal Building, 477 Michigan Ave.
Detroit, MI 48226
(313) 226-4817

Grand Rapids Service Office
HUD
2922 Fuller Ave., N.E.
Grand Rapids, MI 49505
(616) 456-2214

Greensboro Area Office
HUD
415 N. Edgeworth St.
Greensboro, NC 27401
(919) 378-5673

Hartford Area Office
HUD
One Hartford Square West, Suite 204
Hartford, CT 06106
(203) 244-2317

Honolulu Area Office
HUD
300 Ala Moana Blvd.
Honolulu, HI 94830
(808) 546-2137

Houston MF Service Office
HUD
Two Greenway Plaza East, Suite 200
Houston, TX 77046
(713) 226-4352

Indianapolis Area Office
HUD
151 N. Delaware St., P.O. Box 7047
Indianapolis, IN 46207
(317) 269-2087

Jackson Area Office
HUD
U.S. Federal Building, 100 W. Capital St., Room 1016
Jackson, MS 39201
(601) 960-4719

Jacksonville Area Office
HUD
Peninsular Plaza, 661 Riverside Ave.
Jacksonville, FL 32202
(904) 791-2953

Kansas City Area Office
HUD
Professional Building, 1103 Grand St.
Kansas City, MO 64106
(816) 374-6125

Knoxville Area Office
HUD
1 Northshire Building,
1111 Northshire Dr.
Knoxville, TN 37919
(615) 588-1477

Little Rock Area Office
HUD
One Union North Plaza, Suite 1400
Little Rock, AR 72201
(501) 378-6148

Los Angeles Area Office
HUD
2500 Wilshire Blvd.
Los Angeles, CA 90057
(213) 688-5978

Louisville Area Office
HUD
539 River City Mall, P.O. Box 1044
Louisville, KY 40202
(502) 582-6467

Lubbock Service Office
HUD
Federal Building, 1205 Texas Ave.
Lubbock, TX 79408
(806) 762-7275

Manchester MF Service Office
HUD
Norris Cotton Federal Building, 275 Chestnut St.
Manchester, NH 03103
(603) 666-7684

Milwaukee Area Office
HUD
744 N. Fourth St.
Milwaukee, WI 53203
(414) 291-1028

Minneapolis-St. Paul Area Office
HUD
220 South Second St., Bridge Place Building
Minneapolis, MN 55803
(612) 349-2095

Nashville MF Service Office
HUD
1 Commerce Place, Suite 1600
Nashville, TN 37219
(615) 251-5069

New Orleans Area Office
HUD
1001 Howard Plaza Tower
New Orleans, LA 70113
(504) 589-6635

New York Area Office
HUD
26 Federal Plaza
New York, NY 10278
(212) 264-4975

Newark Area Office
HUD
Gateway Building No. 1, Raymond Plaza
Newark, NJ 07102
(201) 645-3230

Omaha Area Office
HUD
Univac Building, 7100 West Center Rd.
Omaha, NE 68106
(402) 229-9428

Philadelphia Area Office
HUD
625 Walnut St.
Philadelphia, PA 19106
(215) 597-3409

Phoenix Service Office
HUD
Arizona Bank Building, 101 N. First Ave., Suite 1800
Phoenix, AZ 85002
(602) 261-4497

Pittsburgh Area Office
HUD
Fort Pitt Commons, 455 Fort Pitt Blvd.
Pittsburgh, PA 15219
(412) 644-3431

Portland Area Office
HUD
Cascade Building, 520 S.W. Sixth Ave.
Portland, OR 97204
(503) 326-3107

Providence MF Service Office
HUD
Room 330, John O. Pastore Federal Building
Providence, RI 02903
(401) 528-4835

Richmond Area Office
HUD
701 E. Franklin St.
Richmond, VA 23219
(804) 771-2001

Sacramento MF Service Office
HUD
545 Downtown Plaza, P.O. Box 1978, Suite 250
Sacramento, CA 95809
(916) 440-2334

San Antonio Area Office
HUD
Washington Square Building, 800 Dolorosa, P.O. Box 9163
San Antonio, TX 78285
(512) 229-6830

San Francisco Area Office
HUD
One Embarcadero Center, Suite 1600
San Francisco, CA 94111
(415) 556-6781

Seattle Area Office
HUD
403 Arcade Plaza Building, 1321 Second Ave.
Seattle, WA 98101
(206) 442-0334

Shreveport Service Office
HUD
50 Fanin St., New Federal Building, 6th Floor
Shreveport, LA 71101
(318) 226-5405

Spokane Service Office
HUD
746 U.S. Courthouse West, 920 Riverside Ave.
Spokane, WA 99201
(509) 456-4571

Springfield Valuation and Endorsement Station
HUD
Lincoln Towers Plaza, 524 S. Second St., Room 600
Springfield, IL 62701
(217) 492-4174

St. Louis Area Office
HUD
270 N. Tucker Blvd.
St. Louis, MO 63101
(314) 425-4777

Tucson Service Office
HUD
33 N. Stone Ave., Arizona Bank Building
Tucson, AZ 85701
(602) 792-6779

Tulsa Service Office
HUD
State Office Building, 440 S. Houston Ave.
Tulsa, Ok 74127
(405) 231-4582

Washington, DC Area Office
HUD
Universal North Building, 1875 Connecticut Ave.
Washington, DC 20009
(202) 673-5839

LOANS—VA REGIONAL OFFICES

Alabama
VA Regional Office
474 South Court St.
Montgomery, AL 36104

Alaska
VA Regional Office
2925 DeBarr Rd.
Anchorage, AK 99508

Arkansas
VA Regional Office
P.O. Box 1280, Building 65, Fort Roots
North Little Rock, AR 72115

California
VA Regional Office
Federal Building, 11000 Wilshire Blvd.
Los Angeles, CA 90024

VA Regional Office
Oakland Federal Building, 1301 Clay St., Room 1300
North Oakland, CA 94612

Colorado
VA Regional Office
P.O. Box 25216, 44 Union Blvd.
Denver, CO 80225

Connecticut
VA Regional Office
Office Center, 450 Main St.
Hartford, CT 06103

Delaware
VA Regional Office
Office Center, 1601 Kirkwood Highway
Wilmington, DE 19805

District of Columbia
VA Regional Office
941 North Capito St., N.E.
Washington, DC 20421

Florida
VA Regional Office
144 First Ave., S.
St. Petersburg, FL 33701

Georgia
VA Regional Office
730 Peachtree St., N.E.
Atlanta, GA 30365

Hawaii
VA Medical and Regional Office Center
P.O. Box 50188, 96850 PJKK Federal Bldg., 300 Ala Moana Blvd.
Honolulu, HI 96813

Idaho
VA Regional Office
Federal Building and U.S. Courthouse, 550 West Fort St., P.O. Box 044
Boise, ID 83724

Illinois
VA Regional Office
536 S. Clark St., P.O. Box 8136
Chicago, IL 60680

Indiana
VA Regional Office
575 North Pennsylvania St.
Indianapolis, IN 46204

Iowa
VA Regional Office
210 Walnut St.
Des Moines, IA 50309

Kansas
VA Medical and Regional Office Center
5500 E. Kellogg
Wichita, KS 67218

Kentucky
VA Regional Office
545 S. 3rd St.
Louisville, KY 40202

Louisiana
VA Regional Office
701 Loyola Ave.
New Orleans, LA 70113

Maine
VA Medical and Regional Office Center
Route 17 East
Togus, ME 04330

Maryland
VA Regional Office
Federal Building, 31 Hopkins Plaza
Baltimore, MD 21201

Massachusetts
VA Regional Office
John Fitzgerald Kennedy Federal Building
Boston, MA 02203

Michigan
VA Regional Office
477 Michigan Ave.
Detroit, MI 48226

Minnesota
VA Regional Office and Insurance Center
Bishop Henry Whipple Federal Building, Fort Snelling
St. Paul, MN 55111

Mississippi
VA Regional Office
100 W. Capitol St.
Jackson, MS 39269

Missouri
VA Regional Office
1520 Market St.
St. Louis, MO 63103

Montana
VA Medical and Regional Office Center
William St. off of Highway 12 West
Fort Harrison, MT 59636

Nebraska
VA Regional Office
5631 South 48th St.
Lincoln, NE 68516

Nevada
VA Regional Office
1201 Terminal Way
Reno, NV 89520

New Hampshire
VA Regional Office
275 Chestnut St.
Manchester, NH 03101

New Jersey
VA Regional Office
20 Washington Place
Newark, NJ 07102

New Mexico
VA Regional Office
500 Gold Ave., S.W.
Albuquerque, NM 87102

New York
VA Regional Office

Federal Building, 111 West Huron St.
Buffalo, NY 14202

VA Regional Office
252 Seventh Ave. at 24th St.
New York, NY 10001

North Carolina
VA Medical and Regional Office Center
2101 North Elm St.
Fargo, ND 58102-2498

Ohio
VA Regional Office
1240 East Ninth St.
Cleveland, OH 44199

Oklahoma
VA Regional Office
125 South Main St.
Muskogee, OK 74401

Oregon
VA Regional Office
Federal Building, 1220 Southwest Third Ave.
Portland, OR 97204

Pennsylvania
VA Regional Office and Insurance Center
P.O. Box 8079, 5000 Wissahickon Ave.
Philadelphia, PA 19101

Puerto Rico
VA Regional Office
GPO Box 4867
San Juan, PR 00936

Rhode Island
VA Regional Office
380 Westminster Mall
Providence, RI 02903

South Carolina
VA Regional Office
1801 Assembly St.
Columbia, SC 29201

South Dakota
VA Regional Office
P.O. Box 5046, 2501 West 22nd St.
Sioux Falls, SD 57117

Tennessee
VA Regional Office
110 Ninth Ave., S.
Nashville, TN 37203

Texas
VA Regional Office
2515 Murworth Dr.
Houston, TX 77054

VA Regional Office
1400 North Valley Mills Dr.
Waco, TX 76799

Utah
VA Regional Office
P.O. Box 11500, 125 South State St.
Salt Lake City, UT 84147

Vermont
VA Medical and Regional Office Center
N. Hartland Rd.
White River Junction, VT 05001

Virginia
VA Regional Office
210 Franklin Rd. S.W.
Roanoke, VA 24011

Washington
VA Regional Office
Federal Building, 915 Second Ave.
Seattle, WA 98174

West Virginia
VA Regional Office
640 4th Ave.
Huntington, WV 25701

Wisconsin
VA Regional Office
Building 6
Milwaukee, WI 53295

Wyoming
VA Medical and Regional Office Center
2360 East Pershing Blvd.
Cheyenne, WY 82001

MATERIAL SUPPLIERS

Electrical Supplies

American Lighting
5211D W. Market St., Suite 803
Greensboro, NC 27265
(800) 741-0571

Eagle Electric Mfg. Co.
45-31 Court Sq.
Long Island City, NY 11101
(718) 937-8000

Fasco Industries
810 Gillespie St., P.O. Box 150
Fayetteville, NC 28302
(919) 483-0421, (800) 334-4126

Fiskars Power Sentry
6271 Bury Dr., Suite A
Eden Prairie, MN 55346
(612) 949-1100, (800) 852-4312

Home Automation
2709 Ridgelake Dr.
Metairie, LA 70002
(504) 833-7256

Honeywell Home & Building Control
1985 Douglas Dr. N.
Minneapolis, MN 55422
(800) 345-6770, ext. 2033

Hunter Fan Co.
2500 Frisco Ave.
Memphis, TN 38114
(901) 743-1360

Lightolier
631 Airport Rd.
Fall River, MA 02720
(508) 679-8131, (800) 215-1068

LiteTouch
3550 S. 700 W.
Salt Lake City, UT 84119
(801) 268-8668

Osram Sylvania Inc.
100 Endicott St.
Danvers, MA 01923
(508) 777-1900

Square D Co.
3201 Nicholasville Rd., Suite 300
Lexington, KY 40503
(800) 392-8781

Sunway Fan Co.
3221 Garden Brook Dr.
Dallas, TX 75234
(214) 241-3778, (800) 484-9096

Sylvan Designs
8921 Quartz Ave.

Northridge, CA 91324
(818) 998-6868

U.S. Gaslight
4658 S. Old Peachtree Rd.
Norcross, GA 30071
(770) 448-1972, (800) 241-4317

WAC Lighting Co.
P.O. Box 560218, 113-25 14th Ave.
College Point, NY 11356
(718) 961-0695, (800) 526-2588

Westinghouse & Cutler-Hammer
1090 W. Thorndale Ave.
Bensenville, IL 60106
(708) 595-1667, (800) 443-3342

Equipment

Ace Hardware Corp.
2200 Kensington Ct.
Oak Brook, IL 60521
(708) 990-6751

Arrow Fastener Co.
271 Mayhill St.
Saddle Brook, NJ 07662
(201) 843-6900

Campbell Hausfeld
100 Production Dr.
Harrison, OH 45030
(513) 367-4811

Caterpillar
100 N.W. Adams St.
Peoria, IL 61629
(309) 675-1000

Cooper Tools
P.O. Box 728
Apex, NC 27502
(919) 387-0099

DeWalt Industrial Tool Co.
626 Hanover Pike
Hampstead, MD 21074
(800) 4-DEWALT

Deere & Co. (John Deere Industrial Equipment Co.)
8000 Jersey Ridge Rd.
Davenport, IA 52807
(309) 765-3195

Fein Power Tools
3019 W. Carson St.
Pittsburgh, PA 15204
(412) 331-2325, (800) 441-9878

Hitachi Koki USA
3950 Steve Reynolds Blvd.
Norcross, GA 30093
(770) 925-1774

Jepson Power Tools
20333 S. Western Ave.
Torrance, CA 90501
(310) 320-3890, (800) 456-8665

Klein Tools
7200 McCormick Blvd., P.O. Box 599033
Chicago, IL 60659-9033
(708) 677-9500

Komatsu Forklift (USA)
5595 Fresca Dr.
La Palma, CA 90623
(714) 228-3877

Kraft Tool Co.
619 E. 19th St.
Kansas City, MO 64108
(816) 474-4555

Laser Alignment
2850 Thornhills
Grand Rapids, MI 49546
(616) 949-7430, (800) 4-LASERS

Makita USA
14930 Northam St.
La Mirada, CA 90638
(714) 522-8088

Milwaukee Electric Tool Corp.
13135 W. Lisbon Rd.
Brookfield, WI 53005
(414) 781-3811

Mitsubishi Caterpillar Forklift
2011 W. Sam Houston Pkwy. N.
Houston, TX 77043-2421
(713) 467-1234

Omaha Industrial Tools
14685 Grover St.
Omaha, NE 68144-5435
(402) 334-8185, (800) 228-2765

Panasonic Power Tool Div.
1 Panasonic Way
Secaucus, NJ 07094
(201) 392-6442

Porter-Cable Corp.
4825 Hwy. 45 N.
Jackson, TN 38302-2468
(901) 668-8600, (800) 4-US-TOOL

Quik Drive USA
7528 Hickory Hills Ct.
Whites Creek, TN 37189
(615) 876-7278

Quikspray
P.O. Box 327
Port Clinton, OH 43452
(419) 732-2601

Ryobi America Corp.
5201 Pearman Dairy Rd.
Anderson, SC 29625
(803) 226-6511, (800) 525-2579

Senco Products
8485 Broadwell Rd.
Cincinnati, OH 45244
(513) 388-2000

Stanley Tools
600 Myrtle St.
New Britain, CT 06053
(203) 225-5111, (800) 648-7654

Stanley-Bostitch
Briggs Dr.
East Greenwich, RI 02818
(401) 884-2500, (800) 556-6696

Wallboard Tool Co.
1697 Seabright Ave.
Long Beach, CA 90813-1146
(310) 437-0701

Exterior Siding Materials

ABC Seamless Siding
3001 Fiechtner Dr.
Fargo, ND 58103
(701) 293-5952

Alcoa Building Products
P.O. Box 716
Sidney, OH 45365
(937) 492-1111

Alsco
3101 Poplarwood Ct., Suite 200

Raleigh, NC 27604
(919) 876-9333

Arriscraft Corp.
875 Speedsville Rd., P.O. Box 3190
Cambridge, ON N3H 4S8 Canada
(800) 265-8123

Cellwood Products (A Div. of Stolle Corp.)
100 Cellwood Pl.
Gaffney, SC 29340-4722
(803) 489-8136

CertainTeed Siding
P.O. Box 860
Valley Forge, PA 19482
(800) 233-8990

Dryvit Systems
1 Energy Way
West Warwick, RI 02893
(401) 822-4100

Georgia-Pacific Corp.
P.O. Box 1763
Norcross, GA 30091
(800) BUILD-GP

Kasko Industries
18 Turtle Creek Dr.
Tequesta, FL 33469
(407) 575-1193

Niagara Fiberboard
10 Stevens St.
Lockport, NY 14095
(716) 434-8881

Royal Building Products
30A Vinyl Ct.
Woodbridge, ON L4L 4A3 Canada
(905) 850-9700

The Pacific Lumber Co.
P.O. Box 565
Scotia, CA 95565
(707) 764-8888

Western Red Cedar Lumber Association
1200-555 Burrard St.
Vancouver, BC V7X 1S7 Canada
(604) 684-0266

Weyerhaeuser Composite Products Div.
CCB-100
Tacoma, WA 98477
(800) 458-7180

Flooring

Buchtal Corp. USA
1325 Northmeadow Pkwy., Suite 114
Roswell, GA 30076
(404) 442-5500

Dalton Georgia Wholesale Carpet
3041 N. Dug Gap Rd.
Dalton, GA 30720
(706) 277-9559, (800) 476-9559

Direct Carpet Mills of America
P.O. Box 2759
Rome, GA 30164-2759
(706) 235-6009, (800) 424-1223

Direct Dalton Carpet Mills
P.O. Box 2645
Dalton, GA 30722
(706) 277-1200, (800) 892-6789

The Georgia Marble Co.
Blue Ridge Ave., P.O. Box 9
Nelson, GA 30151
(404) 735-2591, (800) 334-0122

Mannington Mills
P.O. Box 30
Salem, NJ 08079-0030
(800) 952-1857

PM Cousins Supply Co.
605 Peninsula Blvd.
Hempstead, NY 11550
(516) 485-1300

Quarry Tile Co.
6328 Utah Ave.
Spokane, WA 99212
(509) 924-1466, (800) 423-2608

Rave Carpets
2875 Cleveland Rd.
Dalton, GA 30721
(706) 259-4864, (800) 942-6969

Heating, Ventilation & Air Conditioning

American-Standard Air Conditioning
P.O. Box 9010
Tyler, TX 75711

Bryant
P.O. Box 70
Indianapolis, IN 46206
(317) 243-0851, (800) 468-7253

Cool Attic
Wolters Industrial Park
Mineral Wells, TX 76068
(817) 325-7887, (800) 433-1626

Duralast Products Corp.
P.O. Box 15869
New Orleans, LA 70175
(504) 895-2068

Goodman Manufacturing Co.
1501 Seamist
Houston, TX 77008
(713) 861-2500

Heatway
3131 W. Chestnut Expwy.
Springfield, MO 65802
(417) 864-6108, (800) 255-1996

International Energy Systems
P.O. Box 588
Barrington, IL 60011
(708) 381-0203, (800) 927-0419

Lennox Industries
P.O. Box 799900
Dallas, TX 75379-9900
(214) 497-5000

North American Heating, Refrigeration & Air Conditioning
1389 Dublin Rd., P.O. Box 16790
Columbus, OH 43216-6790
(614) 488-1835

Payne
P.O. Box 70
Indianapolis, IN 46206
(317) 243-0851, (800) 227-4633

Rheem Mfg. Air Conditioning Div.
5600 Old Greenwood Rd.
Fort Smith, AR 72917-7010
(501) 646-4311

Rheem Mfg. Water Heater Div.
P.O. Box 244020
Montgomery, AL 36124
(334) 260-1500

Ruud Air Conditioning Division
P.O. Box 17010
Fort Smith, AR 72917-7010
(501) 646-4311

The Trane Co. Unitary Products Group
6200 Troup Hwy.
Tyler, TX 75707

Vent-Aire Systems
4850 Northpark Dr.
Colorado Springs, CO 80918-3872
(719) 599-9080, (800) 937-9080

WeatherKing Air Conditioning
5600 Old Greenwood Rd.
Fort Smith, AR 72917-7010
(501) 646-4311

Insulation

Amoco Foam Products Co.
2907 Log Cabin Dr.
Smyrna, GA 30080-7013
(800) 241-4402

The Celotex Corp.
4010 Boy Scout Blvd.
Tampa, FL 33607
(813) 873-1700

CertainTeed Corp. Insulation Group
P.O. Box 860
Valley Forge, PA 19482-0860
(610) 341-7739

The Dow Chemical Co. Construction Materials
2020 Dow Center
Midland, MI 48674
(517) 636-2303

Du Pont Tyvek
P.O. Box 80-705, Centre Rd.
Wilmington, DE 19880-0705
(800) 44-TYVEK

Insulated Building Systems
22377 Cedar Green Rd., Suite 2B
Dulles, VA 20166
(703) 450-4886

Owens Corning
Fiberglas Tower
Toledo, OH 43659
(800) GET-PINK

SOLEC (Solar Energy Corp.)
129 Walters Ave.
Ewing Township, NJ 08638-1829
(609) 883-7700

Interior Finish

AJ Stairs
1095 Towbin Ave.
Lakewood, NJ 08701
(908) 905-8500, (800) 425-7824

American Custom Millwork
3904 Newton Rd., P.O. Box 3608
Albany, GA 31706
(912) 888-3303

American Wall Tie Co.
2711 Lake
Chicago, IL 60612
(312) 533-1728

Architectural Paneling
979 Third Ave., Suite 919
New York, NY 10022
(212) 371-9632

Armstrong World Industries
P.O. Box 3001
Lancaster, PA 17604
(717) 397-0611, (800) 233-3823

Builders Edge
P.O. Box 7739
Pittsburgh, PA 15215
(412) 782-4880, (800) 969-7245

The Celotex Corp.
4010 Boy Scout Blvd.
Tampa, FL 33607
(813) 873-1700

Chesapeake Hardwood Products
201 W. Dexter St.
Chesapeake, VA 23324-3023
(804) 543-1601, (800) 446-8162

Classic Architectural Specialties
3223 Canton St.
Dallas, TX 75226
(214) 748-1668, (800) 662-1221

Classic Mouldings
226 Toryork Dr.
Weston, ON M9L 1Y1 Canada
(416) 745-5560

Custom Doors & Drawers
232 St. Amaud St.
Amherstburg, ON N9V 2Z7 Canada
(519) 736-2195

Driwood Moulding Co.
P.O. Box 1729
Florence, SC 29503-1729
(803) 669-2478

Duo-Fast Corp.
3702 River Rd.
Franklin Park, IL 60131
(800) 752-5207

Flexi-Wall Systems (A Div. of Wall & Floor Treatments)
P.O. Box 89
Liberty, SC 29657
(864) 843-3104, (800) 843-5394

Georgia-Pacific Corp.
P.O. Box 1763
Norcross, GA 30091
(800) BUILD-GP

International Cellulose Corp.
P.O. Box 450006
Houston, TX 77245
(713) 433-6701

Kentucky Millwork
P.O. Box 33276
Louisville, KY 40232
(502) 451-3456, (800) 235-5235

Kwikset Corp. (A Sub. of Black & Decker Co.)
1 Park Plaza, Suite 1000
Irvine, CA 92714
(714) 474-8800, (800) 327-LOCK

Masco Corporation
21001 Van Born Rd.
Taylor, MI 48180
(313) 274-7400

Milton W. Bosley Co.
151 Eighth Ave. N.W.
Glen Burnie, MD 21061
(410) 761-7727, (800) 638-5010

Mirrex Corp.
2025 E. Linden Ave.
Linden, NJ 07036
(908) 353-3370

Prest-on Co.
316 N. Point Lookout
Hot Springs, AR 71913
(501) 525-4683, (800) 323-1813

Schlage Lock Co.
2401 Bayshore Blvd.

San Francisco, CA 94134
(415) 467-1100

Spiral Stairs of America
1700 Spiral Ct., Franklin Ave.
Erie, PA 16510
(814) 898-3700, (800) 422-3700

Stanley Hardware
480 Myrtle St.
New Britain, CT 06053
(860) 225-5111

United States Gypsum Co.
P.O. Box 806278
Chicago, IL 60680-4124
(312) 606-4000, (800) USG-4YOU

Weiser Lock (A Div. of Masco Corp.)
6660 S. Broadmoor Rd.
Tucson, AZ 85746
(602) 741-6200, (800) 677-LOCK

Weslock National
13344 S. Main St.
Los Angeles, CA 90061
(310) 327-2770

Yale Locks & Hardware
P.O. Box 25288
Charlotte, NC 28229-8010
(704) 283-2101, (800) 438-1951

Kitchen & Bath

Abet Laminati
60 W. Sheffield Ave.
Englewood, NJ 07631
(201) 541-0700, (800) 228-ABET

Amana Refrigeration
2800 220th Trail, P.O. Box 8901
Amana, IA 52204
(319) 622-5511, (800) 843-0304

American Cabinet
1655 Amity Rd., P.O. Box 640
Conway, AR 72032
(800) 643-8035

American Marazzi Tile
359 Clay Rd.
Sunnyvale, TX 75182
(214) 226-0110

American Olean Tile
1000 Cannon Ave.
Lansdale, PA 19446
(215) 393-2237

American Standard
1 Centennial Plaza
Piscataway, NJ 08855
(908) 980-2400, (800) 524-9797

American Whirlpool
3050 N. 29th Ct.
Hollywood, FL 33020
(305) 921-4400, (800) 327-1394

Aristokraft
P.O. Box 420, 1 Aristokraft Sq.
Jasper, IN 47547-0420
(812) 482-2527

Avonite
1945 S. Hwy. 304
Belen, NM 87002
(505) 864-3800, (800) 428-6648

Block Tops
4770 E. Wesley Dr.
Anaheim, CA 92807
(714) 779-0475

Broan Mfg. Co.
926 W. State St.
Hartford, WI 53092
(414) 673-4340, (800) 692-7626

Cascade Cabinet Corp.
8330 212th St. S.E.
Woodinville, WA 98072-8020
(206) 481-6860, (800) 228-1830

Country Homestead Restoration
Contractor
P.O. Box 188
Kreamer, PA 17833
(717) 374-7122

Craft-Maid Custom Kitchens
501 S. Ninth St., Bldg. C
Reading, PA 19602
(610) 376-8686

Crossville Ceramics
P.O. Box 1168, Industrial Park
Crossville, TN 38557
(615) 484-2110

Custom Wood Products
P.O. Box 4500
Roanoke, VA 24015
(540) 345-8821, (800) 366-2971

Delta Faucet Co.
55 E. 111th St.
Indianapolis, IN 46280-1000
(317) 848-1812, (800) 345-DELTA

Designs In Tile
P.O. Box 358
Mt. Shasta, CA 96067
(916) 926-2629

Du Pont Co. Corian Products
Chestnut Run Plaza, P.O. Box 80702
Wilmington, DE 19880
(800) 4CORIAN

Elkay Mfg. Co.
2222 Camden Ct.
Oak Brook, IL 60521
(708) 574-8484

Elkhart Door
P.O. Box 2177, 1515 Leininger Ave.
Elkhart, IN 46515
(219) 294-5428, (800) 348-7559

Formica Corp.
10155 Reading Rd.
Cincinnati, OH 45241
(513) 786-3400, (800) FORMICA

Frigidaire Co. Frigidaire, Tappan &
White-Westinghouse
6000 Perimeter Dr.
Dublin, OH 43017
(614) 792-4100, (800) 685-6005

GE Appliances
Appliance Park
Louisville, KY 40225-0001
(502) 452-4311, (800) 626-2000

GROHE (A Sub. of Friedrich Grohe,
Germany)
241 Covington Dr.
Bloomingdale, IL 60108
(708) 582-7711

Hearth Kitchens
711 10th Ave.
Hanover, ON N4N 2P7 Canada
(519) 364-1170, (800) 267-4524

Heritage Custom Kitchens
215 Diller Ave.
New Holland, PA 17557
(717) 354-4011

Home Crest Corp.
1002 Eisenhower Dr. N.

Goshen, IN 46526
(219) 533-9571

Hotpoint Appliances
Appliance Park
Louisville, KY 40225
(502) 452-4311, (800) 626-2000

In-Sink-Erator Div. Emerson Electric Co.
4700 21st St.
Racine, WI 53406-5093
(414) 554-5432, (800) 558-5712

Jacuzzi Whirlpool Bath
P.O. Drawer J, 2121 N. California Blvd., Suite 475
Walnut Creek, CA 94596
(510) 938-7070, (800) 678-6889

Jenn-Air
3035 N. Shadeland Ave.
Indianapolis, IN 46226
(317) 545-2271, (800) JENN-AIR

Jensen Industries
1946 E. 46th St.
Los Angeles, CA 90058-2096
(213) 235-6800, (800) 325-8351

King Refrigerator Corp.
7602 Woodhaven Blvd.
Glendale, NY 11385
(718) 897-2200

KitchenAid (A Brand of Whirlpool Corp.)
2000 M-63 N, Mail Drop 4302
Benton Harbor, MI 49022
(616) 923-5000, (800) 253-3977

Kohler Co.
444 Highland Dr.
Kohler, WI 53044
(414) 457-4441

Kolson
653 Middle Neck Rd.
Great Neck, NY 11023
(516) 487-1224, (800) 783-1335

Kountry Kraft
P.O. Box 570
Newmanstown, PA 17073
(610) 589-4575, (800) 628-9061

KraftMaid Cabinetry
P.O. Box 1055, 16052 Industrial Pkwy.
Middlefield, OH 44062
(216) 632-5333

Magic Chef
3035 N. Shadeland Ave.
Indianapolis, IN 46226-0901
(800) 536-6247

Maytag
1 Dependability Sq.
Newton, IA 50208
(515) 792-7000

Merillat Industries
5353 W. US 223
Adrian, MI 49221
(517) 263-0771

Monarch Tile
834 Rickwood Rd.
Florence, AL 35630
(205) 764-6181, (800) 289-8453

Peerless Faucet Co. (A Division of Masco Corp. of Indiana)
55 E. 111th St.
Indianapolis, IN 46280
(317) 848-1812

Price Pfister
13500 Paxton St.
Pacoima, CA 91331
(818) 896-1141, (800) PFAUCET

Quaker Maid
State Rte. 61, Box H
Leesport, PA 19533
(610) 926-3011

Quality Cabinets
515 Big Stone Gap
Duncanville, TX 75137
(800) 298-7020

Rangaire & Co.
P.O. Box 177
Cleburne, TX 76033
(817) 556-6500, (800) 777-7264

Strom Plumbing
Dept RM, 3756 Omec Cir.
Rancho Cordova, CA 95742-7399
(916) 638-2722

Sub-Zero Freezer Co.
4717 Hammersley Rd.
Madison, WI 53711
(608) 271-2233, (800) 222-7820

Tappan (A Brand of Frigidaire Co.)
6000 Perimeter Dr.
Dublin, OH 43017
(614) 792-4100, (800) 685-6005

United Panel
P.O. Box 188, 8 Wildon Dr.
Mt. Bethel, PA 18343
(610) 588-6871, (800) 933-8700

US Brass
17120 Dallas Pkwy.
Dallas, TX 75248
(214) 407-2600, (800) US-BRASS

Vent-A-Hood
P.O. Box 830426
Richardson, TX 75083-0426
(214) 235-5201

Whirlpool (Whirlpool Appliance Group)
2000 M-63
Benton Harbor, MI 49022
(616) 923-5000, (800) 253-1301

White-Westinghouse (A Brand of Frigidaire Co.)
6000 Perimeter Dr.
Dublin, OH 43017
(614) 792-4100, (800) 685-6005

Wilsonart International
2400 Wilson Pl., P.O. Box 6110
Temple, TX 76503-6110
(817) 778-2711, (800) 433-3222

Landscaping Materials

Anchor Wall Systems
6101 Baker Rd., Suite 201
Minnetonka, MN 55345-5973
(612) 933-8855

Belden Brick Co.
700 W. Tuscarawas St.
Canton, OH 44702
(216) 456-0031

Cangelosi Marble & Granite
14021 S. Gessner
Missouri City, TX 77489
(713) 499-7521

Cedarbrook Sauna
21326 Hwy. 9

Woodinville, WA 98072
(509) 782-2447

Creative Building Products
4307 Arden Dr.
Fort Wayne, IN 46804-4400
(219) 459-0456

Dura Art Stone
11010 Live Oak
Fontana, CA 92337
(909) 350-9000

Earthstone Retaining Wall Systems
5850 Canoga Ave., Suite 400
Woodland Hills, CA 91367
(818) 347-4730

Georgia-Pacific Corp.
P.O. Box 1763
Norcross, GA 30091
(800) BUILD-GP

Mobil Chemical Co. Trex Composite Products Division
20 S. Cameron
Winchester, VA 22601
(540) 678-8100

Moultrie Mfg.
P.O. Box 1179
Moultrie, GA 31776-1179
(912) 985-1312

Redland Brick (Cushwa Plant)
P.O. Box 160
Williamsport, MD 21795-0160
(301) 223-7700

Risi Stone Systems
Le Parc Office Tower, 8500 Leslie St., Suite 390
Thornhill, ON L3T 7P1 Canada
(905) 882-5898

Rockwood Retaining Walls
7200 N. Hwy. 63
Rochester, MN 55906
(507) 288-8850

Stone Products Corp.
P.O. Box 270
Napa, CA 94559
(707) 255-1727

UltraGuard
201 E. Palm Dr., Suite 1
Syracuse, IN 46567-1955
(219) 457-4342

VersaDek Ind.
Box 1003
Santa Margarita, CA 93453
(805) 438-3662

Wausau Tile
P.O. Box 1520
Wausau, WI 54402-1520
(715) 359-3121

Manufactured & Kit Homes

AFM Corp.
P.O. Box 246
Excelsior, MN 55331
(612) 474-0809

Acadia Post & Beam
P.O. Box 217, Port Williams
Kings County, NS B0P 1T0 Canada
(902) 542-2298

Advanced Building Systems
P.O. Box 327
Paulding, OH 45879
(419) 399-4412

Algonquin Log Homes
R.R. #1
Norland, ON K0M 2L0 Canada
(705) 454-2311

All Steel Homes
2626 Gribble St.
North Little Rock, AR 72114
(501) 945-8092

Appalachian Log Structures
P.O. Box 614
Ripley, WV 25271
(304) 372-6410

Craft-Bilt Mfg. Co.
53 Souderton-Hatfield Pike
Souderton, PA 18964
(800) 422-8577

Crest Homes Corp. (A Div. of Schult Homes)
30 Industrial Park
Milton, PA 17847
(717) 742-8521

Deck House
930 Main St.
Acton, MA 01720
(617) 259-9450

Deltec Homes
604 College St.
Asheville, NC 28801
(704) 253-0483

Garland Homes
P.O. Box 12
Victor, MI 59875
(406) 642-3095

Insulspan
P.O. Box 120427
Nashville, TN 37212
(800) 726-3510

Lindal Cedar Homes/Sunrooms
4300 S. 104th Pl.
Seattle, WA 98178
(206) 725-0900

Simplex Industries
Keyser Valley Industrial Park, 1 Simplex Dr.
Scranton, PA 18504
(717) 346-5113

Southland Log Homes
P.O. Box 1668
Irmo, SC 29063-1668
(803) 781-5100

Sterling Building Systems
P.O. Box 1967
Wausau, WI 54402-1967
(715) 359-7108

Wausau Homes
P.O. Box 8005
Wausau, WI 54402-8005
(715) 359-7272

Miscellaneous

Abundant Energy
P.O. Box 307
Pine Island, NY 10969
(914) 258-4022, (800) 426-4859

BC Greenhouse Builders
7425 Hedley Ave.
Burnaby, BC V5E 2R1 Canada
(604) 433-4220

Bang & Olufsen of America
1200 Business Center Dr.

Mount Prospect, IL 60056
(708) 299-9380, (800) 323-0378

Carolina Solar Structures
8 Loop Rd.
Arden, NC 28704
(704) 684-9900, (800) 241-9560

Closet Maid
720 S.W. 17th St.
Ocala, FL 32674
(904) 351-6100, (800) 874-0008

Country Stoves
P.O. Box 987
Auburn, WA 98071-0987
(206) 735-1100

First Alert Professional Security Systems
172 Michael Dr.
Syosset, NY 11791
(516) 921-6066

Florian Greenhouse
64 Airport Rd.
West Milford, NJ 07480
(201) 728-7800, (800) FLORIAN

Four Seasons Sunrooms
5005 Veterans Memorial Hwy.
Holbrook, NY 11741
(800) FOUR SEASONS

Hearth Products Association
1601 N. Kent St., Suite 1001
Arlington, VA 22209
(703) 522-0086

Heat-N-Glo
6665 W. Hwy. 13
Savage, MN 55378
(612) 890-8367, (800) 669-HEAT

Heatilator
1915 W. Saunders St.
Mt. Pleasant, IA 52641
(319) 385-9211, (800) 247-6798

IntelliNet
2900 Horseshoe Dr. S.
Naples, FL 33942-3200
(941) 434-5888

Majestic Products Company
1000 E. Market St.
Huntington, IN 46750
(219) 356-8000, (800) 525-1898

Miracle Sealants & Abrasives
12806 Schabarum Ave., Bldg. A
Irwindale, CA 91706
(818) 814-8988, (800) 350-1901

New England Glass Enclosures
7 South End Plaza
New Milford, CT 06776
(203) 355-2299

Rev-A-Shelf
2409 Plantside Dr.
Jeffersontown, KY 40299
(502) 499-5835, (800) 626-1126

Schulte Corp.
12115 Ellington Ct.
Cincinnati, OH 45249
(513) 489-9300, (800) 669-3225

Sears Contract Sales (Sears, Roebuck and Co.)
3333 Beverly Rd., E3-363A
Hoffman Estates, IL 60179
(708) 286-2994, (800) 359-2000

Superior Fireplace Co.
4325 Artesia Ave.
Fullerton, CA 92633
(714) 521-7302

Thomson Consumer Electronics RCA Custom Home Theatre
10330 N. Meridian INH215
Indianapolis, IN 46206-1976
(800) 722-6585

Painting Materials

Benjamin Moore & Co.
51 Chestnut Ridge Rd.
Montvale, NJ 07645
(201) 573-9600

Bondex Intl.
3616 Scarlet Oak Blvd.
St. Louis, MO 63122
(314) 225-5001, (800) 225-7522

Borden
180 E. Broad St.
Columbus, OH 43215-3799
(614) 225-4000, (800) 848-9400

Bostik Tile & Flooring Group
Boston St.
Middleton, MA 01949
(508) 777-0100, (800) 726-7845

DAP
855 N. Third St., P.O. Box 277
Tipp City, OH 45371
(937) 667-4461, (800) 543-3840

Dur-A-Flex
P.O. Box 280166
East Hartford, CT 06128-0166
(860) 528-9838, (800) 253-3539

Dura Seal Div. of Minwax Co.
P.O. Box 438
Montvale, NJ 07645-1814
(201) 391-0253, (800) 526-0495

Dutch Boy Professional Paints
101 Prospect Ave.
Cleveland, OH 44115
(216) 566-2929

Dyco Paints
5850 Ulmerton Rd.
Clearwater, FL 34620
(813) 536-6560, (800) 237-8232

GE Silicones
260 Hudson River Rd.
Waterford, NY 12188
(518) 233-3505, (800) 255-8886

The Glidden Co.
925 Euclid Ave.
Cleveland, OH 44115
(216) 344-8491, (800) 984-5444

Insta-Foam Products
1500 Cedarwood Dr.
Joliet, IL 60435
(815) 741-6800, (800) 800-FOAM

Klean-Strip
P.O. Box 1879
Memphis, TN 38101
(901) 775-0100, (800) 238-2672

Loctite Corp. North American Group
1001 Trout Brook Crossing
Rocky Hill, CT 06067-3910
(203) 571-5100

Macco Adhesives
925 Euclid Ave.
Cleveland, OH 44115
(216) 334-8000, (800) 634-0015

Minwax Co.
50 Chestnut Ridge Rd.

Montvale, NJ 07645
(201) 391-0253, (800) 526-0495

Miracle Adhesives (A Div. of Pratt & Lambert)
75 Tonawanda St., P.O. Box 1505
Buffalo, NY 14240
(716) 873-2770, (800) 876-7005

NPC Sealants
1208 S. Eighth Ave.
Maywood, IL 60153
(708) 681-1040, (800) 654-1042

Olympic Paints and Stains
1 PPG Plaza
Pittsburgh, PA 15272
(412) 434-3900, (800) 621-2024

Osmose Wood Preserving
1016 Everee Inn Rd.
Griffin, GA 30224
(404) 228-8434, (800) 522-9663

PPG Industries
1 PPG Pl.
Pittsburgh, PA 15272
(412) 434-3131, (800) 2-GETPPG

Plasti-Kote Co.
1000 Lake Rd.
Medina, OH 44256
(216) 725-4511, (800) 431-5928

Pratt & Lambert
75 Tonawanda St.
Buffalo, NY 14207
(716) 873-6000

Pro/Cote
101 Prospect Ave.
Cleveland, OH 44115
(216) 566-2929

Red Devil
2400 Vauxhall Rd.
Union, NJ 07083-1933
(908) 688-6900, (800) 4-A-DEVIL

The Sherwin-Williams Co.
101 Prospect Ave. N.W.
Cleveland, OH 44115
(216) 566-2000, (800) 336-1110

Super-Tek Products
25-44 Borough Pl.
Woodside, NY 11377-7899
(718) 278-7900

Surebond
500 E. Remington Rd.
Schaumburg, IL 60173
(708) 843-1818

TEC (A Sub. of HB Fuller Co.)
315 S. Hicks Rd.
Palatine, IL 60067
(847) 358-9500, (800) 323-7407

Tufco
P.O. Box 23500, 3161 Ridge Rd.
Green Bay, WI 54305
(414) 336-0054, (800) 558-8145

Wagner Spray Tech Corp.
1770 Fernbrook Ln.
Plymouth, MN 55447-4663
(612) 553-7000, (800) 328-8251

Wood-Kote Products
8000 N.W. 14th Pl.
Portland, OR 97211
(503) 285-8371

Roofing Materials

Air Vent
3000 W. Commerce
Dallas, TX 75212
(800) 247-8368

Alcoa Building Products
P.O. Box 716
Sidney, OH 45365
(937) 492-1111, (800) 962-6973

Alumax Home Products
450 Richardson Dr., P.O. Box 4515
Lancaster, PA 17604
(717) 299-3711

Atlas Roofing Corp.
1775 The Exchange, Suite 160
Atlanta, GA 30339
(404) 933-4463

Boral Concrete Products
4400 MacArthur Blvd., Suite 500
Newport Beach, CA 92660
(714) 955-4976

Browning Metal Products (A CertainTeed Co.)
4805 N. Prospect Rd.
Peoria Heights, IL 61614
(309) 682-1015

Cal-Shake
P.O. Box 2265, 5355 N. Vincent Ave.
Irwindale, CA 91706-2265
(818) 969-3451, (800) 736-7663

Cedar Plus
P.O. Box 515
Sumas, WA 98295
(800) 963-3388

Celadon
P.O. Box 860
Valley Forge, PA 19482-0860
(800) 782-8777

CertainTeed Corp. Roofing Products Group
P.O. Box 860
Valley Forge, PA 19482-0860
(800) 233-8990

Cobra Ventilation Products GAF Materials Corp.
1361 Alps Rd.
Wayne, NJ 07470
(201) 628-3874

Dura-Loc Roofing Systems
P.O. Box 220
Courtland, ON N0J 1E0 Canada
(519) 688-2200

Everest Roofing Products
2500 Workman Mill Rd.
Whittier, CA 90601
(800) 767-0267

GAF Materials Corp.
1361 Alps Rd.
Wayne, NJ 07470
(201) 628-3000

GS Roofing Co.
5525 MacArthur Blvd., Suite 900
Irving, TX 75038
(214) 580-5600

Georgia-Pacific Corp.
P.O. Box 1763
Norcross, GA 30091
(800) BUILD-GP

Green River
33610 E. Broadway Ave.
Mission, BC V2V 4M4 Canada
(604) 826-9531

IKO Mfg.
120 Hay Rd.
Wilmington, DE 19809
(302) 764-3100

MFM Building Products Corp.
P.O. Box 340
Coshocton, OH 43812
(614) 622-2645

Majestic Forest Products Corp.
P.O. Box 66, 16640 111th Ave.
Edmonton, AB T5J 2G9 Canada
(403) 484-7113

NEI
Pine Rd.
Brentwood, NH 03833
(603) 778-8899

Revere Copper Products
P.O. Box 300
Rome, NY 13440
(315) 338-2022

Schuller Intl.
P.O. Box 5108
Denver, CO 80217
(303) 978-4900

Scott Cedar
P.O. Box 515
Sumas, WA 98295
(800) 963-3388

Southeastern Metals Mfg. Co.
11801 Industry Dr.
Jacksonville, FL 32218
(904) 757-4200

Tamko Roofing Products
220 W. Fourth St.
Joplin, MO 64801
(417) 624-6644, (800) 641-4691

Tri-Ply
1250 Fourteen Mile Rd., Suite 103
Clawson, MI 48017
(810) 288-9780

Universal Marble & Granite
1919 Halethorpe Farms Rd.
Baltimore, MD 21227
(410) 247-2442

Software

ACAD-Plus
9601 Jones Rd., #250
Houston, TX 77065
(713) 890-3300

Autodesk
111 McInnis Pkwy.
San Rafael, CA 94903
(415) 507-5000

BUILDSOFT
P.O. Box 13893
Research Triangle Park, NC 27560
(919) 941-6269

Best Estimate
740 Broadview Ave.
Toronto, ON M4K 2P1 Canada
(416) 463-1108

Cadkey
4 Griffin Rd. N.
Windsor, CT 06095-1511
(203) 298-8888

Construction Data Control (CDCI)
4000 De Kalb Technology Pkwy., Suite 220
Atlanta, GA 30340
(404) 457-7725

Enterprise Computer Systems
1 Independence Pointe, P.O. Box 2383
Greenville, SC 29602-2383
(864) 234-7676

Estimation
805-L Barkwood Ct.
Linthicum Heights, MD 21090
(410) 636-4566

Industry Specific Software
1200 Woodruff Rd., Suite B-20
Greenville, SC 29607
(864) 297-7086

Master Builder Software/Omware
100 Pleasant Hill Ave. N.
Sebastopol, CA 95472
(707) 823-8681

NEBS, Inc.
20 Industrial Park Dr.
Nashua, NH 03062
(603) 880-5100

Peachtree Software
1505 Pavilion Pl.
Norcross, GA 30093
(404) 564-5700

Prosoft
107 N. Armenia Ave.
Tampa, FL 33609
(813) 251-1628

RS Means
100 Construction Plaza, P.O. Box 800
Kingston, MA 02364
(617) 585-7880

Sirius Software
345 W. Second St., Suite 201
Dayton, OH 45402-1443
(937) 228-4849

Small System Design
7464 Arapahoe Ave., Suite A-7
Boulder, CO 80303
(303) 442-9454

Softdesk Retail Products
10725 Ambassador Dr.
Kansas City, MO 64153
(816) 891-1040

Timberline Software
9600 S.W. Nimbus
Beaverton, OR 97008
(503) 626-6775

Virtus Corp.
118 MacKenan Dr., Suite 250
Cary, NC 27511
(919) 467-9700

Win Estimator
8209 S. 222nd St., Suite B
Kent, WA 98032
(206) 395-3631

Windows & Doors

Acadia Windows & Doors
9611 Pulaski Park Dr.
Baltimore, MD 21220-1435
(410) 780-9600, (800) 638-6084

Alcoa Vinyl Windows
725 Pleasant Valley Dr.
Springboro, OH 45066
(513) 746-0488, (800) 238-1866

Andersen Windows
100 Fourth Ave. N.
Bayport, MN 55003
(612) 439-5150, (800) 426-4261

Atlantic Pre-Hung Doors
P.O. Box 1258, 143 Conant St.
West Concord, MA 01742
(508) 369-5600

Better-Bilt
P.O. Box 5700
Norcross, GA 30091
(770) 497-2000, (800) 477-6544

Beveled Glass Designs Larry Robertson Associates
3185 N. Shadeland
Indianapolis, IN 46226
(317) 353-0472, (800) 428-5746

Beveled Glass Works
23715 W. Malibu Rd., Suite 351
Malibu, CA 90265
(310) 456-3201, (800) 421-0518

Caradco (A Sub. of Alcoa)
P.O. Box 920, 201 Evans Rd.
Rantoul, IL 61866
(217) 893-4444, (800) 238-1866

CertainTeed Windows
P.O. Box 860
Valley Forge, PA 19482
(800) 233-8990

Cladwood Div. Smurfit Newsprint
427 Main St.
Oregon City, OR 97045
(503) 650-4274, (800) 547-6633

Clopay Corp. Clopay Building Products Co.
312 Walnut St., Suite 1600
Cincinnati, OH 45202
(513) 381-4800, (800) 225-6729

Crestline/SNE Enterprises (A PLY GEM Co.)
1 Wausau Center, P.O. Box 8007
Wausau, WI 54401
(715) 845-1161, (800) 552-4111

The Genie Co.
22790 Lake Park Blvd.
Alliance, OH 44601-3498
(216) 821-5360, (800) OK-GENIE

Georgia-Pacific Corp.
P.O. Box 1763
Norcross, GA 30091
(800) BUILD-GP

Glass Block Designs
381 11th St.
San Francisco, CA 94103
(415) 626-5770

Glass Blocks Unlimited
126 E. 16th St.
Costa Mesa, CA 92627-3707
(714) 548-8531

Holmes Garage Door Co.
2601 W. Valley Hwy., P.O. Box 1976
Auburn, WA 98071-1976
(206) 931-8900

Karona
3050 44th St., P.O. Box 128
Grandville, MI 49418
(616) 532-5901, (800) 829-9233

Lincoln Wood Products
701 N. State St.
Merrill, WI 54452
(715) 536-2461

Maestro Products
1565 Eastwood Ct.
Riverside, CA 92507
(909) 369-6009

Marvin Windows and Doors
P.O. Box 100
Warroad, MN 56763
(218) 386-1430, (800) 346-5128

Morgan Products
P.O. Box 2446, 601 Oregon St.
Oshkosh, WI 54901-5965
(800) 766-1992

Norco Windows
P.O. Box 140
Hawkins, WI 54530
(715) 585-6311, (800) 826-6793

Nord (A Div. of JELD-WEN)
300 W. Marine View Dr., P.O. Box 1187
Everett, WA 98206-1187
(800) 877-9482

Overhead Door Corp.
6750 LBJ Fwy., Suite 1200
Dallas, TX 75240
(214) 233-6611, (800) 929-DOOR

Peachtree Doors
P.O. Box 5700
Norcross, GA 30091
(770) 497-2000, (800) 477-6544

Pease Industries
7100 Dixie Hwy.
Fairfield, OH 45014
(513) 870-3600, (800) 88-DOORS

Pella Corporation
102 Main St.
Pella, IA 50219
(515) 628-6992, (800) 84-PELLA

Perma-Door
631 N. First St.
West Branch, MI 48661
(517) 345-5110, (800) 842-3667

Pozzi Window Co.
P.O. Box 5249
Bend, OR 97708
(503) 382-4411, (800) 821-1016

Semco Windows & Doors Semling Menke Co.
P.O. Box 378
Merrill, WI 54452
(715) 536-9411, (800) 333-2206

Stanley Door Systems (A Sub. of Stanley Works)
1225 E. Maple Rd.
Troy, MI 48083-5600
(810) 528-1400, (800) 521-2752

Therma-Tru Corp.
1687 Woodlands Dr.
Maumee, OH 43537
(419) 891-7400, (800) 537-8827

Thermal-Gard
400 Walnut St.
Punxsutawney, PA 15767-1368
(814) 938-1408

Weather Shield Mfg.
P.O. Box 309, 1 Weather Shield Plaza
Medford, WI 54451
(715) 748-2100, (800) 477-6808

Weck Glass Block/Glashaus c/o Glashaus
415 W. Golf #13
Arlington Heights, IL 60005-3923
(708) 640-6910

Wood & Structural Materials

Boise Cascade
1111 W. Jefferson St.
Boise, ID 83702
(208) 384-6321, (800) 458-4631

Georgia-Pacific Corp.
P.O. Box 1763
Norcross, GA 30091
(800) BUILD-GP

Homasote Co.
P.O. Box 7240
West Trenton, NJ 08628-0240
(609) 883-3300, (800) 257-9491

International Paper Co.
2 Manhattanville Rd.

Purchase, NY 10577
(914) 397-1500, (800) 223-1268

Louisiana-Pacific Corp.
111 S.W. Fifth Ave.
Portland, OR 97204

Norbord Industries
1 Toronto St., Suite 500
Toronto, ON M5C 2W4 Canada
(416) 365-0710

Trus Joist MacMillan
P.O. Box 60
Boise, ID 83706
(208) 364-1200, (800) 628-3997

Weyerhaeuser Engineered Strand
Products Div.
2000 Frontis Plaza Blvd., Suite 101
Winston-Salem, NC 27103
(910) 760-7151, (800) 648-2566

SECTION III: APPENDICES

Appendix

MASTERPLAN

STEP	DESCRIPTION	Date Started	Date Completed	Date Inspected	NOTES
	PRECONSTRUCTION				
PC1	BEGIN construction project scrapbook				
PC2	DETERMINE size of home				
PC3	EVALUATE alternate house plans				
PC4	SELECT house plan				
PC5	MAKE design changes				
PC6	PERFORM material and labor take-off				
PC7	CUT costs as required				
PC8	UPDATE material and labor take-off				
PC9	ARRANGE temporary housing				
PC10	FORM separate corporation				
PC11	APPLY for business license				
PC12	ORDER business cards				
PC13	OBTAIN free and clear title to land				
PC14	APPLY for construction loan				
PC15	ESTABLISH project checking account				
PC16	CLOSE construction loan				
PC17	OPEN builder accounts				
PC18	OBTAIN purchase order forms				
PC19	OBTAIN building permit				
PC20	OBTAIN builder's risk policy				
PC21	OBTAIN worker's compensation policy				
PC22	ACQUIRE minimum tools				
PC23	VISIT construction sites				
PC24	DETERMINE need for on-site storage				
PC25	NOTIFY all subs and labor				
PC26	OBTAIN compliance bond				
PC27	DETERMINE minimum occupancy requirements				
PC28	CONDUCT spot survey of lot				
PC29	MARK all lot boundaries and corners				
PC30	POST construction signs				
PC31	POST building permit on-site				
PC32	ARRANGE for portable toilet				
PC33	ARRANGE for site telephone				

MASTERPLAN

STEP	DESCRIPTION	Date Started	Date Completed	Date Inspected	NOTES
	EXCAVATION				
EX1	CONDUCT excavation bidding process				
EX2	LOCATE underground utilities				
EX3	DETERMINE exact dwelling location				
EX4	MARK area to be cleared				
EX5	MARK where curb is to be cut				
EX6	TAKE last picture of site area				
EX7	CLEAR site area				
EX8	EXCAVATE trash pit				
EX9	INSPECT cleared site				
EX10	STAKE out foundation area				
EX11	EXCAVATE foundation/basement area				
EX12	INSPECT site and excavated area				
	BATTER BOARDS				
EX13	LAY OUT corner stakes for outer four corners of house				
EX14	DRIVE three batter board stakes at each corner				
EX15	NAIL 1×6 horizontal boards to stakes				
EX16	PULL batter board strings				
EX17	RECHECK distance of each string				
EX18	LAY OUT remaining batter boards and recheck				
	DRAINAGE				
EX19	APPLY crusher run to driveway area				
EX20	SET UP silt fence				
EX21	PAY excavator (initial)				
	BACKFILL				
EX22	PLACE supports on foundation wall				
EX23	INSPECT backfill area				
EX24	BACKFILL foundation				
EX25	CUT driveway area				
EX26	COVER septic tank and septic tank line				
EX27	COVER trash pit				

MASTERPLAN

STEP	DESCRIPTION	Date Started	Date Completed	Date Inspected	NOTES
EX28	PERFORM final grade				
EX29	CONDUCT final inspection				
EX30	PAY excavator (final)				
	PEST CONTROL				
PS1	CONDUCT standard bidding process				
PS2	NOTIFY pest control sub for pretreatment				
PS3	APPLY pesticide to foundation				
PS4	APPLY pesticide to perimeter				
PS5	OBTAIN signed pest control warranty				
PS6	PAY pest control sub				
	CONCRETE				
CN1	CONDUCT bidding process (concrete supplier)				
CN2	CONDUCT bidding process (formwork/finishing)				
CN3	CONDUCT bidding process (block masonry)				
	FOOTINGS				
CN4	INSPECT batter boards				
CN5	DIG footings				
CN6	SET footing forms				
CN7	INSPECT footing forms				
CN8	SCHEDULE and complete footing inspection				
CN9	CALL concrete supplier for concrete				
CN10	POUR footings				
CN11	INSTALL key forms				
CN12	FINISH footings as needed				
CN13	REMOVE footing forms				
CN14	INSPECT footings				
CN15	PAY concrete supplier for footings				
	POURED WALL				
CN16	SET poured-wall concrete forms				
CN17	SCHEDULE concrete supplier				

MASTERPLAN

STEP	DESCRIPTION	Date Started	Date Completed	Date Inspected	NOTES
CN18	INSPECT poured-wall concrete forms				
CN19	POUR foundation walls				
CN20	FINISH poured-foundation walls				
CN21	REMOVE poured-wall concrete forms				
CN22	INSPECT poured walls				
CN23	BREAK OFF tie ends from walls				
CN24	PAY concrete supplier for walls				
	BLOCK WALL				
CN25	ORDER blocks/schedule masons				
CN26	LAY concrete blocks				
CN27	FINISH concrete blocks				
CN28	INSPECT concrete block work				
CN29	PAY concrete block masons				
CN30	CLEAN up after block masons				
	CONCRETE SLAB				
CN31	SET slab forms				
CN32	PACK slab subsoil				
CN33	INSTALL stub plumbing				
CN34	SCHEDULE HVAC sub for slab work				
CN35	INSPECT plumbing and gas lines				
CN36	POUR and spread crushed stone				
CN37	INSTALL vapor barrier				
CN38	LAY reinforcing wire as required				
CN39	INSTALL rigid insulation				
CN40	CALL concrete supplier for slab				
CN41	INSPECT slab forms				
CN42	POUR slab				
CN43	FINISH slab				
CN44	INSPECT slab				
CN45	PAY concrete supplier for slab				
	DRIVEWAY				
CN46	COMPACT drive and walkway soil				

MASTERPLAN

STEP	DESCRIPTION	Date Started	Date Completed	Date Inspected	NOTES
CN47	SET drive and walkway forms				
CN48	CALL concrete supplier				
CN49	INSPECT drive and walkway forms				
CN50	POUR drive and walkway				
CN51	FINISH drive and walkway areas				
CN52	INSPECT drive and walkway areas				
CN53	ROPE OFF drive and walkway areas				
CN54	PAY concrete supplier				
CN55	PAY concrete finishers				
CN56	PAY concrete finisher's retainage				
	WATERPROOFING				
WP1	CONDUCT standard bidding process				
WP2	SEAL footing and wall intersection				
WP3	APPLY first coat of portland cement				
WP4	APPLY second coat of portland cement				
WP5	APPLY waterproofing compound				
WP6	APPLY 6-mil poly vapor barrier				
WP7	INSPECT waterproofing (tar/poly)				
WP8	INSTALL gravel bed for drain tile				
WP9	INSTALL drain tile				
WP10	INSPECT drain tile				
WP11	APPLY top layer of gravel on drain tile				
WP12	PAY waterproofing sub				
WP13	PAY waterproofing sub retainage				
	FRAMING				
FR1	CONDUCT standard bidding process (material)				
FR2	CONDUCT standard bidding process (labor)				
FR3	DISCUSS aspects of framing with crew				
FR4	ORDER special materials				
FR5	ORDER first load of framing lumber				
FR6	INSTALL sill felt				
FR7	ATTACH sill plate with lag bolts				
FR8	INSTALL support columns				
FR9	SUPERVISE framing process				

MASTERPLAN

STEP	DESCRIPTION	Date Started	Date Completed	Date Inspected	NOTES
FR10	FRAME first-floor joists and subfloor				
FR11	FRAME basement stairs				
FR12	POSITION all first-floor large items				
FR13	FRAME exterior walls (first floor)				
FR14	PLUMB and line first floor				
FR15	FRAME second-floor joists and subfloor				
FR16	POSITION all large second-floor items				
FR17	FRAME exterior walls (second floor)				
FR18	PLUMB and line second floor				
FR19	INSTALL second-floor ceiling joists				
FR20	FRAME roof				
FR21	INSTALL roof decking				
FR22	PAY framing first installment				
FR23	INSTALL lapped tar paper				
FR24	FRAME chimney chases				
FR25	INSTALL prefab fireplaces				
FR26	FRAME dormers and skylights				
FR27	FRAME tray ceilings, skylights and bays				
FR28	INSTALL sheathing on exterior walls				
FR29	INSPECT sheathing				
FR30	REMOVE temporary bracing				
FR31	INSTALL exterior windows and doors				
FR32	APPLY dead wood				
FR33	INSTALL roof ventilators				
FR34	FRAME decks				
FR35	INSPECT framing				
FR36	SCHEDULE loan draw inspection				
FR37	CORRECT any framing problems				
FR38	PAY framing labor (second payment)				
FR39	PAY framing retainage				
	ROOFING				
RF1	SELECT shingle style, material and color				
RF2	PERFORM standard bidding process				
RF3	ORDER shingles and roofing felt				
RF4	INSTALL metal drip edge				
RF5	INSTALL roofing felt				

MASTERPLAN

STEP	DESCRIPTION	Date Started	Date Completed	Date Inspected	NOTES
RF6	INSTALL drip edge on rake				
RF7	INSTALL roofing shingles				
RF8	INSPECT roofing				
RF9	PAY roofing subcontractor				
RF10	PAY roofing subcontractor retainage				
	GUTTERS				
GU1	SELECT gutter type and color				
GU2	PERFORM standard bidding process				
GU3	INSTALL underground drainpipe				
GU4	INSTALL gutters and downspouts				
GU5	INSTALL splashblocks				
GU6	INSPECT gutters				
GU7	INSTALL copper awnings				
GU8	PAY gutter subcontractor				
GU9	PAY gutter sub retainage				
	PLUMBING				
PL1	DETERMINE type and quantity of fixtures				
PL2	CONDUCT standard bidding process				
PL3	WALK THROUGH site with plumber				
PL4	ORDER special plumbing fixtures				
PL5	APPLY for water connection and sewer tap				
PL6	INSTALL stub plumbing for slabs				
PL7	MOVE large plumbing fixtures into place				
PL8	MARK location of all fixtures				
PL9	INSTALL rough-in plumbing				
PL10	INSTALL water meter and spigot				
PL11	INSTALL sewer tap				
PL12	SCHEDULE plumbing inspection				
PL13	INSTALL septic tank and line				
PL14	CONDUCT rough-in plumbing inspection				
PL15	CORRECT rough-in plumbing problems				
PL16	PAY plumbing sub for rough-in				

MASTERPLAN

STEP	DESCRIPTION	Date Started	Date Completed	Date Inspected	NOTES
	PLUMBING FINISH				
PL17	INSTALL finish plumbing				
PL18	TAP into water supply				
PL19	CONDUCT finish plumbing inspection				
PL20	CORRECT finish plumbing problems				
PL21	PAY plumbing sub for finish work				
PL22	PAY plumbing retainage				
	HVAC				
HV1	CONDUCT energy audit				
HV2	SHOP for heating and cooling system				
HV3	CONDUCT standard bidding process				
HV4	FINALIZE HVAC design				
HV5	ROUGH-IN heating and air system				
HV6	INSPECT heating and air rough-in				
HV7	CORRECT rough-in problems				
HV8	PAY HVAC sub for rough-in				
HV9	INSTALL finish heating and air				
HV10	INSPECT final HVAC installation				
HV11	CORRECT all HVAC problems				
HV12	CALL gas company to install lines				
HV13	DRAW gas line on plat diagram				
HV14	PAY HVAC sub for finish work				
HV15	PAY HVAC retainage				
	ELECTRICAL				
EL1	DETERMINE electrical requirements				
EL2	SELECT electrical fixtures and appliances				
EL3	DETERMINE phone installation requirements				
EL4	CONDUCT standard bidding process				
EL5	SCHEDULE phone wiring				
EL6	APPLY for temporary electrical hookup				
EL7	INSTALL temporary electric pole				
EL8	PERFORM rough-in electrical				
EL9	INSTALL phone wiring				

MASTERPLAN

STEP	DESCRIPTION	Date Started	Date Completed	Date Inspected	NOTES
EL10	SCHEDULE rough-in electrical inspection				
EL11	INSPECT rough-in electrical				
EL12	CORRECT rough-in electrical problems				
EL13	PAY electrical sub for rough-in work				
EL14	INSTALL garage door for opener installation				
EL15	INSTALL garage door opener				
EL16	PERFORM finish electrical work				
EL17	CALL phone company to connect service				
EL18	INSPECT finish electrical				
EL19	CORRECT finish electrical problems				
EL20	CALL electrical utility to connect service				
EL21	PAY electrical sub for finish work				
EL22	PAY electrical sub retainage				
	MASONRY				
MA1	DETERMINE brick pattern and color				
MA2	DETERMINE stucco pattern and color				
MA3	PERFORM standard bidding process (masonry)				
MA4	PERFORM standard bidding process (stucco)				
MA5	APPLY flashing				
MA6	ORDER brick, stone and angle irons				
MA7	LAY bricks				
MA8	FINISH bricks and mortar				
MA9	INSPECT brickwork				
MA10	CORRECT brickwork deficiencies				
MA11	PAY masons				
MA12	CLEAN UP excess bricks				
MA13	INSTALL base for decorative stucco				
MA14	PREPARE stucco test area				
MA15	APPLY stucco lath				
MA16	APPLY first coat of stucco				
MA17	APPLY second coat of stucco				
MA18	INSPECT stucco work				
MA19	CORRECT stucco deficiencies				
MA20	PAY stucco sub				
MA21	PAY mason sub retainage				

MASTERPLAN

STEP	DESCRIPTION	Date Started	Date Completed	Date Inspected	NOTES
MA22	PAY stucco sub retainage				
	SIDING/CORNICE				
SC1	SELECT siding material and color				
SC2	CONDUCT standard bidding process				
SC3	ORDER windows and doors				
SC4	INSTALL windows and doors				
SC5	INSTALL flashing				
SC6	INSTALL siding				
SC7	INSTALL siding trim around corners				
SC8	CAULK siding joints				
SC9	INSPECT siding				
SC10	PAY sub for siding work				
SC11	INSTALL cornice				
SC12	INSPECT cornice work				
SC13	CORRECT any cornice problems				
SC14	PAY cornice sub				
SC15	CORRECT siding problems				
SC16	ARRANGE for painter to caulk and paint				
SC17	PAY siding retainage				
SC18	PAY cornice retainage				
	INSULATION				
IN1	DETERMINE insulation requirements				
IN2	CONDUCT standard bidding process				
IN3	INSTALL exterior wall insulation				
IN4	INSTALL soundproofing				
IN5	INSTALL floor insulation				
IN6	INSTALL attic insulation				
IN7	INSPECT insulation				
IN8	CORRECT insulation work as needed				
IN9	PAY insulation sub				
IN10	PAY insulation sub retainage				

MASTERPLAN

STEP	DESCRIPTION	Date Started	Date Completed	Date Inspected	NOTES
	DRYWALL				
DR1	PERFORM standard bidding process				
DR2	ORDER and receive drywall				
DR3	HANG drywall on all walls				
DR4	FINISH drywall				
DR5	INSPECT drywall				
DR6	TOUCH UP and repair drywall				
DR7	PAY drywall sub				
DR8	PAY drywall sub retainage				
	TRIM				
TR1	DETERMINE trim requirements				
TR2	SELECT molding and millwork				
TR3	INSTALL windows after framing				
TR4	CONDUCT standard bidding process				
TR5	INSTALL interior doors				
TR6	INSTALL window casings and aprons				
TR7	INSTALL trim around cased openings				
TR8	INSTALL staircase openings				
TR9	INSTALL crown molding				
TR10	INSTALL base and base cap molding				
TR11	INSTALL chair rail molding				
TR12	INSTALL picture molding				
TR13	INSTALL, sand and stain paneling				
TR14	CLEAN all sliding-door tracks				
TR15	INSTALL thresholds and weather stripping				
TR16	INSTALL shoe molding				
TR17	INSTALL door and window hardware				
TR18	INSPECT trimwork				
TR19	CORRECT trim and stain work				
TR20	PAY trim sub				
	PAINTING AND WALLCOVERING				
PT1	SELECT paint schemes				
PT2	PERFORM standard bidding process				

MASTERPLAN

STEP	DESCRIPTION	Date Started	Date Completed	Date Inspected	NOTES
PT3	PURCHASE all painting materials				
	EXTERIOR				
PT4	PRIME and caulk exterior surfaces				
PT5	PAINT exterior siding and trim				
PT6	PAINT cornice work				
PT7	PAINT gutters				
	INTERIOR				
PT8	PREPARE painting surfaces				
PT9	PRIME walls and trim				
PT10	PAINT or stipple ceilings				
PT11	PAINT walls				
PT12	PAINT or stain trim				
PT13	REMOVE paint from windows				
PT14	INSPECT paint job				
PT15	TOUCH UP paint job				
PT16	CLEAN UP				
PT17	PAY painter				
PT18	PAY retainage				
	CABINETRY				
CB1	CONFIRM kitchen design				
CB2	SELECT cabinetry/countertop				
CB3	COMPLETE cabinet layout diagram				
CB4	PERFORM standard bidding process				
CB5	PURCHASE cabinetry and countertops				
CB6	STAIN or paint cabinetry				
CB7	INSTALL bathroom vanities				
CB8	INSTALL kitchen wall cabinets				
CB9	INSTALL kitchen base cabinets				
CB10	INSTALL kitchen and bath countertops				
CB11	INSTALL cabinet hardware				
CB12	INSTALL utility room cabinetry				

MASTERPLAN

STEP	DESCRIPTION	Date Started	Date Completed	Date Inspected	NOTES
CB13	CAULK cabinet joints as needed				
CB14	INSPECT cabinetry and countertops				
CB15	TOUCH UP and repair as needed				
CB16	PAY cabinet sub				
	GLAZING				
GL1	DETERMINE mirror and glass requirements				
GL2	SELECT mirrors and glass				
GL3	CONDUCT standard bidding process				
GL4	INSTALL fixed-pane picture windows				
GL5	INSTALL shower doors				
GL6	INSTALL mirrors				
GL7	INSPECT all glazing installation				
GL8	CORRECT all glazing problems				
GL9	PAY glazier				
	FLOORING AND TILE				
FL1	SELECT carpet color, style and coverage				
FL2	SELECT hardwood floor type				
FL3	SELECT vinyl floor covering				
FL4	SELECT tile type and coverage				
FL5	CONDUCT standard bidding process				
	TILE				
FL6	ORDER tile and grout				
FL7	PREPARE area to be tiled				
FL8	INSTALL tile base in shower stalls				
FL9	APPLY tile adhesive				
FL10	INSTALL tile and marble thresholds				
FL11	APPLY grout over tile				
FL12	INSPECT tile				
FL13	CORRECT tile problems				
FL14	SEAL grout				
FL15	PAY tile sub				

MASTERPLAN

STEP	DESCRIPTION	Date Started	Date Completed	Date Inspected	NOTES
FL16	PAY tile sub retainage				
	HARDWOOD				
FL17	ORDER hardwood flooring				
FL18	PREPARE subfloor for hardwood flooring				
FL19	INSTALL hardwood flooring				
FL20	SAND hardwood flooring				
FL21	INSPECT hardwood flooring				
FL22	CORRECT hardwood flooring problems				
FL23	STAIN hardwood flooring				
FL24	SEAL hardwood flooring				
FL25	INSPECT hardwood flooring (final)				
FL26	PAY hardwood flooring sub				
FL27	PAY hardwood flooring sub retainage				
	VINYL FLOOR COVERING				
FL28	PREPARE subfloor for vinyl floor covering				
FL29	INSTALL vinyl floor covering				
FL30	INSPECT vinyl floor covering				
FL31	CORRECT vinyl floor covering problems				
FL32	PAY vinyl floor covering sub				
FL33	PAY vinyl floor covering sub retainage				
	CARPET				
FL34	PREPARE subfloor for carpeting				
FL35	INSTALL carpet stretcher strips				
FL36	INSTALL carpet padding				
FL37	INSTALL carpet				
FL38	INSPECT carpet				
FL39	CORRECT carpet problems				
FL40	PAY carpet sub				
FL41	PAY carpet sub retainage				

MASTERPLAN

STEP	DESCRIPTION	Date Started	Date Completed	Date Inspected	NOTES
	LANDSCAPING				
	PRECONSTRUCTION				
LD1	EVALUATE lot				
LD2	DETERMINE best house location				
LD3	DEVELOP site plan				
LD4	CONTACT landscape architect				
LD5	FINALIZE site plan				
LD6	SUBMIT site plan				
LD7	DELIVER fill dirt				
LD8	PAY landscape architect				
	AFTER CONSTRUCTION				
LD9	CONDUCT soil test				
LD10	TILL topsoil				
LD11	APPLY soil treatments				
LD12	INSTALL underground sprinkler system				
LD13	PLANT flower bulbs				
LD14	APPLY seed or sod				
LD15	SOAK lawn with water				
LD16	PLANT bushes and trees				
LD17	PREPARE landscaped islands				
LD18	INSTALL mailbox				
LD19	INSPECT landscaping job				
LD20	CORRECT any landscaping problems				
LD21	PAY landscaping sub				
LD22	PAY landscaping sub retainage				
	DECKING				
DE1	EXAMINE building site				
DE2	DRAW complete plan for deck				
DE3	PURCHASE materials for deck				
DE4	FINISH all grading and backfilling				
DE5	LAY OUT deck design				
DE6	DIG footings and pour concrete piers				
DE7	PULL leveling line from corner to corner				
DE8	INSTALL posts on pier anchors				

MASTERPLAN

STEP	DESCRIPTION	Date Started	Date Completed	Date Inspected	NOTES
DE9	ATTACH deck beams to the posts				
DE10	INSTALL flashing on the tops of the beams				
DE11	MOUNT joist header to the house				
DE12	PLACE remaining headers on the tops of the beams				
DE13	INSTALL joists between frame members				
DE14	MOUNT cross bracing between posts				
DE15	APPLY deck boards to joists				
DE16	SAW the deck boards flush with header joist				
DE17	ATTACH stairway stringers to header joist				
DE18	INSTALL treads between the stringers				
DE19	INSTALL post rails around the perimeter				
DE20	TREAT deck with wood preservative				

Project Schedule

HOW TO USE THE PROJECT SCHEDULE

The Project Schedule that follows on the next page gives you a visual representation of your entire project over a span of eighteen months.

The time span of your project may vay considerably from this length, so adjust your timing accordingly. The length of each project bar includes some slack time and starts at the earliest possible date. The **shaded** boxes represent the critical path of the project. In other words, each preceding task must be completed before the next CRITICAL TASK can begin. Step numbers correspond directly to the steps in the Project Management section and in the Master Plan.

MAJOR STEPS		WEEK OF PROJECT																		
		0	1	2	3	4	5	6	7	8	9	10	11	12	13	14	15	16	17	18
PRECONSTRUCTION	PC	1-33																		
EXCAVATION	EX	1	2-17		18-20									21-30						
PEST CONTROL	PS	1-2			3								4-6							
CONCRETE	CN	1-3		4-8	9-45										46-56					
WATERPROOFING	WP	1				2-13														
FRAMING	FR	1-4				5		6-33					34-39							
ROOFING	RF	1-2						3	4-9									10		
PLUMBING	PL	1-5		6				7		8-16									17-22	
HVAC	HV	1-4									5-8									9-15
ELECTRICAL	EL	1-6		7							8-13								14-22	
MASONRY	MA	1-4									5-22									
SIDING & CORNICE	SC	1-2							3-4		5-9	10-18								
INSULATION	IN	1-2											3-5			6-10				
DRYWALL	DR	1												2-8						
TRIM	TR	1-2							3								4-12			13-20
PAINTING	PT	1-3								4				5-6	7		8-18			
CABINETRY	CB	1-4							5									6-16		
FLOORING & TILE	FL	1-5														6-27			28-41	
GLAZING	GL	1-3							4										5-9	
GUTTERS	GU	1-2													3-9					
LANDSCAPING	LD	1-7															8-22			
DECKING	DE																	1-20		

SUPPLIER'S/SUBCONTRACTOR'S REFERENCE SHEET

NAME	PHONE	TYPE PRODUCT
ADDRESS	CALL WHEN	DATE
	PRODUCTS	
	WORKER'S COMP?	BONDED?
REFERENCE NAME	**PHONE**	**COMMENTS**
PRICES/COMMENTS		

NAME	PHONE	TYPE PRODUCT
ADDRESS	CALL WHEN	DATE
	PRODUCTS	
	WORKER'S COMP?	BONDED?
REFERENCE NAME	**PHONE**	**COMMENTS**
PRICES/COMMENTS		

SUPPLIER'S/SUBCONTRACTOR'S REFERENCE SHEET

NAME	PHONE	TYPE PRODUCT
ADDRESS	CALL WHEN	DATE
	PRODUCTS	
	WORKER'S COMP?	BONDED?
REFERENCE NAME	**PHONE**	**COMMENTS**
PRICES/COMMENTS		

NAME	PHONE	TYPE PRODUCT
ADDRESS	CALL WHEN	DATE
	PRODUCTS	
	WORKER'S COMP?	BONDED?
REFERENCE NAME	**PHONE**	**COMMENTS**
PRICES/COMMENTS		

PLAN ANALYSIS CHECKLIST

PROPERTY: ______________________________

PLAN NUMBER: ______________

DESCRIPTION	YES	NO
CIRCULATION: Is there access between the following without going through another room?		
Living room to bathroom		
Family room to bathroom		
Each bedroom to bathroom		
Kitchen to dining room		
Living room to dining room		
Kitchen to outside door		
Can get from auto to kitchen without getting wet if raining		
ROOM SIZE AND SHAPE		
Are all rooms adequate size and reasonable shape based on planned use?		
Can living room wall accomodate sofa and end tables (minimum 10 ft.)?		
Can wall of den take sofa and end tables (minimum 10 ft.)?		
Can wall of master bedroom(s) take double bed and end tables?		
STORAGE AND CLOSETS		
Guest closet near front entry		
Linen closet near baths		
Broom closet near kitchen		
Tool and lawnmower storage		
Space for washer and dryer		
EXTERIOR		
Is the house style "normal" architecture?		
Is the house style compatible with surrounding architecture?		
SITE ANALYSIS		
Does the house conform to the neighborhood?		
Is the house suited to the lot?		
Is the topography good?		
Will water drain away from the house?		
Will grade of driveway be reasonable?		
Is the lot wooded?		
Is the subject lot as good as others in the neighborhood?		
Totals		

LIGHTING AND APPLIANCE ORDER

JOB ______________________ STORE NAME ______________________

DATE ______________________ LIGHTING BUDGET ______________________

DESCRIPTION	STYLE NO.	QTY.	UNIT PRICE	TOTAL PRICE
Front Entrance				
Rear Entrance				
Dining Room				
Living Room				
Den				
Family Room				
Kitchen				
Kitchen Sink				
Breakfast Area				
Dinette				
Utility Room				
Basement				
Master Bath				
Hall Bath 1				
Guest Bath				
Hall 1				
Hall 2				
Stairway				
Master Bedroom				
Bedroom 2				
Bedroom 3				
Bedroom 4				
Closets				
OUTDOOR				
Front Porch				
Rear Porch				
Porch/Patio				
Carport/Garage				
Post/Lantern				
Floodlights				
Chimes				
Pushbuttons				
APPLIANCES				
Ovens				
Hood				

ITEM ESTIMATE WORKSHEET

VENDOR NAME	DESCRIPTION	MEASURING		CONVERSION FACTOR	ORDERING		PRICE EACH	COST	TAX	TOTAL COST	COST TYPE
		QTY.	UNIT		QTY.	UNIT					

EXPENSE CATEGORY	MATERIAL COST	LABOR COST	SUBCONTRACTOR COST	TOTAL COST

COST ESTIMATE SUMMARY

DESCRIPTION	MATERIAL	LABOR	SUBCONTR.	TOTAL
EXCAVATION				
PEST CONTROL				
CONCRETE				
WATERPROOFING				
FRAMING				
ROOFING				
PLUMBING				
HVAC				
ELECTRIC				
MASONRY				
SIDING & CORNICE				
INSULATION				
DRYWALL				
TRIM				
PAINTING				
CABINETRY				
TILE				
CARPET				
OTHER FLOORING				
GLAZING				
GUTTERS				
LANDSCAPING				
SUBTOTAL				
APPLIANCES				
LUMBER				
OTHER MATERIALS				
TOOLS/RENTAL				
LOAN/LEGAL				
INSURANCE				
PERMITS/LICENSE				
SUBTOTAL				
			GRAND TOTAL	

COST ESTIMATE CHECKLIST

CODE NO.	DESCRIPTION	QTY.	MATERIAL		LABOR		SUB-CONTR.		TOTAL
			UNIT PRICE	TOTAL MATL.	UNIT PRICE	TOTAL LABOR	UNIT PRICE	TOTAL SUB	
		SUB-TOTALS							
						GRAND TOTALS			

SUBCONTRACTOR BID CONTROL LOG

SUB/LABOR	NAME	BID	GOOD UNTIL	NAME	BID	GOOD UNTIL	NAME	BID	GOOD UNTIL	BEST BID	NAME
EXCAVATION											
CONCRETE											
PEST CONTROL											
FRAMING											
ROOFING											
PLUMBING											
HVAC											
ELECTRICAL											
MASONRY											
SIDING											
INSULATION											
DRYWALL											
TRIM											
PAINTING											
CABINETRY											
FLOORING											
TILE											
GLASS											
LANDSCAPING											
GUTTERS											
ASPHALT											

PURCHASE ORDER CONTROL LOG

PO NO.	DATE ORDERED	DESCRIPTION	VENDOR NAME	DATE DELIVERED	DISCOUNT		PAYMENTS				TOTAL PAID
					RATE	DUE	CHK#.	AMT.	CHK#.	AMT.	

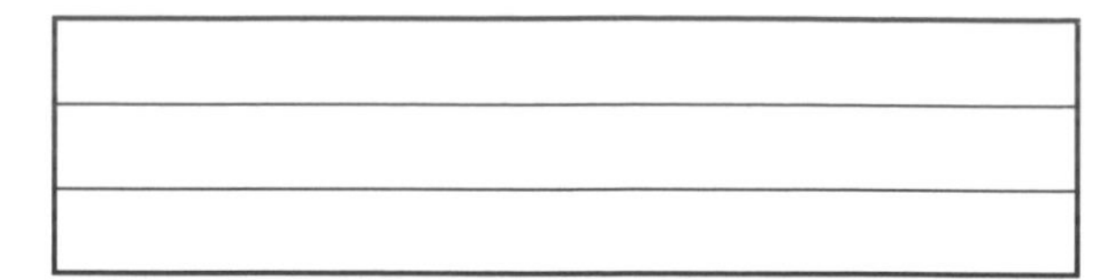

PURCHASE ORDER

TO:

P/O NO.	PAGE NO.
DATE OF ORDER	JOB NO.
SALESMAN	REQ. NO.
DELIVER TO	

REQUESTED BY	DELIVER BY	SHIP VIA	TERMS

B/O	QTY. ORDERED	QTY. RECEIVED	DESCRIPTION	UNIT PRICE	TOTAL

PLEASE SEND A COPY OF THIS PO WITH DELIVERED MATERIAL. CALL IF UNABLE TO MAKE DELIVERY DEADLINE.

PURCHASE ORDER cont.

P/O NO.	PAGE NO.

B/O	QTY. ORDERED	QTY. RECEIVED	DESCRIPTION	UNIT PRICE	TOTAL

PLEASE SEND A COPY OF THIS PO WITH DELIVERED MATERIAL. CALL IF UNABLE TO MAKE DELIVERY DEADLINE.

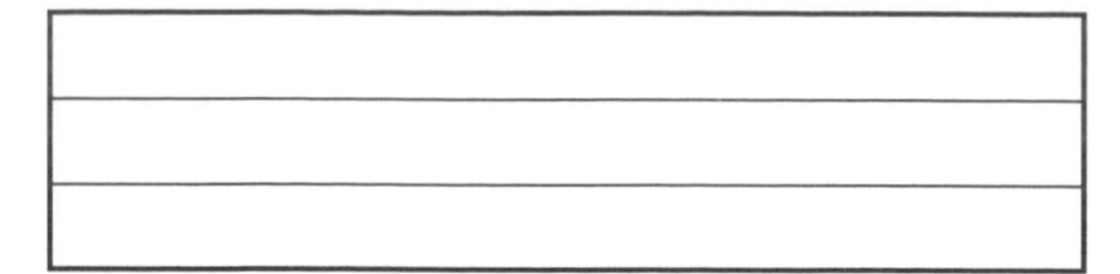

CHANGE ORDER

CHANGE ORDER NO.

DATE OF CHANGE ORDER	JOB NO.
JOB PHONE	CONTRACT DATE
JOB LOCATION	

TO (CONTRACTOR):

You are directed to make the following changes in this Contract:

Contract time will be ☐ increased ☐ decreased by: Days	Original Contract Price	
	Previously Authorized Change Orders	
New Completion Date	New Contract Price	
ACCEPTED—The above prices and specifications are satisfactory and are hereby accepted. All work to be done under same terms and conditions as original contract unless otherwise stipulated.	This Change Order	
	New Authorized Contract Price	
AUTHORIZED BY	SUBCONTRACTOR	
SIGNATURE DATE	SIGNATURE DATE	

SUBCONTRACTOR'S AGREEMENT

Sub. Name ______________________

Address ______________________

____

____

Phone ______________________

Have Worker's Comp? ☐ Yes ☐ No

Worker's Comp. No. ______________________

Expiration Date ______________________

Tax Id No. ______________________

Date ______________________

Job No. ______________________

Job Address ______________________

____

____

Plan No. ______________________

Bid Amount ______________________

Bid Good Until ______________________

Payment to be made as follows:

____

____

____

____

Deduct ________% for worker's compensation

Deduct ________% for retainage.

All material is guaranteed to be as specified. All work to be completed in a timely manner, according to standard practices. Any alteration or deviation from specifications below involving extra costs will be executed only upon written order (refer to Change Order Form).

SPECIFICATIONS: (Attach extra sheets if necessary)

____

____

____

____

____

____

Agreement

____

Subcontractor (date)

PAYMENTS

DATE	CHECK NO.	AMOUNT PAID	AMOUNT DUE

SUBCONTRACTOR'S AFFIDAVIT

STATE OF: ______________________________

COUNTY OF: ______________________________

Personally appeared before the undersigned attesting officer,

____

who being duly sworn, on oath says that he was the contractor in charge of improving the property

owned by ____

located at ____

Contractor says that all work, labor, services and materials used in such improvements were furnished and performed at the contractor's instance; that said contractor has been paid or partially paid the full contract price for such improvements; that all work done or material furnished in making the improvements have been paid for at the agreed price, or reasonable value, and there are no unpaid bills for labor and services performed or materials furnished; and that no person has any claim or lien by reason of said improvements.

____

(Seal)

Sworn to and subscribed before me,

this ________ day of ____________________, 19______.

____

Notary Public

CONSTRUCTION LOAN DRAW SCHEDULE

JOB DESCRIPTION ______________________________ JOB NO. ______________

BANK ______________________________

ADDRESS ______________________________

____

AMOUNT OF CONSTRUCTION LOAN	
CONSTRUCTION LOAN CLOSING COSTS	
TOTAL CONSTRUCTION LOAN COST	

DRAW NO.	DESCRIPTION	DATE	AMOUNT
1			
2			
3			
4			
5			
6			
7			
8			
		TOTAL OF LOAN DRAWS	

BUILDING CONTRACT

(date) ______________________________

1. The undersigned owner authorizes the undersigned builder to construct and deliver to said owner a dwelling in accordance with the plans and specifications which are attached hereto and made a part hereof, on all that tract or parcel of land lying and being in Land:

   Lot ______________________________ of the ______________________________ District

   of ______________________________ County, ______________ (state), and being known as

   ____

2. Except as proved in the paragraph dealing with allowances hereafter, the total cost to the owner (excluding the cost of the land and additional arranged work), shall be:

   ______________________________ DOLLARS ($ ______________).

   Said amount shall be paid as follows:

3. Builder shall commence construction of the house in accordance with the attached plans and specifications after the construction loan has been closed. The construction of the house shall be completed within ______________ days from the time that the construction is commenced.
4. The improvements shall be deemed fully completed when approved by the mortgagee for loan purposes.
5. Builder agrees to give possession of the improvements to the purchaser immediately after the full payment to the builder of any cash balance due. At that time, the builder agrees to furnish the purchaser a notarized affidavit that there are no outstanding liens for unpaid material or labor.
6. It is the understanding of the parties that the purchase price above described includes builder's allowances for items hereinafter described. The amounts are listed at builder's cost. It is the understanding of the parties that the purchase price shall be equal to the price described in Paragraph 2 above, provided the costs of the items below listed do not exceed the amounts shown. In the event that the builder's cost exceeds any item hereinafter listed, the purchase price shall be increased by the amount above the allowance shown. The allowances are as follows:

   ____

   ____

   ____

   ____

7. The cost of any alterations, additions, omissions or deviations at the request of the owner shall be added to the agreed purchase price. No such changes will be made unless stipulated in writing and signed by both parties hereto.
8. Builder shall provide to owner, in writing, a twelve (12) month construction guarantee.
9. Every effort will be exerted on the part of the builder to complete the construction within the period described above. However, builder assumes no responsibilities for delays occasioned by causes beyond his control. In the event of a delay of construction caused by weather, purchaser's selections, etc., the period of delay is to be added to the construction period allotted to builder.
10. This contract shall inure to the benefit of, and shall be binding upon, the parties hereto, their heirs, successors, administrators, executors and assigns.
11. This contract constitutes the sole and entire agreement between the parties hereto and no modification of this contract shall be binding unless attached hereto and signed by all parties to the agreement. No representation, promise, or inducement to this agreement not included in this contract shall be binding upon any party hereto.

The parties do set their hands and seals the day and year first above written.

(Owner)

(Owner)

(Builder)

BUILDING CHECKLIST

JOB ______________________________

DATE STARTED ______________________ ESTIMATED COMPLETION DATE ______________

(**Bold faced** numbered items indicate critical tasks that must be done before continuing.)

		DATE DONE	NOTES
	A. PRE-CLOSING		
☐	1. Take-off		
☐	2. Spec sheet		
☐	3. Contract		
☐	4. Submit loan		
☐	5. Survey		
☐	6. Insurance		
☐	7. Closing		
	B. POST-CLOSING		
☐	1. Business license		
☐	2. Compliance bond		
☐	3. Sewer tap		
☐	4. Water tap		
☐	5. Perk test (if necessary)		
☐	6. Miscellaneous permits		
☐	7. Building permit		
☐	8. Temporary services:		
☐	a. Electrical		
☐	b. Water		
☐	c. Phone		
	C. CONSTRUCTION		
☐	1. Rough layout		
☐	2. Rough grading		
☐	3. Water		
☐	a. Meter		
☐	b. Well		
☐	4. **Set temporary pole**		
☐	5. Crawl or basement		
☐	a. Batter boards		
☐	b. **Footings dug**		
☐	c. **Footings poured**		
☐	d. Lay block/pour		
☐	6. Pre-treat		

		DATE DONE	NOTES
☐	7. Slab		
☐	a. Form boards		
☐	b. **Plumbing**		
☐	c. iscellaneous pipes		
☐	d. Gravel		
☐	e. Poly		
☐	f. **Re-wire**		
☐	g. Pour concrete		
☐	8. Spot survey		
☐	9. Waterproof		
☐	10. Fiberglass tubs		
☐	11. Framing		
☐	a. Basement walls		
☐	b. First floor		
☐	c. Walls—1st floor		
☐	d. Second floor		
☐	e. Walls—2nd floor		
☐	f. Sheathing		
☐	g. Ceiling joists		
☐	h. Rafters/trussed		
☐	i. Decking		
☐	j. Felt		
☐	12. Roofing		
☐	13. Set fireplace		
☐	14. Set doors		
☐	15. Set windows		
☐	16. Siding and cornice		
☐	17. **Rough plumbing**		
☐	18. **Rough HVAC**		
☐	19. Rough electrical		
☐	20. Measure cabinets		
☐	21. Rough phone/cable		
☐	22. Rough miscellaneous		
☐	23. Insulation		
☐	24. Framing inspection		
☐	25. Clean interior		
☐	26. Drywall		
☐	27. Clean interior		

	DATE DONE	NOTES
☐ 28. Stucco		
☐ 29. Brick		
☐ 30. Clean exterior		
☐ 31. Septic tank		
☐ 32. Utility lines		
☐ a. Gas line		
☐ b. Sewer line		
☐ c. Phone line		
☐ d. Electric line		
☐ 33. Backfill		
☐ 34. Pour garage		
☐ 35. Drives and walks		
☐ 36. Landscape		
☐ 37. Gutters		
☐ 38. Garage door		
☐ 39. Paint/stain		
☐ 40. Cabinets		
☐ 41. Trim		
☐ 42. Tile		
☐ 43. Rock		
☐ 44. Hardwood floors		
☐ 45. Wallpaper		
☐ 46. **Plumbing (final)**		
☐ 47. **HVAC (final)**		
☐ 48. **Electrical (final)**		
☐ 49. Mirrors		
☐ 50. Clean interior		
☐ 51. Vinyl/carpet		
☐ 52. Final trim		
☐ 53. Bath accessories		
☐ 54. Paint touch up		
☐ 55. Final touch up		
☐ 56. Connect utilities		
☐ **DATE COMPLETED**		

Glossary

Acid. Cleaning agent used to clean brick.

Acrylic resin. A thermoplastic resin used in latex coatings. *See* Latex paint.

Aggregate. Irregular-shaped gravel suspended in cement.

Air chamber. Pipe appendage with trapped air added to a line to serve as a shock absorber to retard or eliminate air hammer.

Air-dried lumber. Dried by exposure to air, usually in a yard, without artificial heat. For the United States, the minimum moisture content of thoroughly air-dried lumber is 12 percent to 15 percent and average is somewhat higher.

Alkyd resin. One of a large group of synthetic resins used in making latex paints.

Amperage. The amount of current flow in a wire. Similar to the amount of water flowing in a pipe.

Anchor. Irons of special form used to fasten together timbers or masonry.

Anchor bolt. Bolt that secures a wooden sill plate to concrete or masonry floor or foundation wall.

Apron. Trim used at base of windows. Also used as base to build out crown molding.

Asphalt. Base ingredient of asphalt shingles and roofing paper (felt composition saturated in asphalt base). Used widely as a waterproofing agent in the manufacture of waterproof roof coverings of many types, exterior wall coverings, flooring tile and the like. Most native asphalt is a residue from evaporated petroleum. It is insoluble in water but soluble in gasoline and melts when heated.

Attic ventilators. In houses, screened openings provided to ventilate an attic space. They are located in the soffit area as inlet ventilators and in the gable ends or along the ridge as outlet ventilators. They can also consist of power-driven fans used as an exhaust system. *See also* Louver.

BTU. British thermal unit. A standard unit of hot or cold air output.

Backfill. Process of placing soil up against foundation after all necessary foundation treatments have been performed. In many instances, foundation wall must have temporary interior bracing or house framed in to support weight of dirt until it has had time to settle.

Backflow. The flow of water or other fluids or materials into the distributing source of potable water from any source other than its intended source.

Backhoe. A machine that digs narrow, deep trenches for foundations, drainpipe, cable, etc.

Backing. The bevel on the top edge of a hip rafter that allows the roofing board to fit the top of the rafter without leaving a triangular space between it and the lower side of the roof covering.

Backsplash. A small strip (normally 3″ or 4″) placed against the wall and resting on the back of the countertop. This is normally used to protect the wall from water and stains.

Balloon frame. The lightest and most economical form of construction, in which the studding and corner posts are set up in continuous lengths from first floor line or sill to the roof plate.

Baluster. A vertical member in a railing used on the edge of stairs, balconies and porches. A small pillar or column used to support a rail.

Balustrade. A railing made up of balusters, top rail and sometimes bottom rail.

Band. A low, flat decorative trim around windows and doors and horizontal relief; normally 2×4 or 2×6.

Band joist. *See under* Joist, band.

Base. The bottom of a column; the finish of a room at the junction of the walls and floor.

Base or baseboard. A board placed against the wall around a room next to the floor to finish properly the area between the floor and wall.

Base cabinet. A cabinet resting on the floor at waist level, supporting a countertop or an appliance.

Base coat. First coat put on drywall. Normally a very light color. Should even be put on prior to wallpaper so that the paper can be easily removed.

Base molding. Molding applied at the intersection of the wall and floor. Molding used to trim the upper edge of baseboard.

Base shoe. Molding used next to the floor on interior baseboards. Sometimes called a carpet strip.

Batten. Narrow strips of wood used to cover joints or as decorative vertical members over plywood or wide boards.

Batter board. A pair of horizontal boards nailed to vertical posts set at the corners of an excavation area used to indicate the desired level of excavation. Also used for fastening taut strings to indicate outlines of the foundation walls.

Batter pile. Pile driven at an angle to brace a structure against lateral thrust.

Bay window. Any window that

projects out from the walls of the structure.

Bazooka. Term for an automated drywall application tool that tapes, applies mud and feathers in one pass. Gets its name because it looks like a bazooka. Expensive to buy or rent—only used on large-scale operations.

Bead. Any corner or edge that must be finished off with stucco.

Beam. A long piece of lumber or metal used to support a load placed at right angles to the beam—usually floor joists. An inclusive term for joists, girders, rafters and purlins.

Bearing partition. A partition that supports any vertical load in addition to its own weight.

Bearing wall. A wall that supports any vertical load in addition to its own weight.

Bedding. A layer of mortar into which brick or stone is set.

Bed molding. The trim piece (molding) that covers the intersection of the vertical wall and any overhanging horizontal surface, such as the soffit. In a series of moldings, the lowest one.

Belt course. A horizontal board across or around a building, usually made of a flat member and a molding.

Bend. Any change in direction of a line.

Bending strength. The resistance of a member when loaded like a beam.

Bent. A single vertical framework consisting of horizontal and vertical members supporting the deck of a bridge or pier.

Berm. Built-up mound of dirt for drainage and landscaping. A raised area of earth such as earth pushed against a wall.

Bevel board (pitch board). A board used in framing a roof or stairway to lay out bevels.

Bibb. Special cover used where lines are stubbed through an exterior wall.

Blind-nailing. Nailing in such a way that the nail heads are not visible on the face of the work—usually at the tongue of matched boards.

Board. Lumber less than 2 inches thick.

Board foot. A unit of measurement equal to a 1″ thick piece of wood one foot square. Thickness × Length × Width equals board feet.

Boarding in. The process of nailing boards on the outside studding of a house.

Bolster. A short horizontal wood or steel beam on top of a column to support and decrease the span of beams or girders.

Boston ridge. A method of applying asphalt or wood shingles at the ridge or at the hips of a roof as a finish.

Brace. Diagonally framed member used to temporarily hold wall in place during framing. An inclined piece of framing lumber applied to wall or floor to stiffen the structure.

Bracket. A projecting support for a shelf or other structure.

Break joints. To arrange joints so that they do not come directly under or over the joints of adjoining pieces, as in shingling, siding, etc.

Breather paper. A paper that lets water vapor pass through, often used on the outer face of walls to stop wind and rain while not trapping water vapor.

Brick veneer. A facing of brick laid against and fastened to the sheathing of a frame wall or tile wall construction.

Bridging. Diagonal metal or wood cross braces installed between joists to prevent twisting and to spread the load to adjoining joists.

Btu. British thermal unit. A standard unit of hot or cold air output.

Builder's level. A surveying tool consisting of an optical siting scope and a measuring stick. It is used to check the level of batter boards and foundation.

Building drain. The common artery of the drainage system receiving the discharge from other drainage pipes inside the building and conveying it to the sewer system outside the building.

Building paper. Cheap, thick paper, used to insulate a building before the siding or roofing is put on; sometimes placed between double floors.

Building sewer. That part of the drainage system extending from the building drain to a public or private sewer system.

Building supply. The pipe carrying potable water from the water meter or other water source to points of distribution throughout the building and lot.

Built-up member. A single structural component made from several pieces fastened together.

Built-up roof. A roofing composed of three to five layers of asphalt felt laminated with coal tar, pitch or asphalt. The top is finished with crushed slag or gravel. Generally used on flat or low-pitched roofs.

Bulkhead. Vertical drop in footing when changing from one depth to another.

Bull. A covering for a wide soil stack so that rain will not enter.

Butt joint. Junction of the ends of two framing members such as on a sill. Normally a square-cut joint.

CPVC. Chlorinated Polyvinyl Chloride. A flexible form of water line recently introduced. No soldering involved. Plastic.

Can. Recessed lighting fixture.

Cant strip. A piece of lumber triangular in cross section, used at the junction of a flat deck and a wall to avoid a sharp bend and possible cracking of the covering which is applied over it.

Cantilever. A horizontal structural component that projects beyond its support, such as a second-story floor that projects out from the wall of the first floor.

Cap. Hardware used to terminate any plumbing line. The upper member of a column, pilaster, door cornice or molding.

Cap molding. Trim applied to the top of base molding.

Carriage. The supports for the steps

and risers of a flight of stairs. *See* Stringer.

Casement. A window in which the sash opens upon hinges.

Casement frame-and-sash. A frame of wood or metal enclosing part or all of a sash, which can be opened by means of hinges affixed to the vertical edge.

Casement window. A window that swings out to the side on hinges.

Casing. Molding of various widths used around door and window openings.

Casing nails. Used to apply finish trim and millwork. Nail head is small and is set below the surface of the wood to hide it.

Caulk. To fill or close a joint with a seal to make it watertight and airtight. The material used to seal a joint.

Cement mortar. A mixture of cement with sand and water used as a bonding agent between bricks or stones.

Center set. Standard holes for a standard faucet set.

Center-hung sash. A sash hung on its centers so that it swings on a horizontal axis.

Chair rail. Molding applied to the walls, normally at hip level.

Chalking string. A string covered with chalk that, when stretched between two points and snapped, marks a straight line between the points.

Chamfer. The beveled edge of a board.

Check rails. Also called meeting rails. The upper rail of the lower sash and the lower rail of the upper sash of a double-hung window. Meeting rails are made sufficiently thicker than the rest of the sash frame to close the opening between the two sashes. Check rails are usually beveled to ensure a tight fit between the two sashes.

Checking. Fissures that appear with age in many exterior paint coats. They are at first superficial, but in time they may penetrate entirely through the coating.

Checks. Splits or cracks in a board, ordinarily caused by seasoning.

Chock. Heavy timber fitted between fender piles along wheel guard of a pier or wharf.

Chord. The principal member of a truss on either the top or bottom.

Circuit breaker. A device to ensure that current overloads do not occur. A circuit breaker breaks the circuit when a dangerous overload or short circuit occurs.

Clamp. A mechanical device used to hold two or more pieces together.

Clapboards. A special form of outside covering of a house; siding.

Clean-out. A sealed opening in a pipe which can be screwed off to unclog the line if necessary.

Clear wood. Wood that has no knots.

Cleat. A length of wood fixed to a surface, as a ramp, to give a firm foothold or to maintain an object in place.

Collar beam. Nominal 1″- or 2″-thick members connecting opposite roof rafters at or near the ridge board. They serve to stiffen the roof structure.

Column. In architecture, a perpendicular supporting member, circular or rectangular in section, usually consisting of a base, shaft and capital. In engineering, a vertical structural compression member that supports loads acting in the direction of its longitudinal axis.

Combination frame. A combination of the principal features of the full and balloon frames.

Compressor. Component of the central air-conditioning system that sits outside of dwelling.

Concrete. An artificial building material made by mixing cement and sand with gravel, broken stone, or other aggregate, and sufficient water to cause the cement to set and bind the entire mass.

Condensation. In a building, beads or drops of water (and frequently frost in extremely cold weather) that accumulate on the inside of the exterior covering of a building when warm, moisture-laden air from the interior reaches a point where the temperature no longer permits the air to sustain the moisture it holds. Use of louvers or attic ventilators will reduce moisture condensation in attics. A vapor barrier under the gypsum lath or drywall on exposed walls will reduce condensation in them.

Conductors. Pipes for conducting water from a roof to the ground or to a receptacle or downspout.

Conduit. Metal pipe used to run wiring through when extra protection of wiring is needed or wiring is to be exposed.

Construction, frame. A type of construction in which the structural parts are wood or depend upon a wood frame for support. In codes, if masonry veneer is applied to the exterior walls, the classification of this type of construction is usually unchanged.

Control joint. A joint that penetrates only partially through a concrete slab or wall so that if cracking occurs it will be a straight line at that joint.

Coped joint. *See* Scribing.

Corbel. Extending a course or courses of bricks beyond the face of a wall. No course should extend more than 2″ beyond the course below it. Total corbeling projection should not exceed wall thickness.

Corian. An artificial material simulating marble. Manufactured by Corning.

Corner bead. A strip of formed sheet metal, sometimes combined with a strip of metal lath, placed on corners before plastering to reinforce them. Also, a strip of wood finish three-quarters-round or angular placed over a plastered corner for protection.

Corner board. A board used as trim for the external corner of a house or other frame structure, against which the ends of siding are butted.

Corner brace. A diagonal brace placed at the corner of a frame structure to stiffen and strengthen the wall.

Cornice. When used on the outside of

the house, the trimwork that finishes off the intersection of the roof and siding. The cornice may be flush with the siding or may overhang the siding by as much as 2′. It usually consists of a fascia board, a soffit for a closed cornice and appropriate moldings.

Cornice return. The underside of the cornice at the corner of the roof where the walls meet the gable end roofline. The cornice return serves as trim rather than as a structural element, providing a transition from the horizontal eave line to the sloped roofline of the gable.

Counterflashing. A flashing usually used on chimneys at the roofline to cover shingle flashing and to prevent moisture entry. These also allow expansion and contraction without danger of breaking the flashing.

Countersink. To set the head of a nail or screw at or below the surface.

Course. A horizontally laid set of bricks. A 32-course wall with ⅜″ mortar joints stands 7′ tall. A horizontal row of shingles.

Cove molding. A molding with a concave face used as trim or to finish interior corners.

Coverage. The maximum number of shingles that overlap in any one spot. This determines the degree of weather performance a roof has.

Crawl space. A shallow space below the living quarters of a house with no basement, normally enclosed by the foundation wall.

Creosote. A distillate of coal tar produced by high temperature carbonization of bituminous coal; it consists principally of liquid and solid aromatic hydrocarbons used as a wood preservative.

Cricket. A sloped area at the intersection of a vertical surface and the roof, such as a chimney. Used to channel off water that might otherwise get trapped behind the vertical structure. *Also see* Saddle.

Crimp. A crease formed in sheet metal for fastening purposes or to make the material less flexible.

Cripple. A short stud used as bracing under windows and other structural framing.

Cross brace. Bracing with two intersecting diagonals.

Crown molding. The trim piece that tops off the trim on a vertical structure. Usually refers to the more ornamental pieces of cornice trim. Molding applied at the intersection of the ceiling and walls. Normally associated with more formal areas. Can also refer to the ornamental trim applied between the fascia and the roof. If a molding has a concave face, it is called a cove molding.

Crusher run. Crushed stone, normally up to 2″ or 3″ in size. Sharp edges. Used for drive and foundation support as a base. Very stable surface, as opposed to gravel, which is not very stable.

Cube. A standard ordering unit for masonry block units (6×6×8).

Curing. Process of maintaining proper moisture level and temperature (about 73° Fahrenheit) until the design strength is achieved. Curing methods include adding moisture (sprinkling with water, applying wet coverings such as burlap or straw) and retaining moisture (covering with waterproof materials such as polyurethane).

d. *See* Penny.

Dado. A rectangular groove across the width of a board or plank. In interior decoration, a special type of wall treatment.

Damper. A metal flap controlling the flow of conditioned air through ductwork.

Dead load. The weight, expressed in pounds per square foot, of elements that are part of the structure.

Deadening. Construction intended to prevent the passage of sound.

Deadman timber. A large buried timber used as an anchor as for anchoring a retaining wall.

Deadwood. Wood used as backing for drywall.

Decay. Disintegration of wood or other substance through the action of fungi, as opposed to insect damage.

Deck paint. An enamel with a high degree of resistance to mechanical wear, designed for use on such surfaces as porch floors.

Decking. Heavy plank floor of a pier or bridge.

Deformed shank nail. A nail with ridges on the shank to provide better withdrawal resistance. *See under* Nail.

Den paneling. Paneling composed normally of sections of stain-grade materials, joined with a number of vertical and/or horizontal strips of molding.

Density. The mass of a substance in unit volume. When expressed in the metric system, it is numerically equal to the specific gravity of the same substance.

Dentil molding. A special type of crown molding with an even pattern of "teeth."

Dewpoint. Temperature at which a vapor begins to deposit as a liquid. Applies especially to water in the atmosphere.

Diagonal. Inclined member of a truss or bracing system used for stiffening and wind bracing.

Dimension. *See* Lumber dimension.

Direct nailing. To nail perpendicular to the initial surface or to the junction of the pieces joined. Also termed "face nailing."

Door jamb, interior. The surrounding case into which and out of which a door closes and opens. It consists of two upright pieces, called side jambs, and a horizontal head jamb.

Dormer. An opening in a sloping roof, the framing of which projects out to form a vertical wall suitable for windows or other openings. *See* Eye dormer and Shed dormer.

Dovetail. Joint made by cutting pins the shape of dove tails which fit between dovetails upon another piece.

Downspout. A pipe, usually of metal, for carrying rainwater from roof gutters.

Drag time. Time required to haul

heavy excavation equipment to and from the site.

Drawboard. A mortise-and-tenon joint with holes so bored that when a pin is driven through, the joint becomes tighter.

Dressed and matched (tongued and grooved). Boards or planks machined in such a manner that there is a groove on one edge and a corresponding tongue on the other.

Dressed size. Dimensions of lumber after planing smooth.

Dried in. Term describing the framed structure after the roof deck and protective tar paper have been installed.

Drip. (1) A structural member of a cornice or other horizontal exterior-finish course that has a projection for water runoff. (2) A groove in the underside of a sill or drip cap to cause water to run off on the outer edge.

Drip cap. A molding placed on the exterior top side of a door or window frame to cause water to drip beyond the outside of the frame.

Drip edge. Metal flashing normally 3″ wide that goes on the eaves and rakes to provide a precise point for water to drip from so that the cornice does not rot.

Drywall. Interior covering material that is applied in large sheets or panels. The term has become basically synonymous with gypsum wallboard.

Ducts. In a house, usually round or rectangular metal pipes for distributing warm air from the heating plant to rooms, or air from a conditioning device or as cold air returns.

EER. (Energy Efficiency Rating) A national rating required to be displayed on appliances measuring their efficient use of electrical power.

Eaves. The portions of the roof that extend beyond the outside walls of the house. The main function of an overhanging eave is to provide visual separation of the roof and wall and to shelter the siding and windows from rain.

Edge nailing. Nailing into the edge of a board. *See* Nail.

Elastomeric. Having elastic, rubber-like properties.

Elbow. Trough corner extending outward from roof. Ductwork joint used to turn supply or return at any angle. A section of line which is used to change directions. Normally at right angles.

Enamel. Oil-base paint used in high soil areas such as trim and doors.

End nailing. Nailing into the end of a board, which results in very poor withdrawal resistance. *See* Nail.

Exhaust. Air saturated with carbon monoxide. The by-product of natural gas combustion in a forced-air gas system. This exhaust gas should be vented directly out the top of the roof. It is dangerous to breathe this gas.

Expansion joint. A bituminous fiber strip used to separate blocks or units of concrete to prevent cracking caused by expansion as a result of temperature changes. Also used on concrete slabs.

Exposure. The vertical length of exposed shingle (portion not lapped by the shingle above).

Eye dormer. A dormer that has a gable roof.

Face nailing. Nailing applied perpendicular to the members. Also known as direct nailing.

Fascia. A flat board, band or face, used by itself or, more often, in combination with moldings, that covers the end of the rafter or the board that connects the top of the siding to the bottom of the soffit. The board of the cornice to which the gutter is fastened.

Fascia backer. The main structural support member to which the fascia is nailed.

Feathering. Successive coats of drywall compound applied to joints. Each successive pass should widen the compound joint.

Felt. Typical shingle underlayment. Also known as roofing felt or tar paper. *See* Asphalt.

Ferrule. Aluminum sleeve used in attaching trough to gutter spike.

Fill dirt. Loose dirt. Normally, dirt brought in from another location to fill a void. Sturdier than topsoil. Used under slabs, drives and sidewalks.

Filler. Putty or other pasty material used to fill nail holes prior to painting or staining. A heavily pigmented preparation used for filling and leveling off the pores in open-pored woods.

Finger joint. Trim composed of many small scrap pieces by a joint resembling two sets of interlocking fingers. This trim is often used to reduce costs where such trim will be painted.

Finish grade. Process of leveling and smoothing topsoil into final position prior to landscaping.

Fire stop. A solid, tight closure of a concealed space, placed to prevent the spread of fire and smoke through such a space. In a frame wall, this will usually consist of 2×4 cross blocking between studs.

Fished. An end butt splice strengthened by pieces nailed on the sides.

Fishplate. A wood or plywood piece used to fasten the ends of two members together at a butt joint with nails or bolts. Sometimes used at the junction of opposite rafters near the ridge line.

Fixture. Any end point in a plumbing system used as a source of potable water. Fixtures normally include sinks, tubs, showers, spigots, sprinkler systems, washer connections and other related items.

Flagstone (flagging or flags). Flat stones, from 1″ to 4″ thick, used for rustic walks, steps and floors.

Flashing. Galvanized sheet metal used as a lining around joints between shingles and chimneys, exhaust and ventilation vents and other protrusions in the roof deck. Flashing helps prevent water from seeping under the shingles.

Flat paint. An interior paint that contains a high proportion of pigment and dries to a flat or lusterless finish.

Float. To spread drywall compound smooth.

Floating. A process used after screeding to provide a smoother surface. The process normally involves embedding larger aggregate below the surface by vibrating; removing imperfections, high and low spots; and compacting the surface concrete.

Flue. The space or passage in a chimney through which smoke, gas or fumes ascend. Each passage is called a flue, which together with any others and the surrounding masonry, make up the chimney.

Flue lining. Fire clay or terra-cotta pipe, round or square, usually made in all ordinary flue sizes and in 2′ lengths, used for the inner lining of chimneys with the brick or masonry work around the outside. Flue lining in chimney runs from about a foot below the flue connection to the top of the chimney.

Flush. Adjacent surfaces even, or in same plane (with reference to two structural pieces).

Fly rafters. End rafters of the gable overhang supported by roof sheathing and lookouts.

Footing. Lowest perimeter portion of a structure resting on firm soil or rock that supports the weight of the structure. With a pressure-treated wood foundation, a gravel footing may be used in place of concrete.

Footing ditch. Trough area dug to accommodate concrete or footing forms.

Footing form. A wooden or steel structure placed around the footing that will hold the concrete to the desired shape and size.

Formwork. A temporary mold for giving a desired shape to poured concrete.

Foundation. The supporting portion of a structure below the first-floor construction, or below grade, including the footings.

Frame. The surrounding or enclosing woodwork of windows, doors, etc., and the timber skeleton of a building.

Framing. The rough timber structure of a building, including interior and exterior walls, floor, roof and ceilings.

Framing, balloon. A system of framing a building in which all vertical structural elements of the bearing walls and partitions consist of single pieces extending from the top of the foundation sill plate to the roofplate, and to which all floor joists are fastened.

Framing, ladder. Framing for the roof overhang at a gable. Crosspieces are used similar to a ladder to support the overhang.

Framing, platform. A system of framing a building in which floor joists of each story rest on top plates of the story below or on the foundation sill for the first story, and the bearing walls and partitions rest on the subfloor of each story.

Freon. A special liquid used by an air-conditioning compressor to move heat in or out of the dwelling. This fluid circulates in a closed system, and is also used in refrigerators and freezers.

Frieze. A vertical piece of wood used with or without molding to top off the intersection of the siding and the cornice. Frieze boards may be anywhere from 4″ to 12″ wide. A horizontal member connecting the top of the siding with the soffit of the cornice.

Frostline. The depth to which frost penetrates the soil. Footings should always be poured below this line to prevent cracking.

Full frame. The old-fashioned mortised-and-tenoned frame, in which every joint was mortised and tenoned. Rarely used at the present time.

Fungi, wood. Microscopic plants that live in damp wood and cause mold, stain and decay.

Fungicide. A chemical that is poisonous to fungi.

Furring. Long strips of wood attached to walls or ceilings to allow attachment of drywall or ceiling tiles. Furring out refers to adding furring strips to a wall to bring it out further into a room. Furring down refers to using furring strips to lower a ceiling.

GFCI. (Ground-Fault Circuit Interrupter) An extra sensitive circuit breaker usually installed in outlets in bathrooms and exterior locations to provide additional protection against shock. Required now by most building codes.

Gable. The vertical part of the exterior wall that extends from the eaves upward to the peak or ridge of the roof. The gable may be covered with the same siding material as the exterior wall or may be trimmed with gable trim material.

Gable end. An end wall having a gable.

Gage. A tool used by carpenters to strike a line parallel to the edge of a board.

Gambrel. A roof that slopes steeply at the edge of the building, but changes to a shallower slope across the center of the building. This allows the attic to be used as a second story.

Gem box. A metal box installed in electrical rough-in that holds outlets, receptacles and other electrical units.

Girder. A large or principal beam of wood or steel used to support concentrated loads at isolated points along its length.

Girt (ribband). The horizontal member which supports the floor joists or is flush with the top of the joists. A horizontal member used as a stiffener between studs.

Gloss enamel. A finishing material made of varnish and sufficient pigments to provide opacity and color, but little or no pigment of low opacity. Such an enamel forms a hard coating with maximum smoothness of surface and a high degree of gloss.

Glue. A joint held together with glue.

Glueline, exterior. Waterproof glue at the interface of two veneers of plywood.

Gooseneck. Section of staircase trim with a curve in it.

Grade. The ground level around a building. The natural grade is the original level. Finished grade is the level after the building is complete and final grading is done.

Grading. Process of shaping the surface of a lot to give it the desired contours. *See* Finish grade.

Grain. The direction of the fibers in the wood. Edge grain wood has been sawed parallel to the growth rings. Flat grain lumber is sawed perpendicular to the growth rings. Studs are flat grain lumber. The direction, size, arrangement, appearance or quality of the fibers in wood.

Grain, edge (vertical). Edge-grain lumber has been sawed parallel to the pitch of the log and approximately at right angles to the growth rings; i.e., the rings form an angle of 45° or more with the surface of the piece.

Grain, flat. Flat-grain lumber has been sawed parallel to the pitch of the log and approximately tangent to the growth rings; i.e., the rings form an angle of less than 45° with the surface of the piece.

Grain, quartersawn. Another term for edge grain.

Groove. A long hollow channel cut by a tool, into which a piece fits or in which it works. Two special types of grooves are the dado, a rectangular groove cut across the full width of a piece, and the housing, a groove cut at any angle with the grain and partway across a piece. Dadoes are used in sliding doors, window frames, etc.; housings are used for framing stair risers and treads in a string.

Grounds. Guides used around openings and at the floor line to strike off plaster. They can consist of narrow strips of wood or of wide subjambs at interior doorways. They provide a level plaster line for installation of casing and other trim.

Grout. Mortar made of such consistency (by adding water) that it will just flow into the joints and cavities of the masonry work and fill them solid.

Grouted. Filled with a mortar thin enough to fill the spaces in the concrete or ground around the object being set.

Gusset. A flat wood, plywood or similar type member used to provide a connection at intersections of wood members. Most commonly used in joints of wood trusses.

Gutter or eave trough. A shallow channel or conduit of metal or wood set below and along the eaves of a house to catch and carry off rainwater from the roof.

Gypsum board. Same as drywall.

Gypsum plaster. Gypsum formulated to be used with the addition of sand and water for base-coat plaster.

H-clip. A metal clip into which edges of adjacent plywood sheets are inserted to hold edges in alignment.

Halved. A joint made by cutting half the wood away from each piece so as to bring the sides flush.

Hang. To hang drywall is to nail it in place.

Hanger. Vertical-tension member supporting a load.

Head lap. Length (inches) of the amount of overlap between shingles (measured vertically).

Header. One or two pieces of lumber installed over doors and windows to support the load above the opening. A beam placed perpendicular to joists, to which joists are nailed in framing for chimneys, stairways or other openings. A wood lintel.

Headroom. The clear space between floor line and ceiling, as in a stairway.

Hearth. The inner or outer floor of a fireplace, usually made of brick, tile or stone.

Heartwood. Older wood from the central portion of the tree. As this wood dies, it undergoes chemical changes that often impart a resistance to decay and a darkening in color.

Heel of a rafter. The end or foot that rests on the wall plate.

Heel wedges. Triangular-shaped pieces of wood that can be driven into gaps between rough framing and finished items, such as window frames, to provide a solid backing for these items.

Hip. The external angle formed by the meeting of two sloping sides of a roof.

Hip roof. A roof which slopes up toward the center from all sides, necessitating a hip rafter at each corner.

Hopper window. A window that is hinged at the bottom to swing inward.

Hot. A wire carrying current.

Housed. A joint in which a piece is grooved to receive the piece which is to form the other part of the joint.

Humidifier. A device designed to increase the humidity within a room or a house by means of the discharge of water vapor. It may consist of individual room-size units or larger units attached to the heating plant to condition the entire house.

Hydration. The chemical process wherein portland cement becomes a bonding agent as water is slowly removed from the mixture. The rate of hydration determines the strength of the bond and hence the strength of the concrete. Hydration stops when all the water has been removed. Once hydration stops, it cannot be restarted.

I-beam. A steel beam with a cross section resembling the letter I. I-beams are used for long spans as basement beams or over wide wall openings, such as a double garage door, when wall and roof loads are imposed on the opening.

Inside miter. Trough corner extending toward roof.

Insulation board, rigid. A structural building board made of foam or coarse wood or cane fiber impregnated with asphalt or given other treatment to provide water resistance. It can be obtained in various size sheets, in various thicknesses and in various densities.

Insulation, thermal. Any material high in resistance to heat transmission that, when placed in the walls, ceiling or floors of a structure, will reduce the rate of heat flow.

Interior finish. Material used to cover the interior framed areas or materials of walls and ceilings.

Involute. Curved portion of trim used to terminate a piece of staircase railing. Normally used on traditional homes.

Isolation joint. A joint in which two incompatible materials are isolated from each other to prevent chemical action between the two.

Jack post. A hollow metal post with a jack screw in one end so that it can be adjusted to the desired height.

Jack rafter. A rafter that spans the distance from the wall plate to a hip, or from a valley to a ridge.

Jack stud. A short stud that does not extend from floor to ceiling; for example, a stud that extends from the floor to a window.

Jamb. Exterior frame of a door. The side and head lining of a doorway, window or other opening.

Joint. Mortar in between bricks or blocks. The space between the adjacent surfaces of two members or components that are held together by nails, glue, cement, mortar or other means. *See* Control joint, Expansion joint and Isolation joint.

Joint butt. Squared ends or ends and edges adjoining each other.

Joint cement. A powder usually mixed with water that is used for joint treatment in gypsum-wallboard finish. Often called "Spackle."

Joist. A long piece of lumber used to support the load of a floor or ceiling and supported in turn by larger beams, girders or bearing walls. Joists are always positioned on edge. *See* Band joist, Header, Tail beam and Trimmer.

Kerf. The area of a board removed by the saw when cutting. Vertical notch or cut made in a batter board where a string is fastened tightly.

Key. Fancy decorative lintel above window made of brick, normally placed on various angles for a flared effect. Also known as a keystone.

Keyways. A tongue-and-groove type connection where perpendicular planes of concrete meet to prevent relative movement between the two components.

Kiln-dried. Dried in a kiln with the use of artificial heat.

Kiln-dried lumber. Lumber that has been dried by means of controlled heat and humidity, in ovens or kilns, to specified ranges of moisture content. *See* also Air-dried lumber and Lumber, moisture content.

Knee brace. A corner brace, fastened at an angle from wall stud to rafter, stiffening a wood or steel frame to prevent angular movement.

Knee wall. A short wall extending from the floor to the roof in the second story of a multi-story.

Knot. In lumber, the portion of a branch or limb of a tree that appears on the edge or face of the piece.

Lag screws. Large screws with heads designed to be turned with a wrench.

Laminate. Any thin material, such as plastic or fine wood, glued to the exterior of the cabinet.

Laminated beam. A very strong beam created from several smaller pieces of wood that have been glued together under heat and pressure.

Landing. A platform between flights of stairs or at the termination of a flight of stairs.

Lap. A joint of two pieces lapping over each other.

Latex paint. A coating in which the vehicle is a water emulsion of rubber or synthetic resin. Water soluble paint. Normally recommended because of ease of use for interior work.

Lath. A grid of some sort (normally metal or fiberglass) applied to exterior sheathing as a base for stucco. Expanded metal or wood strips are commonly used.

Lattice. Crossed wood, iron plate or bars.

Lay up. To place materials together in the relative positions they will have in the finished building.

Ledger strip. A strip of lumber nailed along the bottom of the side of a girder on which joists rest.

Ledgerboard. The support for the second-floor joists of a balloon-frame house.

Let-in brace. Nominal 1″ thick boards that are applied into notched studs diagonally.

Level. A term describing the position of a line or plane when parallel to the surface of still water; an instrument or tool used in testing for horizontal and vertical surfaces, and in determining differences of elevation.

Light. Space in a window sash for a single pane of glass. Also, a pane of glass.

Line. Any section of plumbing, whether it is copper, PVC, CPVC or cast iron. A string pulled tight. Normally used to check or establish straightness.

Lineal foot. A measure of lumber based on the actual length of the piece.

Lintel. A structural member placed above doors and window openings to support the weight of the bricks above the opening, usually precast concrete. Also called a header.

Liquefied gas. A carrier of wood preservatives, this is a hydrocarbon that is a gas at atmospheric pressure but one that can be liquefied at moderate pressures (similar to propane).

Live load. The load, expressed in pounds per square foot, of people, furniture, snow, etc., that are in addition to the weight of the structure itself.

Load-bearing wall. Any wall that supports the weight of other structural members.

Lookout. The horizontal board (usually a 2×4 or 1×4) that connects the ends of the rafters to the siding. This board becomes the base for nailing on the soffit covering. The end of a rafter, or the construction which projects beyond the sides of a house to support the eaves; also, the projecting timbers at the gables that support the verge boards.

Louver. An opening with a series of horizontal slats so arranged as to permit ventilation, but to exclude rain, sunlight or vision. *See also* Attic ventilators.

Lumber timbers. Yard lumber 5″ or more in least dimension. Includes beams, stringers, posts, caps, sills, girders and purlins.

Lumber, boards. Lumber less than 2″ thick and 2″ or more wide.

Lumber, dimension. Yard lumber from 2″ to, but not including, 5″ thick, and 2″ or more wide. Includes joists, rafters, studs, planks and small timbers.

Lumber, dressed size. The dimension of lumber after shrinking from green dimension and after machining to size or pattern.

Lumber, matched. Lumber that is dressed and shaped on one edge in a grooved pattern and on the other in a tongued pattern. *See* Tongue and groove.

Lumber, moisture content. The weight of water contained in wood, expressed as a percentage of the total weight of the wood. *See also* Air-dried lumber and Kiln-dried lumber.

Lumber, pressure-treated. Lumber that has had a preservative chemical forced into the wood under pressure to resist decay and insect attack.

Lumber, shiplap. Lumber that has been milled along the edge to make a close rabbet or lap joint.

Lumber, yard. Lumber of those grades, sizes and patterns which are generally intended for ordinary construction, such as framework and rough coverage of houses.

Mansard. A type of roof that slopes very steeply around the perimeter of the building to full wall height, providing space for a complete story. The center portion of the roof is either flat or very low sloped.

Mantel. The shelf above a fireplace. Also used in referring to the decorative trim around a fireplace opening.

Masonry. Stone, brick, concrete, hollow tile, concrete block, gypsum block or other similar building units or materials or a combination of the same, bonded together with mortar to form a wall, pier, buttress or similar mass.

Mastic. Any pasty material used as a cement in such applications as setting tile or as a protective coating for thermal insulation or waterproofing. Usually comes in caulking tubes or five-gallon cans.

Matching, or tonguing and grooving. The method used in cutting the edges of a board to make a tongue on one edge and a groove on the other.

Meeting rail. The bottom rail of the upper sash of a double-hung window. Sometimes called the check rail.

Member. A single piece in a structure, complete in itself.

Metal lath. Sheets of metal that are slit and drawn out to form openings. Used as a plaster base for walls and ceilings and as reinforcing over other forms of plaster base.

Mildewcide. A chemical that is poisonous specifically to mildew fungi. A specific type of fungicide.

Millwork. Generally, all building materials made of finished wood and manufactured in millwork plants and planing mills are included under the term "millwork." It includes such items as inside and outside doors, window and door frames, blinds, porchwork, mantels, panelwork, stairways, moldings and interior trim. It normally does not include flooring, ceilings or siding.

Miter. The joint formed by two abutting pieces meeting at an angle.

Miter box. Special guide and saw used to cut trim lumber at precise angles.

Miter joint. A diagonal joint formed at the intersection of two pieces of molding. For example, the miter joint at the side and head casing at a door opening is made at a 45° angle.

Moisture content of wood. Weight of the water contained in the wood, usually expressed as a percentage of the weight of oven dry wood. *See* Lumber, moisture content.

Molding. Decorative strips of wood or other material applied to wall joints and surfaces as a decorative accent. Molding does not have any structural value.

Molding base. The molding on the top of a baseboard.

Mortar. *See* Cement mortar.

Mortise. A slot cut into a board, plank or timber, usually edgewise, to receive a tenon of another board, plank or timber to form a joint.

Mud. Slang for Spackle, or drywall compound. Used to seal joints and hide nail head dimples.

Mullion. A vertical bar or divider in the frame between windows, doors or other openings. The construction between the openings of a window frame to accommodate two or more windows.

Muntin. A short bar, horizontal or vertical, separating panes of glass in a window sash. The vertical member between two panels of the same piece of panel work.

Natural finish. A transparent finish which does not seriously alter the original color or grain of the natural wood. Natural finishes are usually provided by sealers, oils, varnishes, water-repellent preservatives and other similar materials.

Neoprene. A synthetic rubber characterized by superior resistance to oils, gasoline and sunlight.

Newel. A post to which the end of a stair railing or balustrade is fastened. Also, any post to which a railing or balustrade is fastened.

Nominal. Of wood dimension, the approximate size of a sawn wood section before it is planed.

Nominal size. Original size of lumber when cut.

Nonleachable. Not dissolved and removed by the action of rain or other water.

Nonbearing wall. A wall supporting no load other than its own weight.

Nosing. The projecting edge of a molding or strip. Usually applied to the projecting molding on the edge of a stair tread.

Notch. A crosswise rabbet at the end of a board.

O.C., on center. The measurement of spacing for studs, rafters, joists and the like in a building from the center of one member to the center of the next. Normally refers to wall or joist framing of 12″, 16″ or 18″ O.C.

Oriented strand board (OSB). A type of structural flakeboard composed of layers, with each layer consisting of compressed strand-like wood particles in one direction, and with layers oriented at right angles to each other. The layers are bonded together with a phenolic resin.

Outrigger. An extension of a rafter beyond the wall line. Usually a smaller member nailed to a larger rafter to form a cornice or roof overhang.

Paint. A combination of pigments with suitable thinners or oils to provide decorative and protective coatings.

Paint grade. Millwork of quality intended for a painted finish. Not as fine as stain grade.

Panel. In house construction, a thin, flat piece of wood, plywood or similar material, framed by stiles and rails as in a door, or fitted into grooves of thicker material with molded edges for decorative wall treatment. A sheet of plywood, fiberboard, structural flakeboard or similar material.

Panelized. Construction where framing is assembled in "panels" for easy and space-efficient shipping.

Paper, building. A general term, without reference to properties or uses, for papers, felts and similar sheet materials used in buildings.

Paper, sheathing. A building material, generally paper or felt, used in wall and roof construction as a protection against the passage of air and sometimes moisture.

Parging. Thin coatings (¼″) of mortar applied to the exterior face of concrete block where block wall and footing meet; serves as a waterproofing mechanism.

Parquet. A floor with inlaid design. For wood flooring, it is often laid in blocks with boards at angles to each other to form patterns.

Particleboard. Panels composed of small wood particles usually arranged in layers without a particular orientation and bonded together with a phenolic resin. Some particleboards are structurally rated. *See also* Structural flakeboard.

Parting stop or strip. A small wood piece used in the side and head jambs of double-hung windows to separate upper and lower sashes.

Partition. An interior wall in a framed structure dividing two spaces.

Penny. As applied to nails, it originally indicated the price per hundred. The term now serves as a measure of nail length and is abbreviated by the letter "d."

Penta grease. A penta-petroleum emulsion system suspended in water by the use of emulsifiers and dispersing agents.

Pentachlorophenol (penta). A chlorinated phenol, usually in petroleum oil, used as a wood preservative.

Perm. A measure of water vapor movement through a material (grains per square foot per hour per inch of mercury difference in vapor pressure).

Picture. A molding shaped to form a support for picture hooks, often placed at some distance from the ceiling upon the wall to form the lower edge of the frieze.

Pier. A column of masonry, used to support other structural members, such as concrete supports for a floor beam.

Pigment. A powdered solid in suitable degrees of subdivision for use in paint or enamel.

Pilaster. A projection from a wall forming a column to support the end of a beam framing into the wall.

Piles. Long posts driven into the soil in swampy locations or whenever it is difficult to secure a firm foundation, upon which the footing course of masonry or other timbers are laid.

Piling. Large timbers or poles driven into the ground or the bed of a stream to make a firm foundation.

Pitch. The measure of the steepness of the slope of a roof, expressed as the ratio of the rise of the slope over a corresponding horizontal distance. Roof slope is expressed in the inches of rise per foot of run, such as 4 in 12.

Pitch board. Board sawed to the exact shape formed by the stair tread, riser and slope of the stairs, and used to lay out the carriage and stringers.

Pith. The small, soft core at the original center of a tree around which wood formation takes place.

Plan. A horizontal geometrical section of a building, showing the walls, doors, windows, stairs, chimneys, columns, etc.

Plank. A wide piece of sawed timber, usually ½″ to 4″ thick and 6″ or more wide.

Plaster. A mixture of lime, air and sand, or of lime, cement and sand, used to cover outside and inside wall surfaces.

Plaster grounds. Strips of wood used as guides or strike-off edges around window and door openings and at base of walls.

Plastic. Term interchangeable with "wet," as in "plastic cement."

Plate. A horizontal member used to anchor studs to the floor or ceiling. Sill plate: a horizontal member anchored to a masonry wall. Sole plate: bottom horizontal member of a frame wall. Top plate: top horizontal member of a frame wall supporting ceiling joists, rafters or other members.

Plate cut. The cut in a rafter which rests upon the plate; sometimes called the seat cut.

Plenum. Chamber immediately outside of the HVAC unit where conditioned air feeds into all of the supplies. A space in which air is contained under slightly greater than atmospheric pressure. In a house, it is used to distribute heated or cooled air.

Plow. To cut a groove running in the same direction as the grain of the wood.

Plugged exterior. A grade of plywood used for subfloor underlayment. The knot holes in the face plies are plugged and the surface is touch-sanded.

Plumb. The condition when something is exactly vertical to the ground, such as the wall of a house.

Plumb bob. A weight attached to a string used to indicate a plumb (vertical) condition.

Plumb cut. Any cut made in a vertical plane; the vertical cut at the top end of a rafter.

Ply. A term to denote the number of thicknesses or layers of roofing felt, veneer in plywood or layers in built-up materials, in any finished piece of such material.

Plywood. A piece of wood made of three or more layers of veneer joined with glue, and usually laid with the grain of adjoining plies at right angles. Almost always, an odd number of plies are used to provide balanced construction.

Poly. Polyethylene. A heavy-gauge plastic used for vapor barriers and material protection. The accepted thickness is 6 mil.

Porch. An ornamental entranceway.

Post. A timber set on end to support a wall, girder or other member of the structure.

Post-and-beam roof. A roof consisting of thick planks spanning beams that are supported on posts. This construction has no attic or air space between the ceiling and roof.

Potable water. Water satisfactory for human consumption and domestic use, meeting the local health authority requirements.

Preservative. Any substance that, for a reasonable length of time, will prevent the action of wood-destroying fungi, borers of various kinds and similar destructive agents when the wood has been properly coated or impregnated with it.

Primer or prime coat. The first coat in a paint job that consists of two or more coats. The primer may have special properties that provide an improved base for the finish coat.

Pulley stile. The member of a window frame which contains the pulleys and between which the edges of the sash slide.

Purlin. A horizontal board that supports a roof rafter or stud to prevent bowing of the member by weight.

Putty. A type of cement usually made of whiting and boiled linseed oil, beaten or kneaded to the consistency of dough, and used in sealing glass in sash, filling small holes and crevices in wood and for similar purposes.

Quarter round. A small strip of molding whose cross section is similar to a quarter of a circle. Used with or without base molding. May be applied elsewhere.

Quoin. Fancy edging on outside corners made of brick veneer or stucco.

Rabbet. A rectangular longitudinal groove cut in the corner edge of a board or plank.

Racking resistance. A resistance to forces in the plane of a structure that tend to force it out of shape.

Radiant heating. A method of heating, usually consisting of a forced hot water system with pipes placed in the floor, wall or ceiling; or with electrically heated panels.

Rafter. One of a series of structural members of a roof designed to support roof loads. The rafters of a flat roof are sometimes called roof joists. *See also* Fly rafter and Jack rafter.

Rafter, hip. A rafter that forms the intersection of an external roof angle.

Rafter, valley. A rafter that forms the intersection of an internal roof angle. The valley rafter is normally made of double 2″ thick members.

Rafters, common. Those which run square with the plate and extend to the ridge. Cripple—those which cut between valley and hip rafters. Hip—those extending from the outside angle of the plates toward the apex of the roof. Jacks—those square with the plate and intersecting the hip rafter.

Rail. Cross members of panel doors or of sashes. Also, the upper and lower members of a balustrade or staircase extending from one vertical support, such as a post, to another.

Rake. The angled edge of a roof located at the end of a roof that extends past gable. Trim members that run parallel to the roof slope and form the finish between the wall and a gable roof extension.

Re-bar. Metal rods used to improve the strength of concrete structures.

Reflective insulation. Sheet material with one or both surfaces of comparatively low heat emissivity, such as aluminum foil. When used in building construction, the surfaces face air spaces, reducing the radiation across the air space.

Register. Metal facing plate on wall where supply air is released into room and where air enters returns. Registers can be used to direct the flow of air.

Reinforcing. Steel rods or metal fabric placed in concrete slabs, beams or columns to increase their strength.

Relative humidity. The amount of water vapor in the atmosphere, expressed as a percentage of the maximum quantity that could be present at a given temperature. The actual amount of water vapor that can be held in space increases with the temperature.

Resorcinol. An adhesive that is high in both wet and dry strength and resistant to high temperatures. It is used for gluing lumber or assembly joints that must withstand severe conditions.

Retainage. A percent of payment retained to ensure completion of a job.

Return. Ductwork leading back to the HVAC unit to be reconditioned. The continuation of a molding or finish of any kind in a different direction.

Reverse board and batten. Siding in which narrow battens are nailed vertically to wall framing and wider boards are nailed over these so that the edges of boards lap battens. A slight space is left between adjacent boards. This pattern is simulated with plywood by cutting wide vertical grooves in the face ply at uniform spacing.

Ribband. *See* Ledgerboard.

Ribbon (Girt). Normally a 1×4 board let into the studs horizontally to support ceiling or second-floor joists.

Ridge. Intersection of any two roofing planes where water drains away from the intersection. Special shingles are applied to ridges.

Ridge board. The board placed on edge at the ridge of the roof into which the upper ends of the rafters are fastened.

Ridge cut. *See* Plumb cut.

Ridge vent. Opening at the point where roof decking normally intersects along the highest point on a roof where air is allowed to flow from the attic. A small cap covers this opening to prevent rain from entering. When these are long, they are normally known as continuous ridge vents.

Ring shank nail. A nail with ridges forming rings around the shank to provide better withdrawal resistance.

Ripping. Cutting lumber parallel to the grain.

Rise. In stairs, the vertical height of a step or flight of stairs. The vertical distance through which anything rises, as the rise of a roof or stair.

Riser. The vertical board between two treads of a flight of stairs.

Rolled roofing. Roofing material composed of fiber and saturated with asphalt, that is supplied in 36-inch-wide rolls with 108 square feet of material. Weights are generally 45 to 90 lbs. per roll.

Roof, built-up. *See* Built-up roof.

Roof sheathing. The boards or sheet material fastened to the rafters on which shingles or other roof covering is laid.

Roof valley. *See* Valley.

Roofing. The material put on a roof to make it wind and waterproof.

Rottenstone. A slightly abrasive stone used to rub a transparent interior finish to achieve a smooth surface.

Rough grade. First grading effort used to level terrain to approximate shape for drainage and landscaping.

Rout. The removal of material by cutting, milling or gouging to form a groove.

Row lock. Intersecting bricks that overlap on outside corners.

Rubble masonry. Uncut stone, used for rough work, foundations, backing and the like.

Run. In stairs, the net front-to-back width of a step or the horizontal distance covered by a flight of stairs. The length of the horizontal projection of a piece such as a rafter when in position.

STC (Sound Transmission Class). A numerical measure of the ability of a material or assembly to resist the passage of sound. Materials with higher STC numbers have greater resistance to sound transmission.

Saddle. A sloped area at the intersection of a vertical surface, such as chimney, and the roof. Used to channel off water that might otherwise get trapped behind the vertical structure. *Also see* Cricket.

Saddle board. The finish of the ridge of a pitch-roof house. Sometimes called comb board.

Sagging. Slow dripping of excessively heavy coats of paint.

Sapwood. The outer zone of wood, next to the bark. In the living tree it contains some living cells (the heartwood contains none), as well as dead and dying cells. In most species, it is lighter colored than the heartwood. In all species, it is lacking in decay resistance.

Sash. A single light frame containing one or more lights of glass.

Saturated felt. A felt which is impregnated with tar or asphalt.

Sawkerf. *See* Kerf.

Scab. A short length of board nailed over the joint of two boards butted end to end to transfer tensile stresses between the two boards.

Scaffold or staging. A temporary structure or platform enabling workmen to reach high places.

Scale. A short measurement used as a proportionate part of a larger dimension. The scale of a drawing is expressed as ¼″ = 1′.

Scaling. Loss of smooth surface of concrete as a result of flaking or scaling.

Scantling. Lumber with a cross section ranging from 2″ by 4″ to 4″ by 4″.

Scarfing. A joint between two pieces of wood which allows them to be spliced lengthwise.

Scotia. A hollow molding used as a part of a cornice, and often under the nosing of a stair tread.

Scratch coat. The first coat of plaster, which is scratched to form a bond for the second coat.

Screed. A small strip of wood, usually the thickness of the plaster coat, used as a guide for plastering.

Screeding. The process of running a straightedge over the top of forms to produce a smooth surface on wet (plastic) cement. The screeding proceeds in one direction, normally in a sawing motion.

Screen. Metallic or vinyl grid used to keep troughs free of leaves and other debris. Some screens are hinged.

Scribing. Fitting woodwork to an irregular surface. With moldings, scribing means cutting the end of one piece to fit the molded face of the other at an interior angle, in place of a miter joint.

Scuttle hole. An opening in the ceiling to provide access to the attic. It is covered by a closure panel when not in use.

Sealant. *See* Caulk.

Sealer. A finishing material, either clear or pigmented, that is usually applied directly over uncoated wood for the purpose of sealing the surface.

Seam, standing. A joint between two adjacent sheets of metal roofing in which the edges are bent up to prevent leakage and the joint between the raised edges is covered.

Seasoning. Removing moisture from green wood to improve its serviceability.

Seat cut or plate cut. The cut at the bottom end of a rafter to allow it to fit upon the plate.

Section. A drawing showing the kind, arrangement and proportions of the various parts of a structure. It is assumed that the structure is cut by a plane, and the section is the view gained by looking in one direction.

Self-rimming. Term used to describe a type of sink that has a heavy rim around the edge that automatically seals the sink against the Formica top.

Semigloss paint or enamel. A paint or enamel made with a slight insufficiency of nonvolatile vehicle so that its coating, when dry, has some luster but is not very glossy.

Septic tank. A receptacle that receives the discharge from a sewage system and is designed to separate liquids from solids and digest the organic matter through bacteria, discharging the liquid portion into the soil through the subsurface, disposal fields or seepage pits.

Service panel. Junction where main electrical service to the home is split among the many circuits internal to the home. Circuit breakers should exist on each internal circuit.

Settling. Movement of unstable dirt over time. Fill dirt normally settles downward as it is compacted by its own weight or a structure above it.

Shake. A thick hand-split shingle, resawed to form two shakes, usually edge-grained.

Shakes. Imperfections in timber caused during the growth of the timber by high winds or imperfect conditions of growth.

Sheathing. The structural covering, usually wood boards or plywood, used over studs or rafters of a structure. Structural building board is normally used only as wall sheathing.

Sheathing paper. *See* Paper, sheathing. A building material, generally paper or felt, used in wall and roof construction as a protection against the passage of air and water.

Shed dormer. A dormer that has a roof sloping in only one direction at a much shallower slope than the main roof of the house.

Sheet metal work. All components of a house employing sheet metal, such as flashing, gutters and downspouts.

Sheetrock. A term commonly applied to gypsum board.

Shellac. A transparent coating made by dissolving in alcohol "lac," a resinous secretion of the lac bug (a scale insect that thrives in tropical countries, especially India).

Shim. A thin wedge of wood for driving into crevices to bring parts into alignment.

Shingle butt. The lower, exposed side of a shingle.

Shingles. Roof covering of asphalt, fiberglass, asbestos, wood, tile, slate or other material, or combinations of materials, such as asphalt and felt, cut to stock lengths, widths and thicknesses.

Shingles, siding. Various kinds of shingles, such as wood shingles or shakes and nonwood shingles, that are used over sheathing for exterior wall covering of a structure.

Shiplap. *See* Lumber, shiplap.

Shutter. A lightweight, louvered, flush wood or nonwood frame in the form of a door, located at each side of a window. Some are made to close over the window for protection; others are fastened to the wall for decorative purposes.

Side lap. Length (inches) of the amount of overlap between two horizontally adjoining shingles.

Siding. The finish covering of the outside wall of a frame building, whether made of horizontal weatherboards, vertical boards with battens, shingles or other material. The outside finish between the casings.

Siding, Dolly Varden. Beveled wood siding that is rabbeted on the bottom edge.

Siding, bevel (lap siding). Wedge-shaped boards used as horizontal siding in a lapped pattern. This siding varies in butt thickness from ½″ to ¾″ and in widths up to 12″. It is normally used over some type of sheathing.

Siding, drop. Siding that is usually ¼″ thick and 6″ or 8″ wide, with tongue-and-groove or shiplap edges. Often used as siding without sheathing in secondary buildings.

Silicone. One of a large group of polymerized organic siloxanes that are available as resins, coatings, sealants, etc., with excellent waterproofing characteristics.

Sill. (1) The lowest member of the frame of a structure, resting on the foundation and supporting the floor joists or the uprights of the wall. (2) The member forming the lower side of an opening—such as a doorsill or windowsill.

Sill caulk. Mastic placed between top of foundation wall and sill studs to make an airtight seal.

Silt fence. A barrier constructed of burlap, plastic or bales of hay used to prevent the washing away of mud and silt from a cleared lot onto street or adjacent lots.

Sizing. Working material to the desired size. Also, a coating of glue, shellac or other substance applied to a surface to prepare it for painting or other method of finish.

Slab. A concrete floor poured on the ground.

Sleeper. A wood member embedded in or resting directly on concrete, as in a floor, that serves to support and to fasten subfloor or flooring.

Slip tongue. A spline used to connect two adjacent boards that have grooves facing each other.

Smokepipe thimble. *See* Thimble.

Soffit. The underside of the cornice or any part of the roof that overhangs the siding.

Soil cover or ground cover. A light covering of plastic film, roll roofing, or similar material used over the soil in crawl spaces of buildings to minimize movement of moisture from the soil into the crawl space.

Soil stack. A vent opening out to the roof that allows the plumbing system to equalize with external air pressure and allows the sewer system to "breathe."

Sole or sole plate. *See* Plate.

Solid bridging. A solid member placed between adjacent floor joists near the center of the span to prevent joists from twisting.

Solids. Solid bricks used for fireplace hearths, stoops, patios or driveways.

Spackle. Soft puttylike compound used for drywall patching and touch-up; does not shrink as much as regular joint compound. Allows painting immediately after application. *See* Joint cement.

Spalling. Chips or splinters breaking loose from the surface of concrete because of moisture moving through from the reverse side.

Span. The distance between two supporting members of a joist or beam. The longest unsupported distance along a joist.

Specifications. The written or printed directions regarding the details of a building or other construction.

Spike. A long nail used to attach the trough to the roof. Used in conjunction with ferrule. *See* Ferrule.

Splashblock. A small masonry block laid with the top close to the ground surface, to receive roof drainage from downspouts and carry it away from the building.

Splice. Joining of two similar members in a straight line.

Spline. A long, narrow, thin strip of wood or metal often inserted into the edges of adjacent boards to form a tight joint.

Spread set. Faucet set that requires three holes to be cut wider than normal.

Square. A unit of measure usually applied to roofing material, denoting a sufficient quantity to cover 100 square feet of surface. A tool used by mechanics to obtain accuracy.

Stack. Also known as vent stack. A ventilation pipe coming out of the roofing deck.

Stain grade. Millwork of finest quality, intended for a stain finish. Capable of receiving and absorbing stain easily.

Stain, shingle. A form of oil paint, very thin in consistency, intended for coloring wood with rough surfaces, such as shingles, without forming a coating of significant thickness or gloss.

Stair carriage. Supporting member for stair treads. Usually a 2″ plank notched to receive the treads; sometimes called a "rough horse."

Stair landing. *See* Landing.

Stair rise. *See* Rise.

Stairs, box. Those built between walls, and usually with no support except the wall.

Staking. To lay out the position of a home, the batter boards, excavation lines and depth(s).

Starter strip. A continuous strip of asphalt roofing used as the first course, applied to hang over the eave.

Stick built. A house built "one stick at a time" on-site vs. in a factory.

Stiffness. Resistance to deformation by loads that cause bending stresses.

Stile. An upright framing member in a panel door.

Stipple. Rough and textured coatings applied to ceilings. This process makes it easier to finish ceilings since they do not have to be taped and sanded as many times as the walls.

Stool. A flat molding fitted over the windowsill between jambs and contacting the bottom rail of the lower sash.

Stoop arms. Section of foundation wall extending out perpendicular to exterior wall used to support a masonry or stone stoop.

Stoop iron. Corrugated iron sheeting used as a base for the top of a brick or stone stoop. This is to eliminate the need for completely filling the stoop area with fill dirt (which may settle) or concrete.

Stop valve. A shutoff valve allowing water to be cut off at a particular point in the system.

Stop, gravel. A raised ridge of metal at the edge of a tar-and-gravel roof that keeps the gravel from falling off the roof.

Stop, trim. The trim member on the jambs of an opening that a door or window closes against.

Storm sash or storm window. An extra window usually placed on the outside of an existing one as additional protection against cold weather.

Story. That part of a building between any floor and the floor or roof next above.

Strike plate. A metal plate mortised into or fastened to the face of a door frame side jamb to receive the latch or bolt when the door is closed.

Stringer. A timber or other support for cross members in floors or ceilings. In stairs, the stringer (or stair carriage) supports the stair treads.

Strip flooring. Wood flooring consisting of narrow, matched strips.

Stub-in. A term applied to the process of installing rough plumbing "stubs" to simplify the installation of plumbing at a later date.

Structural flakeboard. A panel material made of specially produced flakes that are compressed and bonded together with phenolic resin. Popular types include waferboard and OSB (oriented strand board). Structural flakeboards are used for many of the same applications as plywood.

Stucco. A fine plaster used for interior decoration and fine work; also for rough outside wall coverings.

Stud. Standard 2×4 lumber normally cut to 8′ or 10′ nominal lengths, used for framing walls. (Plural = studs or studding.)

Studding. The framework of a partition or the wall of a house; usually referred to as 2×4s.

Studwall. A wall consisting of spaced vertical structural members with thin facing material applied to each side.

Subfloor. Boards or plywood laid on joists over which a finish floor is to be laid.

Superstructure. The structural part of the deck above the posts or supports.

Supply. Ductwork leading from the HVAC unit to the registers.

Suspended ceiling. A ceiling system supported by hanging it from the overhead structural framing.

3-way switch. Type of switch that allows control of a lighting fixture from two switch locations, such as both ends of a hallway.

T&G. A type of wood joint machined with a tongue or protusion on one side and a groove on the other. This allows the two pieces to be joined snugly together. This term is commonly used to refer to tongue-and-groove flooring or roofing material.

Tail beam. A relatively short beam or joist supported by a wall at one end and by a header at the other.

Tape. Paper used to cover the joints between sheets of gypsum. Tape joints are then sealed with mud (Spackle).

Tenon. A projection at the end of a board, plank or timber for insertion into a mortise.

Termite shield. A shield, normally of galvanized sheet metal, placed between footing and foundation wall to prevent the passage of termites.

Termites. Insects that superficially resemble ants in size, general appearance and habit of living in colonies; hence, they are frequently called "white ants." Subterranean termites establish themselves in buildings not by being carried in with lumber, but by entering from ground nests after the building has been constructed. If unmolested, they eat out the woodwork, leaving a shell of sound wood to conceal their activities. Damage may proceed so far as to cause collapse of parts of a structure before discovery. There are about fifty six species of termites known in the United States. But the two major ones, classified by the manner in which they attack wood, are ground-inhabiting or subterranean termites (the most common) and dry wood termites, which are found almost exclusively along the extreme southern border and the Gulf of Mexico in the United States.

Thimble. The section of a vitreous clay flue that passes through a wall.

Thinwall. Thin, flexible conduit used between outlet boxes.

Threshold. A strip of wood or metal with beveled edges used over the finish floor and the sill of an exterior door.

Tie beam (collar beam). A beam so situated that it ties the principal rafters of a roof together.

Tieback member. A timber, oriented perpendicular to a retaining wall, that ties the wall to a deadman buried behind it.

Tin shingle. A small piece of tin used in flashing and repairing a shingle roof.

To the weather. A term applied to the projecting of shingles or siding beyond the course above.

Toenailing. Driving a nail at a slant to the initial surface to permit it to penetrate into a second member.

Ton. An industry standard measure used to express a quantity of cold air that is produced by an air-conditioning system.

Tongue and groove. Boards that join on edge with a groove on one unit and a corresponding tongue on the other to interlock. Certain plywoods and hardwood flooring are tongue and grooved. "Dressed and matched" is an alternative term with the same meaning.

Top plate. Piece of lumber supporting ends of rafters.

Topsoil. Two- or three-inch layer of rich, loose soil. This must be removed from areas to be cleared or excavated and replaced in other areas later. Not for load bearing areas.

Touch sanding. Very light sanding of prime paint coat.

Transit. Similiar to a builder's level except that the instrument can be adjusted vertically. Used for testing walls for plumb and laying out batter board, and establishing degree of a slope.

Trap. A device providing a liquid seal that prevents the backflow of air without materially affecting the flow of sewage or waste water. "S"-shaped drain traps are required in most building codes.

Tray ceiling. Raised area in a ceiling. Looks like a small vaulted ceiling.

Tray molding. Special type of crown molding where a large portion of the molding is applied to the ceiling as opposed to the wall.

Tread. The horizontal board in a stairway on which the foot is placed.

Trig. A string support for guide lines to prevent sagging and wind disturbances on long expanses of wall.

Trim. The finish materials in a building, such as moldings, applied around openings (window trim, door trim) or at the floor and ceiling of rooms (baseboard, cornice and other moldings).

Trimmer. A beam or joist to which a header is nailed in framing for a chimney, stairway or other opening.

Trimming. Putting the inside and outside finish and hardware upon a building.

Troweling. A process used after floating to provide an even smoother concrete surface.

Truss. A frame or jointed structure designed to act as a beam of long span, while each member is usually subjected to longitudinal stress only, either tension or compression.

Truss plate. A heavy-gauge, pronged metal plate that is pressed into the sides of a wood truss at the point where two or more members are to be joined together.

Turpentine. A volatile oil used as a thinner in paints and as a solvent in varnishes. Chemically, it is a mixture of terpenes.

Undercoat. A coating applied prior to the finishing or top coats of a paint job. It may be the first of two or the second of three coats. In some uses of the word, it may become synonymous with priming coat.

Underlayment. Any paper or felt composition used to separate the roofing deck from the shingles. Underlayments such as asphalt felt (tar paper) are common. A material placed under flexible flooring materials such as carpet, vinyl tile, or linoleum to provide a smooth base over which to lay such materials.

Underlayment exterior. *See* Plugged exterior.

Valley. The internal angle formed by the junction of two sloping sides of a roof where water drains at the intersection.

Vapor barrier. Material used to retard the movement of water vapor into walls and prevent condensation in them. Usually considered as having a perm value of less than 1.0. Applied separately over the warm side of exposed walls or as a part of batt or blanket insulation.

Varnish. A thickened preparation of drying oil or drying oil and resin suitable for spreading on surfaces to form continuous, transparent coatings, or for mixing with pigments to make enamels.

Vehicle. The liquid portion of a finishing material; it consist of the binder (nonvolatile) and volatile thinners.

Veneer. Thin sheets of wood made by rotary cutting or slicing of a log. Veneer is glued in plies or on top of other wood to improve appearance or strength.

Vent. A pipe or duct, or a screened or louvered opening that provides an inlet or outlet for the flow of air. Common types of roof vents include ridge vents, soffit vents and gable end vents.

Vent system. A pipe or network of pipes providing a flow of air to or from a drainage system to protect trap seals from siphonage or back pressure.

Vestibule. An entrance to a house; usually enclosed.

Volatile thinner. A liquid that evaporates readily, used to thin or reduce the consistency of finishes without altering the relative volumes of pigments and nonvolatile vehicles.

Voltage. The "force" of electrical potential. Similar to the pressure of water in a pipe.

Waferboard. A type of structural flakeboard made of compressed, waferlike wood particles or flakes bonded together with a phenolic resin. The flakes may vary in size and thickness and may be either randomly or directionally oriented.

Wainscoting. Paneling and trim applied from the floor to a height of about 3′. Used in dining areas to protect against marks from dining chairs. Popular decorative touch in modern homes.

Wale. A horizontal beam.

Wallplate. The cover over an electrical outlet or switch on the wall.

Wane. Bark, or lack of wood from any cause, on the edge or corner of a piece of wood. Hence, waney.

Wash. The slant upon a sill, capping, etc., to allow the water to run off easily.

Water line. Decorative relief line around foundation approximately 3′ from the ground.

Water table. The finish at the bottom of a house that carries water away from the foundation.

Water-repellent preservative. A liquid designed to penetrate wood and impart water repellency and a moderate preservative protection. It is used for millwork, such as sash and frames, and is usually applied by dipping.

Wattage. The product of the amperage times the voltage. A good indicator of the amount of electrical power needed to run the particular appliance. The higher the wattage, the more electricity used per hour.

Weather stripping. Strips of thin metal or other material that prevent infiltration of air and moisture around windows and doors. Compression weather stripping on single- and double-hung windows performs the additional function of holding the windows in place in any position.

Web. The thin center portion of a beam that connects the wider top and bottom flanges.

Weep hole. Small gap in brick wall, normally on garage, that allows water to drain.

Whaler. A large structural member placed horizontally against foundation forms to which braces are temporarily attached to prevent forms from moving horizontally under the pressure of concrete.

Wind. ("i" pronounced as in "kind") A term used to describe the surface of a board when twisted (winding) or when resting upon two diagonally opposite corners if laid upon a perfectly flat surface.

Wire mesh. A heavy-gauge steel mesh sold in rolls for reinforcing concrete slabs.

Withe. A vertical layer of bricks, one brick thick.

Wood mold. A brickmaking process in which bricks are actually molded instead of extruded. Fancy shapes are the result of this process.

Wooden brick. Piece of seasoned wood made the size of a brick and laid where it is necessary to provide a nailing space in masonry walls.

Index

L

M

N

P

R

S